The Coconut

Botany, Production and Uses

The Coconut

Botany, Production and Uses

Edited by

Steve W. Adkins[1]

Julianne M. Biddle[1,2]

Amirhossein Bazrafshan[1]

Sundaravelpandian Kalaipandian[1]

[1]The University of Queensland, School of Agriculture and Food Sustainability, Gatton Campus, Gatton, QLD 4343, Australia
[2]Australian Centre for International Agricultural Research, Canberra, ACT, Australia

CABI

CABI is a trading name of CAB International

CABI
Nosworthy Way
Wallingford
Oxfordshire OX10 8DE
UK

Tel: +44 (0)1491 832111
E-mail: info@cabi.org
Website: www.cabi.org

CABI
200 Portland Street
Boston
MA 02114
USA

Tel: +1 (617)682-9015
E-mail: cabi-nao@cabi.org

A catalogue record for this book is available from the British Library, London, UK.

Library of Congress Cataloging-in-Publication Data

Names: Adkins, Steve W., editor. | Kalaipandian, Sundaravelpandian, editor. | Biddle, Julianne M., editor. | Bazrafshan, Amirhossein, editor.
Title: The coconut : botany, production and uses / edited by Stephen W Adkins, Julianne M. Biddle, Amirhossein Bazrafshan, Sundaravelpandian Kalaipandian
Other titles: Botany, production and uses.
Description: Boston, MA : CAB International, [2023] | Series: Botany, production and uses | Includes bibliographical references and index. | Summary: "Comprehensive book covering all aspects of coconut from origins to cultivation, ecology and physiology, pest and disease management, harvest and post harvesting management, breeding and biotechnology as well as the current and future status of the coconut as one of the most important palm species"-- Provided by publisher.
Identifiers: LCCN 2023032711 (print) | LCCN 2023032712 (ebook) | ISBN 9781789249712 (hardback) | ISBN 9781789249729 (ebk)
Subjects: LCSH: Coconut. | Coconut--Ecophysiology. | Coconut--Climatic factors. | Coconut--Biotechnology. | Coconut--Diseases and pests. | Coconut--Propagation.
Classification: LCC SB401.C6 C544 2023 (print) | LCC SB401.C6 (ebook) | DDC 634/.61--dc23/eng/20231011
LC record available at https://lccn.loc.gov/2023032711
LC ebook record available at https://lccn.loc.gov/2023032712

ISBN-13: 9781789249712 (hardback)
9781789249729 (ePDF)
9781789249736 (ePub)

DOI: 10.1079/9781789249736.0000

Commissioning Editor: Rebecca Stubbs
Editorial Assistant: Emma McCann
Production Editor: Shankari Wilford

Typeset by Straive Pondicherry, India
Printed and bound in the UK by CPI Group (UK) Ltd, Croydon, CR0 4YY

Contents

Contributors

P. Abhin Sukumar, Division of Physiology, Biochemistry and Post-Harvest Technology, Indian Council of Agricultural Research (ICAR) Central Plantation Crops Research Institute (CPCRI), Kasaragod, Kerala, India. Email: abhinsukumarp@gmail.com

S.W. Adkins, School of Agriculture and Food Sustainability, The University of Queensland, Gatton Campus, Gatton, QLD 4343 Australia. Email: s.adkins@uq.edu.au

E.A. Aguilar, Institute of Crop Science, College of Agriculture and Food Science, University of the Philippines, Los Baños, Philippines. Email: eaaguilar1@up.edu.ph

J. Alouw, International Coconut Community (ICC), Jakarta, Indonesia. Email: jelfina@coconutcommunity.org; Horticulture and Estate Research Center, National Research and Innovation Agency, Indonesia, Cibinong Science Center, Jl. Raya Jakarta – Bogor, Cibinong Kabupaten Bogor 16915, Indonesia. Email: jelf001@brin.go.id

A. Bazrafshan, The University of Queensland, School of Agriculture and Food Sustainability, Gatton Campus, Gatton, QLD 4343 Australia. Email: a.bazrafshan@uq.edu.au

F.C. Beveridge, School of Agriculture and Food Sustainability, The University of Queensland, Gatton Campus, Gatton, QLD 4343, Australia. Email: fernanda.carobeveridge@uq.edu.au

M.A. Bezerra, Embrapa Tropical Agroindustry, Rua Doutora Sara Mesquita 2270, Fortaleza – CE, Brazil. Email: marlos.bezerra@embrapa.br

R. Bhat, Division of Crop Production, Indian Council of Agricultural Research (ICAR)-Central Plantation Crops Research Institute (CPCRI), Kasaragod, 671124, Kerala, India. Email: ravi.bhat@icar.gov.in

J.M. Biddle, Australian Centre for International Agricultural Research, Canberra, ACT, Australia. Email: julianne.biddle@aciar.gov.au; School of Agriculture and Food Sustainability, The University of Queensland, Gatton Campus, Gatton, QLD 4343, Australia. Email: Julianne.biddle@uq.edu.au

R. Cave, School of Agriculture and Food Sustainability, The University of Queensland, Gatton Campus, Gatton, QLD 4343, Australia. Email: r.cave@uq.edu.au

K. Chapman, InterAgConsult, Buderim, QLD, Australia. Email: keith.buderim@gmail.com

R. Dakshayani, Indian Institute of Food Processing Technology, Thanjavur, Tamil Nadu, India.

F.M. Dayrit, Ateneo de Manila University, Loyola Heights, Quezon City, Manila, Philippines. Email: fdayrit@ateneo.edu

C.F. de Lacerda, Federal University of Ceará, Av. da Universidade, 2853 – Benfica, Fortaleza – CE, 60020-181, Brazil. Email: cfeitosa@ufc.br

P.D. Dissanayake, Coconut Research Institute, 61150 Lunuwila, Sri Lanka. Email: pavanidissanayake@ymail.com

M. Ferreira-Neto, Federal Rural University of the Semi-arid Region, R. Francisco Mota, 572 – Pres. Costa e Silva, Mossoró – RN, 59625-900, Brazil. Email: miguel@ufersa.edu.br

H.R. Gheyi, Federal University of Campina Grande, Campina Grande, PB, Brazil. Email: hgheyi@gmail.com

K.B. Hebbar, Division of Physiology, Biochemistry, and Post-Harvest Technology, Indian Council of Agricultural Research (ICAR), Central Plantation Crops Research Institute (CPCRI), Kasaragod, Kerala, India. Email: hebbar.kb@icar.gov.in

V.B. Johnson, GreenQuills International (GQI), 34730 Prades-le-Lez, France. Email: vinbjohnson@yahoo.com; Alliance of Bioversity International and CIAT, Montpellier, France. Email: v.johnson@cgiar.org

S. Kalaipandian, School of Agriculture and Food Sustainability, The University of Queensland, Gatton Campus, Gatton, QLD 4343, Australia. Email: s.kalaipandian@uq.edu.au

S. Karouw, Indonesian Board of Coconut Research, Manado, North Sulawesi, Indonesia.

A.C. Khanashyam, Department of Food Science and Technology, Kasetsart University, 50 Ngamwongwan Road, Ladyao, Chatuchak, Bangkok 10900, Thailand. Email: khanashyamac@gmail.com

E.Y.Y. Kong, School of Agriculture and Food Sustainability, The University of Queensland, Gatton Campus, Gatton, QLD 4343, Australia. Email: e.kong@uq.edu.au

J.C. Lacsina, Independent Consultancy, Tarlac City, Philippines. Email: juca@goedenph.com

M.R. Manikantan, Indian Council of Agricultural Research – Central Plantation Crops Research Institute, Kasaragod, Kerala, India.

E.C. Manohar, International Coconut Genetic Resources (COGENT), ICC, Quezon Province, Philippines. Email: erlenemanohar2020@gmail.com

Z. Mu, School of Agriculture and Food Sustainability, The University of Queensland, Gatton Campus, Gatton, QLD 4343 Australia. Email: zhihua.mu@uq.net.au

M. Narváez, Centro de Investigación Científica de Yucatán, 97205 Mérida, Yucatán, México. Email: maryn@cicy.mx

S. Neenu, Division of Crop Production, Indian Council of Agricultural Research (ICAR)-Central Plantation Crops Research Institute (CPCRI), Kasaragod, 671124, Kerala, India. Email: neenu.s@icar.gov.in

Q.T. Nguyen, School of Agriculture and Food Sustainability, The University of Queensland, Gatton Campus, Gatton, QLD 4343, Australia. Email: q.nguyen212@uq.edu.au

C. Oropeza, Centro de Investigación Científica de Yucatán, 97205 Mérida, Yucatán, México. Email: cos@cicy.mx

C.F. Ortiz, Colegio de Postgraduados. 86500 Cárdenas, Tabasco, Mexico. Email: cfortiz@colpos.mx

R. Pandiselvam, Indian Council of Agricultural Research – Central Plantation Crops Research Institute, Kasaragod, Kerala, India. Email: anbupandi1989@yahoo.co.in

B. Panis, Dept of Biosystems, KU Leuven, Willem Decroylaan 42, 3001 Leuven, Belgium. Email: bart.panis@kuleuven.be; The Alliance of Bioversity International and CIAT, Willem Decroylaan 42, 3001 Leuven, Belgium. Email: b.panis@cgiar.org

A. Peña, Centro de Investigación Científica de Yucatán, 97205 Mérida, Yucatán, México. Email: augustojavierpena@gmail.com

L. Perera, Coconut Research Institute, 61150 Lunuwila, Sri Lanka. Email: lalithperera1234@yahoo.com

M. Ramaswamy, Indian Institute of Agricultural Research, Sugarcane Breeding Institute, Coimbatore, India. Email: rmanimekalaiicar@gmail.com

S.V. Ramesh, Division of Physiology, Biochemistry, and Post-Harvest Technology, Indian Council of Agricultural Research (ICAR), Central Plantation Crops Research Institute (CPCRI), Kasaragod, Kerala, India. Email: ramesh.sv@icar.gov.in

E. Ramos, Instituto Nacional de Investigaciones Forestales, Agrícolas y Pecuarias, Campo Experimental Huimanguillo, México. Email: ramos.eder@inifap.go

C. Silverio, Instituto Nacional de Investigaciones Forestales, Agrícolas y Pecuarias. Campo Experimental Huimanguillo, México. Email: silverio.maria@inifap.gob.mx

S. Sisunandar, Biology Education Department, Muhammadiyah University of Purwokerto, Kampus Dukuhwaluh, Purwokerto 53182, Indonesia. Email: sisunandar@ump.ac.id

P. Subramanian, Division of Crop Production, Indian Council of Agricultural Research (ICAR)-Central Plantation Crops Research Institute (CPCRI), Kasaragod, 671124, Kerala, India. Email: subramanian.p@icar.gov.in

R. Surekha, Division of Crop Production, Indian Council of Agricultural Research (ICAR)-Central Plantation Crops Research Institute (CPCRI), Kasaragod, 671124, Kerala, India. Email: surekha.r@icar.gov.in

V.R.M. Vidhanaarachchi, Tissue Culture Division, Coconut Research Institute, Lunuwila 61150, Sri Lanka. Email: vijitharma@yahoo.com

H. Wilms, KU Leuven, Oude Markt 13, 3000 Leuven, Belgium; Research Institute for Nature and Forest (INBO), Genetic Diversity, Geraardsbergen, Belgium. Email: Hannes.wilms@inbo.be

C. Yang, School of Agriculture and Food Sustainability, The University of Queensland, Gatton Campus, Gatton, QLD 4343, Australia. Email: chongxi.yang@uq.net.au

Preface

In the vast realm of botanical treasures, few captivate our imagination like the coconut. Its attraction spans continents, cultures and industries, making it a subject of fascination for many scientists, growers and enthusiasts worldwide. We proudly present to you *The Coconut: Botany, Production and Uses*, a thoughtful compilation that delves deep into the captivating world of this remarkable plant.

This comprehensive book comprises 14 chapters, each offering a unique perspective on the multifaceted coconut. The book systematically reports from the origins to production, threats, opportunities and the future of the coconut industry. This is the first book that considers such a thorough range of topics for coconut. A diverse array of experts have contributed their expertise to the chapters of this book, hailing from many countries including Australia, Belgium, Brazil, Chile, China, France, India, Indonesia, Iran, Malaysia, Mexico, the Philippines, Sri Lanka, Thailand, the UK and Vietnam. Their collective knowledge and experiences enrich every chapter, providing a truly global exploration of the coconut's botany, production and versatile uses. The subject matter ranges from the most basic to the applied and is designed to be a compilation that will remain highly relevant for researchers and growers for many years to come.

The account begins with 'An Introduction: Botany, Origin and Diversity', where we embark on an exploration of the coconut's evolutionary history, its diverse origins and the astonishing range of forms that exist. From there, we explore the subject of 'Coconut Ecophysiology', examining the intricate interplay between the coconut palm and its environments. The authors shed light on the physiological mechanisms that enable coconut palms to flourish in numerous ecological settings. For those involved in coconut cultivation, 'Management Practices for Coconut Production' provides invaluable insights and practical guidelines for optimizing yield and ensuring sustainable farming practices. As our climate rapidly changes, 'Coconut Production under a Changing Climate' takes centre stage, exploring the impacts of environmental shifts on coconut cultivation and offering strategies for adaptation. Numerous biotic factors pose significant challenges to coconut growers, and 'Pests and Diseases Affecting Coconut' offers a comprehensive examination of these threats, including detailed insights into their identification, prevention and control. Similarly, 'Physiological and Nutritional Disorders of Coconut' unravels the complexities of maintaining coconut palm health, addressing common disorders and emphasizing the importance of appropriate nutrition. Effective management of harvest and postharvest steps is vital to preserve the quality and value of fruit, and 'Harvest and Postharvest Management of Coconut' provides essential guidance in this regard. The potential for enhancing coconut varieties through genetic advancements is explored in 'Breeding and Genetics for Coconut Improvement', unveiling the innovative approaches employed to enhance the productivity and resilience. Recognizing the importance of conservation of genetic diversity, 'Coconut Conservation and Propagation' sheds light on the efforts to preserve coconut germplasm and to ensure this is available for the industry to thrive. The cutting-edge field of 'Coconut Biotechnology' unveils the techniques and advancements guiding coconut research to new frontiers. Beyond its agricultural significance, the coconut has long been cherished for its role in nutrition and health. 'Coconut Oil: Nutrition and Health' unravels the myriad of benefits and applications of this versatile fruit, unveiling its potential value for human well-being. Meanwhile, 'Traditional Coconut Products' pays tribute to the rich cultural heritage surrounding the coconut, showcasing the traditional practices and artisanal creations deeply intertwined with its history. Understanding the global context is crucial, and 'Global Coconut Trade, Economy and the Value Chain' offers a comprehensive analysis of the interplay between coconut, international commerce and the socioeconomic dynamics of different regions. Lastly, 'The Future of Coconut as an Economic Crop Worldwide' presents an inspiring outlook, exploring the potential of coconut to shape the global economy and drive sustainable development in the years to come.

We would like to express our deep appreciation and thanks to all the authors for their persistence during the 3-year preparation period. We now invite you to embark on a journey through the pages of *The Coconut: Botany, Production and Uses*. May it deepen your appreciation for the coconut's boundless potential and inspire further exploration and discovery of this captivating plant.

1 An Introduction: Botany, Origin and Diversity

Steve W. Adkins*, Robyn Cave and Fernanda Caro Beveridge
School of Agriculture and Food Sustainability, The University of Queensland, Gatton, QLD, Australia

Abstract

Coconut is a monoecious perennial palm and the only species in the genus *Cocos* within the Arecaceae, a family that first appeared 100 million years ago on the Gondwana supercontinent. It is likely that the modern coconut evolved around 85 million years ago, when Gondwana fragmented and newly formed land masses drifted through the Tethys Sea. Two development stages probably followed, the first running through geological time and dependent upon natural events, and the second a response to human activities around 100,000 years ago, during colonization of the tropical coasts. In the tropics, the palm has become a principal source of nourishment and was the first vegetable oil to be traded. While food products are the main use, other products such as timber and coir ensure that all parts of the palm are utilized. New processing methods have led to coconut oil, sugar and water becoming more valuable, highlighting the need for conservation and utilization of germplasm.

1.1 Introduction

The coconut palm (*Cocos nucifera* L.) is one of the world's most important palms, currently cultivated on 12 million ha of tropical coastal lowlands, contributing significantly to the income and livelihood of millions of farmers and their families in tropical countries (FAOSTAT, 2013). Widely referred to as the 'Tree of Life', the fruit's kernel is converted locally into food products with high nutritional and therapeutic value (Foale, 2003; Perera *et al.*, 2009; Dayrit and Dayrit, 2013), being rich in lipids, edible fibre, proteins and inorganic nutrients (Lim, 2012). Commercially refined products, including virgin coconut oil, shell charcoal and husk fibre (coir), are manufactured regionally and exported. Virgin coconut oil, which has emerged recently as a high-quality edible product compared with traditional copra-derived oil, is extracted at low temperature. It possesses powerful antioxidant and antibacterial properties (Chakraborty and Mitra, 2008; Marina *et al.*, 2009), and possibly has anticancer (Koschek *et al.*, 2007; Enos *et al.*, 2016) and anti-Alzheimer's potential (see Chapter 11, this volume). Components found in several coconut products have been used to prevent and treat a wide range of diseases, including hypertension, cardiovascular disease, diabetes, obesity, ulcers and other gut ailments, as well as hormonal imbalance in postmenopausal women (Ross, 2005; Lim, 2012). As well as the products derived from the fruit, coconut sugar is produced from the sap of cut inflorescence buds. Moreover, coconut wood, cut from the older portions of the stem, is used as a robust timber component for building construction, furniture and handicrafts. This chapter summarizes the current knowledge on

* Email: s.adkins@uq.edu.au

DOI: 10.1079/9781789249736.0001

coconut taxonomy, its origins and distribution, and its botanical features, and provides a description of some of the most important coconut varieties.

1.2 Taxonomy

Coconut is a perennial monoecious palm from the monocotyledon group. It is the only species in the genus *Cocos* within the family Arecaceae and the subfamily Arecoideae, which is made up of 27 genera containing approximately 600 species (Dransfield *et al.*, 2008). Previously, there were over 30 species included in the genus *Cocos*, distributed mainly in Central and South America (Niral and Jerard, 2018). However, later analysis placed *Cocos* as a single-species genus, despite its pantropical distribution, with *Cocos nucifera* as the sole member (Beccari, 1916). Based on detailed anatomical examination of all genera, Tomlinson (1961) justified this systematic classification of *Cocos* (Table 1.1). Subspecific descriptions of *typica*, *nana*, *javanica*, etc., have been proposed but have had limited use.

Different names have been used for *C. nucifera* around the world. In the Federated States of Micronesia it is called *nu*, in the Mariana Islands *niyog* or *nizok*, in the Republic of Kiribati *te ni*, in the Republic of Palau *iru*, in the Marshall Islands *ni*, in the Society Islands *ha'ari*, and in Samoa, Tonga, French Polynesia, the Cook Islands, Papua New Guinea and Fiji *niu*. In other regions of the world, in English it is called the coconut or coconut palm, in Dutch *coco*, *cocos*, *cocospalm* or *klapperboom*, in French *coco*, *cocoyer*, *coq au lait*, *cocotier* or *noix de coco*, in German *Kokospalme*, in Malaysian/Indonesian *kelapa* or *nyior*, in Filipino *niyog*, in Portuguese *coco da Bahia*, *coco da India* or *coqueiro de Bahia*, and in Spanish *coco*, *cocotero*, *palma de coco*, *coco de agua* or *palmera de coco*.

Table 1.1. Systematic classification of the genus *Cocos* as justified by Tomlinson (1961), based on detailed anatomical examination of all genera.

Kingdom	Plantae
Subkingdom	Tracheobionta
Super division	Spermatophyta
Division	Magnoliophyta
Class	Liliopsida
Subclass	Arecidae
Order	Arecales
Family	Arecaceae
Genus	*Cocos*
Species	*Cocus nucifera* L.

1.3 Origins and Distribution

The first members of the Arecaceae appeared around 100 million years ago on the Gondwana supercontinent, as shown by fossil records. Fragmentation of Gondwanaland approximately 85 million years ago, along with continental drift, promoted diversification through isolation and climate change (Harries, 1978). It has been hypothesized that the modern coconut species evolved during this time, when Gondwana's land fragments, east of the forming African continent, drifted through the warm ocean region, known as Tethys Sea (Fig. 1.1).

The coconut fruit evolved adaptive traits enabling it to survive long-distance dissemination by ocean currents. It produced an extraordinarily resilient and buoyant fruit that allowed it to be dispersed by floating in sea water and to form populations on other landmasses (Harries and Clement, 2014). Therefore, ocean drift of fruit could be the most important contributor to natural dispersal in the present-day Indo-Pacific regions (Bourdeix *et al.*, 2020). It has been estimated that coconut fruit can survive for up to 110 days floating in the ocean, which is assessed to represent a distance of around 4800 km (Edmondson, 1941). By floating, coconut colonized many scattered islands including coral atolls and recently forming volcanic islands (Harries and Clement, 2014), where it could establish without herbivorous attack and competition from other plant species (Foale and Harries, 2011).

A two-stage developmental history has been described for coconut: the first stage was a progression through geological time and was dependent upon natural phenomena, while the second was as a response to human activities (Gunn *et al.*, 2011). Most present-day coconut fruit have a thick fibrous mesocarp (husk) with a hard stony endocarp. Having a high proportion of husk provides a high degree of buoyancy, and being able to float is crucial for embryo protection and viability preservation during long-distance

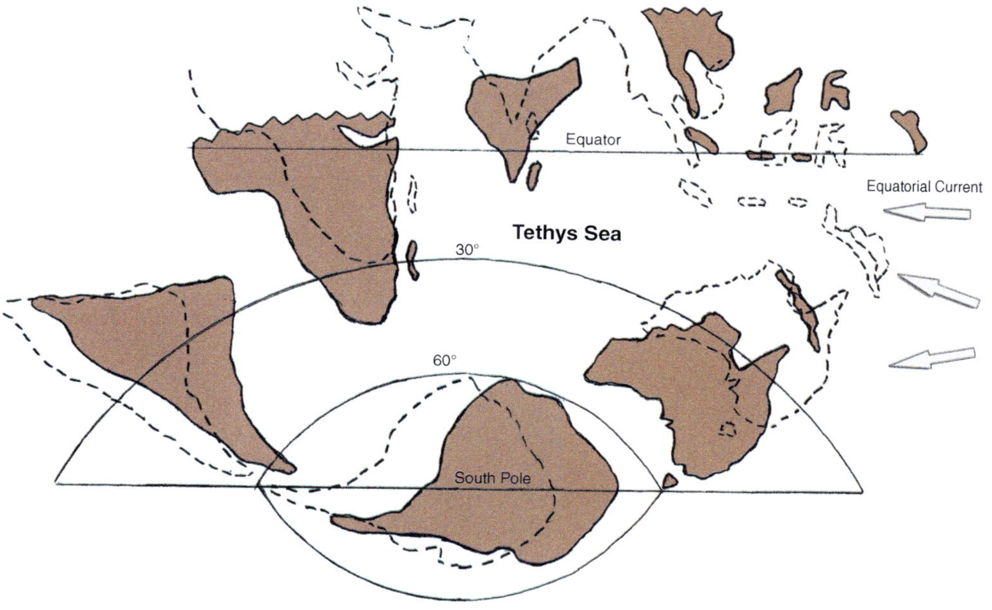

Fig. 1.1. The modern coconut species probably evolved during the fragmentation of the supercontinent Gondwanaland to form the continents of Australia, South America, Africa and southern India, and were drifting through the warm ocean region, known as Tethys Sea. From Foale, 2003.

travel on ocean currents (Harries, 2012; Harries and Clement, 2014). Moreover, dormancy can be present within the coconut seed for several months, allowing the fruit to travel long distances within ocean currents to remote locations, such as islands or more substantial land masses. After arrival at a new tropical beach, establishment success could be achieved partly through the relatively large and nutritious endosperm, when compared with most other fruits. This substantial food reserve enabled roots to grow beyond the usual reach of saltwater brought by the tide to the freshwater table beneath every well-watered beach. Humid tropical environments, such as those found on the east coast of Africa, Southeastern Asia and the tropical islands of the Pacific and the west coast of Mexico, would have allowed the coconut to establish naturally (Sauer, 1971). The length of this wild coconut prosperity period must have continued over several million years.

The beginning of the modern phase of the coconut population expansion, in response to human activities, is less certain. However, it probably dates to around 100,000 years ago, to a time when human colonization of the tropical and subtropical coasts is thought to have begun (Ingicco *et al.*, 2018). As the coconut fruit's kernel provides not only an important amount of high-energy food but also drinking water, the coconut fruit was probably carried on sea voyages as a secure food and water reserve. After arriving at a potentially suitable coastline to colonize, which might have already hosted a wild coconut population, the remaining seed from the provisions used by the early voyagers would have been propagated. Once the newly established coconut population grew, it would become the parent of subsequent introgressed populations due to cross-pollination with the wild palms, for which there are numerous examples around the tropical world (Harries, 1978). As a staple food for all coastal communities, the coconut was propagated inland as the human population increased in the tropics and way beyond its original coastline location. Coconut was cultivated successfully, up to altitudes of 1000 m in some cases, by human settlement that spread into the lowland interior of sizeable islands and the tropical regions of the Asian subcontinent (Child, 1974).

Besides the strong connection between coconuts and human settlement, during the last two centuries the coconut has gained a far greater role, as a response to areas outside the tropics demanding its oil, which has been used since the 1830s for manufacturing soap. Coconut demand expanded to edible coconut oil within a few decades. It was used to supplement animal lipid supply, which could not meet the requirements of the increasing population of industrializing countries. Coconut oil had such a high market value at that time that investment by industrialists in plantations thrived throughout the tropics, particularly in South-east Asia, in nearby islands and throughout the South Pacific, together with East Africa, Central America and the Caribbean. Plantation development began around 1880, peaked between 1900 and 1925, but then declined due to the Great Depression of the 1930s, with a further decline during the Second World War. The expansion of other plant oil plantations caused the final collapse of large coconut oil plantations.

None the less, despite the coconut oil decline in the late 20th century, the widely expanded distribution of the coconut palm during the plantation era remained in place. As the palm continued to be important for the life and culture of the regions where it is grown, the traditional landowners took over former plantation lands, while other palms were made available to tenants who integrated them into a semi-subsistence lifestyle. Currently, the coconut palm seems to exclusively dominate most suitable beaches throughout the tropics, together with significant plantings where the original vegetation had been cleared. At present, Brazil has the most significant expansion of coconut on a plantation scale, in response to a strong coconut water demand from the USA (FAO, 2018).

1.4 Botanical Features

The coconut palm belongs to the monocotyledons, a group of flowering plants having an embryo with a single large cotyledon, flower parts in groups of three and usually leaves with parallel veins. A single, erect stem is produced that can reach 24–30 m for tall varieties (Paull and Duarte, 2012; Naresh Kumar *et al.*, 2018). As with most monocotyledonous species, the stem lacks a vascular cambium and secondary thickening is absent. A single-shoot apical meristem produces a crown that consists of 30–35 green pinnate leaves at various stages of development (Ranasinghe *et al.*, 2015). Depending on the growing conditions, one leaf is produced each month over the life of the plant. Floral buds are produced in the axil of each leaf after 3–7 years of growth (Niral and Jerard, 2018). The inflorescence bears separate male and female flowers (Fig. 1.2), with many more male flowers produced than female. The adventitious roots are produced continuously throughout the life of the palm from the base or bole of the stem. The long-lived roots have a structure that is typical of most plants, but they lack root hairs for water and nutrient absorption.

1.4.1 Vegetative features

Stem

The single stem produced by the coconut palm lacks true bark and vascular cambium, and does not undergo secondary growth as seen in dicotyledonous species. Hence, once formed, the stem neither increases nor decreases in diameter. However, diurnal expansion and contraction have been reported and also general shrinkage as the palm ages (Carr and Lockwood, 2011). In common with all palms, the stem is formed at its full diameter beneath the apical meristem. The stem becomes visible after 3–4 years as the outer leaves senesce and detach. The stem diameter of a tall variety can be up to 50 cm at the base (Fig. 1.3). The dwarf variety has a trunk diameter of up to 30 cm and reaches its full diameter after 4 years. The rate of vertical growth is rapid in young palms, up to 1.5 m year^{-1} (Niral and Jerard, 2018). This slows once the reproductive phase has been reached and continues to decline as the palm ages to only 10–20 cm year^{-1} around the age of 40 years. However, planting palms close together, and/or in shade, will cause the stem to elongate rapidly. The developing portion of the stem is protected and strengthened by the overlapping leaf bases that wrap around the stem. When the leaves senesce and detach, a scar remains that can be used to determine the

Fig. 1.2. The coconut inflorescence bears separate male and female flowers, with many more male flowers produced than female. A fully opened inflorescence of the coconut showing (A) shedding male flowers at the midpoint on the rachilla and (B) a female flower with stigma. Photos courtesy of Dr Sisunandar, Indonesia.

age of the palm (Naresh Kumar *et al.*, 2018). The stem of the coconut palm consists of many vascular bundles, each enclosed in a fibrous sheath, which strengthens the stem and enables it to withstand very strong winds (Paull and Duarte, 2012). The vascular bundles are scattered throughout the stem with a higher concentration towards the periphery (Tomlinson and Huggett, 2012). This pattern is maintained along the length of the stem, although at the base of the stem, the vascular bundles are wider. The vascular tissues are alive at maturity and remain functional for the life of the palm. The size and strength of the trunk has been observed to be greater when soil chloride is readily available (Ochs *et al.*, 1991). The density of the trunk increases with age so that the lower region is always denser than the upper region. The density of the newly formed stem tissue is approximately 400 kg m^{-3} and increases to around 800 kg m^{-3} during the next 30 years. With the absence of bark, the stem is protected with a type of cork cell layer (Niral and Jerard, 2018).

Root system

The adventitious roots produced by coconut palms arise from the stem base and are produced throughout the life of the palm (Niral and Jerard, 2018). They consist of uniformly sized primary and secondary roots that also produce pneumatophores. The main adventitious roots are around 8 mm in diameter and capable of extending approximately 8 m horizontally and 2 m or more vertically downwards, although most roots occur within 1 m of the stem and to a depth of 0.5–1.0 m (Paull and Duarte, 2012). Avilan *et al.* (1984) reported that root systems of tall and dwarf varieties that were 4–12 years old reached similar depths, and that root distribution was influenced by certain features of the soil, and by the fertilizer and irrigation regimes used. Extending from the upper surface of the main and secondary roots are small, whitish pneumatophores that are entry points for oxygen. The main roots are long lived, whereas the secondary roots, which produce multiple branches and are responsible for most of the water and nutrient uptake, are relatively short lived (Niral and Jerard, 2018). The absorbing region of the root is immediately behind the protective root cap and is hairless. Young, actively growing roots are yellow to white in colour and turn red to brown as they age. Mature palms can have between 1500 and 7000 roots (Paull and Duarte, 2012; Niral and Jerard, 2018).

Fig. 1.3. The coconut palm can have a stem base up to 50 cm in diameter. This swollen base, termed the 'bole', is present in all tall varieties and in some hybrids. Photo courtesy of Dr S.W. Adkins, Australia.

Heart, crown and leaves

The apical meristem of the palm comprises a broad, cone-shaped mass of soft, undifferentiated tissue, referred to as the heart, on the 'pedestal' of the extending stem. Around the periphery of the meristem, leaf and inflorescence primordia form and expand, while growth in the central part extends and forms the stem.

The assemblage of fronds on a mature palm is known as the crown. It consists of three main sets of pinnate leaves, each with about 10–12 or 14 leaves. The oldest set of leaves is those from which fruit has been produced, the middle set comprises those that are providing photosynthates for the palm and fruit, and the youngest set is those that are not fully expanded and have rudimentary inflorescences in their axils (Niral and Jerard, 2018). The coconut palm produces leaves year-round, with dwarf varieties producing about 17 leaves year^{-1} and tall varieties about 12 leaves year^{-1} (Paull and Duarte, 2012). After about 30 years, leaf production starts to decline and the size and lifespan of the leaf decreases. The entire life cycle of a typical leaf is about 5 years from the time the leaf primordia are produced by the shoot apical meristem to the time the leaf senesces and falls from the palm (Niral and Jerard, 2018). Leaf opening tends to occur more consistently in regions where there is little seasonality and particularly when rainfall is consistent. The leaves are produced in five-merous spirals with the bunch of fruit either hanging on the left-hand side of the petiole or on the right. In mature palms, leaves can reach 4 m in length for dwarf varieties and 7 m for tall varieties, and consist of a short petiole and around 200–250 leaflets. Each leaflet has 20–25 parallel veins and caries 170–220 stomata mm^{-2}, mostly on the lower leaflet surface, although dwarf varieties tend to have more stomata than tall varieties.

Productivity

Tall varieties sustain peak productivity from 15 to about 35 years of age, although it is quite common that there will be alternate years of high and low productivity. Older palms exhibit a gradual reduction in the length of the frond and the diameter of the crown. The accompanying reduction in light interception brings about a steady downward trend in biomass production, causing stem extension to become very small (6 cm $year^{-1}$) beyond 60 years of age. This low rate of productivity still leaves sufficient biomass to support the production of a small number of fruit, even in 90-year-old palms, although such longevity is rare due to the increased likelihood of breakage of the now-fragile upper stem by cyclonic winds (Foale and Ashburner, 2004).

1.4.2 Reproductive features

Fruit

Coconut palms produce separate male and female flowers. Typically, there are thousands of male flowers in contrast to about 50 female flowers (Paull and Duarte, 2012). Following pollination, either by wind or insects (Niral and Jerard, 2018), a period of shedding occurs where about 55–95% of female flowers are lost (Paull and Duarte, 2012; Solangi, 2014). Further shedding occurs before the endosperm tissues have fully developed and usually coincides with periods of drought (Paull and Duarte, 2012). Most bunches of fruit produced on an average palm growing in an ideal environment comprise about ten fruit of equal size, shape and colour. After fertilization, the developing fruit turn from a yellowish colour to green, yellow or red-bronze, depending on the variety (Fig. 1.4), after exposure to light. Bronze or green of varying intensities is the most common colour of the immature fruit of tall varieties, while yellow or orange immature fruit are generally found in the self-pollinating dwarf varieties. The coconut fruit is a drupe that reaches its full volume 6–8 months after fertilization and then turns brown at maturity after 10–12 months (Niral and Jerard, 2018). Dwarf varieties take a slightly shorter time, maturing at 10–11 months. After turning brown, the fruit of some varieties will naturally fall. Those that do not readily fall, which is more common in dwarf varieties, need to be harvested.

At 6 months of age, the fruit comprises a soft husk, an inner hard shell and a large cavity filled with liquid endosperm or coconut water. The husk consists of many vascular fibres that have the role of conducting biomass to the developing seed. The endosperm or kernel layer on the inside of the shell is very soft and gelatinous at first, but by month 10 it firms, and by the month

Fig. 1.4. After fertilization, the developing fruit turn from a yellowish colour to green, yellow or red-bronze, depending on the variety. (A) A yellow dwarf and (B) an orange dwarf. Photos courtesy of Mr Shyam, ICAR-Central Plantation Crops Research Institute, Kasaragod, Kerala, India.

12 it is solid. This kernel at maturity contains 33% oil, 50% water and 5% protein with the remainder comprising carbohydrates, including a significant fibre content. It is this kernel that has become the highly profitable commodity, traded in its dry form as copra to industrialized countries from about 1880 until the late 20th century, with some exportation continuing in many regions to the present day.

At maturity, a medium-sized coconut fruit from a tall variety (18 cm long × 15 cm wide) consists of a low-density resilient covering that contains a spherical nut 12 cm in diameter. The shell of the nut is 3 mm thick, and its inner surface is occupied by the kernel, which can be up to 14 mm thick. The inner cavity (approx. 9 cm in diameter with a volume of around 450 mL) is filled with liquid endosperm. At maturity, and following fruit abscission from the palm, the water begins to vaporize, allowing air to enter the cavity. The embryo is embedded in the kernel directly beneath the 'soft eye', the region in the shell through which the embryo will protrude during germination (Naresh Kumar *et al.*, 2018).

Germination and growth

The fruit are usually sown horizontally to maximize contact of the embryo with the soil (Naresh Kumar *et al.*, 2018). Germination usually commences when sufficient moisture is provided for a sustained period. After 10–20 days, enlargement of the embryo occurs, and growth commences in two directions. As the shoot extends through the husk, a spongy haustorium expands from the opposite end of the embryo into the vacuole, absorbing the liquid endosperm and secreting enzymes that hydrolyse the kernel to supply energy-rich sugars and nutrients to the developing seedling. The haustorium continues to draw sugars and nutrients from the kernel for up to 13–16 weeks, which are transferred directly to the growing palm (Foale, 1968; Naresh Kumar *et al.*, 2018). The first leaves start to unfold about 2 months after germination, with the leaf blades of the first few leaves being fused (Niral and Jerard, 2018). For the first 4 months of seedling growth, the leaves are very small, and the plant remains dependent on the endosperm reserves for growth. From 6 to 12 months after germination, the leaves gradually increase in size. This coincides with the gradual transition of the palm to be fully autotropic 4–15 months after germination (Foale, 1968).

1.5 Important Types and Unique Varieties

1.5.1 Dwarf and tall

Early coconut farmers described two basic forms of coconut, tall and dwarf (Fig. 1.5), and called them *Cocos typica* and *C. nana* (Niral and Jerard, 2018). More recently, however, coconut agronomists have acknowledged that within each of these two forms there are several botanical varieties that, in turn, can be subdivided into more named forms based on their geographical origin, growth characteristics, and fruit size and number (Niral and Jerard, 2018). Today, these names often combine an origin locator with the palm's stature, such as Rennell Island Tall or Malayan Dwarf, or may indicate a hybrid ancestry, such as Maypan (Malayan Yellow Dwarf × Panama Tall) or the fruit colour (e.g. Malayan Orange Dwarf), or another fruit attribute such as a unique endosperm (e.g. Makapuno, Kopyor or Aromatic). It is believed that tall varieties are principally cross-pollinated and outbred, while the dwarf varieties are dedicated self-pollinators and are therefore inbred. Self-pollination in dwarf varieties results from features of their floral architecture, although dwarf varieties can cross-pollinate, especially when surrounded by tall palms. Historically, the tall varieties have been more extensively cultivated than dwarfs. Tall varieties are typified by having long, stout stems with a swollen base or bole, grow up to a height of 30 m and can live for 90 years. They form their first fruit at the age of about 7 years and yield high-quality copra, oil and fibre. Dwarf varieties have shorter, more slender stems without a swollen base, grow up to a height of around 15 m and live for up to 60 years. The palms come to bearing at the age of about 3 years but yield only poor-quality copra. However, at an immature stage, they produce around 170–225 g of sweet drinking water rich in potassium and other nutrients. Due to their shorter stature, dwarf palms are often cultivated for sugar production, which is made from the sap regularly collected from the cut flower bud stem. As dwarf

Fig. 1.5. Early coconut farmers described two basic forms of coconut, tall and dwarf, and called them *Cocos typica* and *C. nana*. Mature tall coconut palms are to be seen on the left-hand side next to a plantation of dwarf coconut palms. Photo courtesy of Dr Sisunandar, Indonesia.

plants under favourable conditions come into fruit production 2 years earlier than tall varieties, and because the demand for fresh drinking fruit, coconut water and sugar production is rapidly increasing, the planting of the dwarf varieties in many regions is becoming common.

1.5.2 Soft endosperm types

The kernel of a mature coconut is normally hard, but in some types of coconut, the fruit have a thicker but creamier kernel with a buttery texture, with little or no liquid endosperm. Such coconut types with this kind of endosperm are generally referred to as Makapuno and Kopyor types, based on the names given in the Philippines and Indonesia, respectively, where such fruit were first identified (Fig. 1.6). These types of fruit attract a premium price at market and are consumed fresh or used in the preparation of specific coconut delicacies. Forms such as Makapuno and Kopyor have also been described from most other coconut-growing countries, and are known as 'Garuk' in Papua New Guinea, 'Mapharao Khati' in Thailand, 'Thairu Thengai' in India, 'Dikiri Pol' in Sri Lanka and 'Dua Sap' in Vietnam (Niral and Jerard, 2018). Studies on these soft endosperm types have shown that the endosperm has less

Fig. 1.6. In some types of coconut, the fruit have a thicker but creamier kernel with a buttery texture, with little or no liquid endosperm (A–C), compared with the normally hard kernel of a mature coconut (D). Coconut types with this kind of endosperm are generally referred to as Kopyor (A, B) or Makapuno (C) types based on the names given in Indonesia and the Philippines, respectively, where such fruit were first identified. Photos courtesy of Dr Sisunandar, Indonesia, and Dr Adkins, Australia.

fibre, a lower oil content and a greater sweetness (Niral and Jerard, 2018). The embryos from all of these types have a normal morphology but will not germinate from within the fruit. Instead, they must be germinated *in vitro* when cultured on an appropriate tissue-culture medium. Based on biochemical studies (Mujer *et al.*, 1983, 1984a,b), these soft endosperm types arise due to the low activity of the enzyme α-d-galactosidase, which results in an inadequately formed but tasty endosperm. Soft endosperm types have been found among tall, dwarf and hybrid coconut types.

1.5.3 Aromatic types

The fragrance given off when a young coconut is opened can vary among types. The aromatic coconut types not only give off a pleasant smell but produce a flavoursome liquid endosperm (Niral and Jerard, 2018). Such coconuts are found in several coconut-producing countries, including Thailand, the Philippines and Vietnam. The liquid and solid endosperm from these kinds of coconuts is characterized by an agreeable pandan-like aroma, presumably caused by the presence of 2-acetyl-1-pyrroline, the major constituent of the aroma in the fragrant leaves of the palm-like *Pandanus* species or from aromatic rice. The flavour of the coconut endosperm seems to be a qualitative trait controlled by a single recessive gene (Saensuk *et al.*, 2016). Such flavours add tastiness to the consumption of fresh coconut water, and this results in the fruit attracting a higher market price than other varieties.

1.5.4 Sweet and soft husk types

Certain coconut palms produce young fruit with a mesocarp containing a low tannin content, soft fibres and a higher sugar content. Normally, due to the high tannin content and sturdy fibres, the sweetness of the young pericarp is concealed, resulting in an indigestible and unremarkable taste. In the sweet husk forms of coconut, the husk of the young fruit can be chewed like sugarcane or eaten like any tropical fruit. These types of fruit have been found in several coconut-growing countries including Indonesia, Sri Lanka, Fiji, Malaysia, the Philippines, Thailand, Vietnam and more recently in French Polynesia. Soft-husked varieties with lower levels of fibre, which are not sweet, have been recorded from several countries. Such forms have a weakly attached husk, especially when the fruit is dry, and the husk tends to be easily removed (Niral and Jerard, 2018). The economic value of such coconut varieties has yet to be determined.

1.6 Conclusion

Humans have long recognized the value of the coconut palm and have cultivated it for thousands of years. The genus *Cocos* contains only one species, with a few wild types among its closest relatives. Fossil records demonstrate how variable the palm was with fruit ranging from 3 to 25 cm in size depending on the region in which it was growing. The coconut palm has become established all around the tropical world, most likely due to ocean currents transporting its fruit and the movement of fruit by people during the practice of trading goods. Consequently, the exact origin of the palm remains uncertain.

Coconut remains an important crop for many countries in the tropics, as many communities rely on its products for their livelihood. While food products are the main reason for growing this crop, other products such as timber and coir ensure that all parts of the palm are utilized. The coconut palm is distinctive, with a single, slender stem that can reach up to 10–30 m in height, depending on whether they are a dwarf or tall variety, and a crown of 30–35 leaves. Both types have their advantages, such as self-pollination and ease of harvest for dwarf palms, and excellent longevity and yield for tall varieties. Breeders have been developing hybrids that combine the desirable traits from both types. The use of genetic resources from wild types would be ideal to provide additional traits and hybrid vigour, but given there are few wild populations left, preserving the current genetic diversity that is present within the tall and dwarf varieties should be a high priority.

The vegetative and reproductive organs of coconut are relatively slow to mature. For example, the life cycle of a leaf is about 5 years and the fruit take just under 12 months to fully mature. Likewise, seedlings rely on the seed endosperm for nutrients for at least 15 months, although their reliance is reduced after 4 months. Typically, plants that grow slowly require a stable environment to prevent the onset of decline. With more extreme climatic events, perennial plants are experiencing an increased frequency of environmental stress. To prevent pathogen attack, slow-growing plants typically allocate substantial resources to defence. However, being a monocotyledonous species, the coconut does not have secondary growth and cannot adequately protect itself from pathogen attack. Combined, these botanical characteristics suggest that coconut palms could be highly vulnerable to changes in the environment and to further pest and disease outbreaks.

1.7 References

Avilan, L.A., Rivas, N. and Sucre, R. (1984) Estudio del sistema radical del cocotero (*Cocos nucifera* L.). *Oléagineux* 39, 13–23.

Beccari, O. (1916) *Cocos nucifera* Linn forma *palmyrensis* Becc. In: Rock, J.F (ed.) *Palmyra Island with a Description of its Flora*. College of Hawaii, Honolulu, pp. 1–58.

Bourdeix, R., Adkins, S., Johnson, V., Perera, L. and Sisunandar. (2020) *In situ* and *ex situ* conservation of coconut genetic resources. In: Adkins, S., Foale, M., Bourdeix, R., Nguyen, Q. and Biddle, J. (eds)

Coconut Biotechnology: Towards the Sustainability of the 'Tree of Life'. Springer, Cham, Switzerland, pp. 51–75.

Carr, M.K.V. and Lockwood, G. (2011) The water relations and irrigation requirements of cocoa (*Theobroma cacao* L.): a review. *Experimental Agriculture* 47, 653–676.

Chakraborty, M. and Mitra, A. (2008) The antioxidant and antimicrobial properties of the methanolic extract *from Cocos nucifera* mesocarp. *Food Chemistry* 107, 994–999.

Child, R. (1974) *Coconuts*. Longmans, London.

Dayrit, C.S. and Dayrit, F.M. (2013) *Coconut Oil: From Diet to Therapy*. Anvil Publishing, Mandaluyong, The Philippines.

Dransfield, J., Uhl, N.M., Asmussen, C.B., Baker, W.J., Harley, M.M. *et al.* (2008) *Genera Palmarum: The Evolution and Classification of Palms*. Royal Botanic Gardens, London.

Edmondson, C.H. (1941) Viability of coconut seeds after floating in sea. *Occasional Papers of the Bernice Pauahi Bishop Museum of Polynesian Ethnology and Natural History, Hawaii* 16, 293–304.

Enos, R.T., Velázquez, K.T., McClellan, J.L., Cranford, T.L., Nagarkatti, M. *et al.* (2016) High-fat diets rich in saturated fat protect against azoxymethane/dextran sulphate sodium-induced colon cancer. *American Journal of Physiology – Gastrointestinal* and *Liver Physiology* 310, G906–G919.

FAO (2018) Statistics. Food and Agriculture Organization of the United Nations, Rome. Available at: www.fao.org/statistics/en/ (accessed 10 August 2023).

FAOSTAT (2013) World Coconut Harvested Areas in 2013. Food and Agriculture Organization of the United Nations, Rome. Available at: www.fao.org/faostat/ (accessed 10 August 2023).

Foale, M.A. (1968) The growth of the young coconut palm (*Cocos nucifera* L.). I. The role of the seed and of photosynthesis in seedling growth up to 17 months of age. *Australian Journal of Agricultural Research* 19, 781–789.

Foale, M.A. (2003) *The Coconut Odyssey: The Bounteous Possibilities of the Tree of Life*. Australian Centre for International Agricultural Research, Canberra.

Foale, M.A. and Ashburner, R. (2004) The coconut palm. In: Chopra, V.L. and Peter, K.V. (eds) *Handbook of Industrial Crops*. Indian Council of Agricultural Research, New Delhi.

Foale, M. and Harries, H. (2011) Coconut. In: Elevitch, C. (ed.) *Specialty Crops for Pacific Islands*. Permanent Agricultural Resources. Holualoa, Hawaii, pp. 113–134.

Gunn, B.F., Baudouin, L. and Olsen, K.M. (2011) Independent origins of cultivated coconut (*Cocos nucifera* L.) in the Old World tropics. *PLoS One* 6: e21143.

Harries, H.C. (1978) The evolution, dissemination and classification of *Cocos nucifera* L. *Botanical Review* 44, 266–319.

Harries, H.C. (2012) Germination rate is the significant characteristic determining coconut palm diversity. *AoB Plants* 2012: pls045.

Harries, H.C. and Clement, C.R. (2014) Long-distance dispersal of the coconut palm by migration within the coral atoll ecosystem. *Annals of Botany* 113, 565–570.

Ingicco, T., van den Berg, G. and Jago-On, C. (2018) Earliest known hominin activity in the Philippines by 709 thousand years ago. *Nature* 557, 233–237.

Koschek, P.R., Alviano, D.S., Alviano, C.S. and Gattass, C.R. (2007) The husk fiber of *Cocos nucifera* L. (Palmae) is a source of anti-neoplastic activity. *Brazilian Journal of Medical and Biological Research* 40, 1339–1343.

Lim, T.K. (2012) *Cocos nucifera*. In: Lim, T.K (ed.) *Edible Medicinal and Non-medicinal Plants*. Springer, Berlin, pp. 301–334.

Marina, A.M., Man, Y.B.C., Nazimah, S.A.H. and Amin, I. (2009) Antioxidant capacity and phenolic acids of virgin coconut oil. *International Journal of Food Science and Nutrition* 60, 114–123.

Mujer, C.V., Arambulo, A.S., Mendoza, E.M.T. and Ramirez, D.A. (1983) The viscous component of the mutant (Makapuno) coconut endosperm. I. Isolation and characterization. *Philippine Journal of Biology* 12, 42–50.

Mujer, C.V., Ramirez, D.A. and Mendoza, E.M.T. (1984a) α-d-galactosidase deficiency in coconut endosperm: its possible pleiotropic effects in makapuno. *Phytochemistry* 23, 893–894.

Mujer, C.V., Ramirez, D.A. and Mendoza, E.M.T. (1984b) Coconut α-d-galactosidase isoenzymes: isolation, purification and characterization. *Phytochemistry* 23, 1251–1254.

Naresh Kumar, S., Hebbar, K.B., Kasturi Bai, K.V. and Rajagopal, V. (2018) Physiology and biochemistry. In: Nampoothiri, K.U.K., Krishnakumar, V., Thampan, P.K. and Nair, M.A. (eds) *The Coconut Palm (*Cocos nucifera *L.) – Research and Development Perspectives*. Springer, Singapore, pp. 443–488.

Niral, V. and Jerard, B.A. (2018) Botany, origin and genetic resources of coconut. In: Nampoothiri, K.U.K., Krishnakumar, V., Thampan, P.K. and Nair, M.A. (eds) *The Coconut Palm (*Cocos nucifera *L.) – Research and Development Perspectives*. Springer, Singapore, pp. 57–111.
Ochs, R., Olivin, J., Quencez, P. and Hornus, P. (1991) Response to potassium fertilizer on acid sands of tertiary sediments. *Oléagineux* 46, 1–11.
Paull, R.E. and Duarte, O. (2012) Mangosteen. In: *Tropical Fruits*. CAB International, Wallingford, UK, pp. 123–138.
Perera, L., Perera, S.A.C.N., Bandaranayake, C.K. and Harries, H.C. (2009) Coconut. In: Vollmann, J. and Rajcan, I. (eds) *Oil Crops*. Springer, New York, pp. 369–396.
Ranasinghe, C.S., Kumarathunge, M.D.P., Jayaranjini, S. and de Costa, W.A.J.M. (2015) Photosynthetic irradiance response, canopy photosynthesis and their variation with canopy strata in tall and dwarf×tall coconut cultivars (*Cocos nucifera* L.). *Scientia Horticulturae* 189, 175–183.
Ross, I.A. (2005) *Cocos nucifera* L. In: Ross, I.A. (ed.) *Medicinal Plants of the World*. Humana Press, Totowa, New Jersey, pp. 117–154.
Saensuk, C., Wanchana, S., Choowongkomon, K., Wongpornchai, S., Kraithong, T. *et al.* (2016) *De novo* transcriptome assembly and identification of the gene conferring a "pandan-like" aroma in coconut (*Cocos nucifera* L.). *Plant Science* 252, 324–334.
Sauer, J.D. (1971) A reevaluation of the coconut as an indicator of human dispersal. In: Riley, C.L. *et al.* (eds) *Man Across the Sea*. Texas University Press, Austin, Texas, pp. 309–319.
Solangi, A.H. (2014) Effect of organic and inorganic fertilizers amendments on inflorescence and flower morphology and yield of coconut (*Cocos nucifera* L.). *Cord* 30, 10–18.
Tomlinson, P.B. (1961) *Anatomy of the Monocotyledons: II. Palmae*. Oxford University Press, Oxford.
Tomlinson, P.B. and Huggett, B.A. (2012) Cell longevity and sustained primary growth in palm stems. *American Journal of Botany* 99, 1891–1902.

2 Coconut Ecophysiology

Claudivan Feitosa de Lacerda[1]*, Eveline Y.Y. Kong[2], Miguel Ferreira-Neto[3], Robyn Cave[2], Marlos Alves Bezerra[4] and Hans Raj Gheyi[5]

[1]Federal University of Ceará, Fortaleza, CE, Brazil; [2]School of Agriculture and Food Sustainability, The University of Queensland, Gatton, QLD, Australia; [3]Federal Rural University of the Semi-arid Region, Mossoró, RN, Brazil; [4]Embrapa Tropical Agroindustry, Fortaleza, CE, Brazil; [5]Federal University of Campina Grande, PB, Brazil

Abstract

The different types of coconut palm, tall, dwarf and hybrid, grow best in coastal areas with a humid tropical climate. In these regions, many biotic and abiotic constraints are serious threats that affect the sustainability of coconut production, and these problems are likely to be exacerbated by global climate change. Thus, ecophysiology studies and the search for coconut genotypes tolerant to different environmental constraints are an important focus to help deal with future challenges. This chapter describes the vegetative and reproductive development of the coconut palm and explores the physiological and productive responses of this crop to the main abiotic and biotic constraints, such as salinity, waterlogging, water deficit, temperature stress, nutrient deficiency, diseases and pests. This text highlights the need to improve water- and fertilizer-use efficiency in coconut production systems worldwide to achieve maximum productivity from this permanent crop.

2.1 Introduction

Coconut (*Cocos nucifera* L.) is a crop grown in tropical coastal regions, widespread in practically all continents, between the latitudes 23°N and 23°S, and at altitudes of up to 600 m above sea level (Foale and Harries, 2009; Gunn *et al.*, 2011; Naresh Kumar and Aggarwal, 2013). Three types of coconut are found around the world. Tall types comprise palms that reach more than 20 m height and are generally used to supply copra for the coconut derivatives industry. Such palms start fruit production after 5–7 years, with a peak in production between 30 and 40 years. Dwarf types, comprising palms of smaller stature, reaching about 10 m in height, are generally used to supply fresh drinking water. These types start fruit production after 2–3 years, producing 120–250 fruit per palm every year, under irrigated systems. Hybrid types comprise palms of intermediate stature and are often used for the dual purpose of supplying copra and drinking water. These types have an annual productivity of 120–150 fruit per plant, with fruit production beginning after 3–4 years (Child, 1974; Aragão *et al.*, 1999).

The coconut palm grows best under humid tropical climates with an average annual temperature of 27–29°C, a low range of daily temperatures (<10°C is ideal) and an annual precipitation of 1100–2300 mm, with at least 2000 h of sunshine per year, and a minimum of 120 h month^{-1} (Azevedo *et al.*, 2006; Rajagopal *et al.*, 2006; Ribeiro *et al.*, 2010; Naresh Kumar and Aggarwal, 2013). The biotic and abiotic constraints that impact coconut physiology depend on soil

* Email: cfeitosa@ufc.br

 The Coconut: Botany, Production and Uses (eds S. Adkins *et al.*)

DOI: 10.1079/9781789249736.0002

and climate conditions, the type of cultivation system and the occurrence of natural phenomena (e.g. tsunamis and tropical storms). These impacts on physiology result in productivity losses that are likely to be exacerbated by global climate change, which tends to increase drought events in some regions and causes sea-level rises in coastal areas. Thus, it is important to understand the growth and physiological responses of coconut to the main abiotic and biotic constraints, which will be covered in this chapter.

2.2 Vegetative and Reproductive Growth

The coconut palm belongs to the group of flowering plants known as monocotyledons. The genus *Cocos* is typified by long pinnate leaves, a single straight stem, paniculate inflorescences and fruit that are botanically called drupes. Coconut palms are characterized by their fasciculate root systems, which explore the upper layers of the soil. Kushwah *et al.* (1973) observed that the highest concentration of roots is found at depths of 0.3–1.2 m and that 74% of the root system does not extend beyond 2.0 m from the stem base. Cintra *et al.* (1992) evaluated a Green Dwarf coconut variety in the production phase and found that the highest concentration of its roots was at a depth of 0.2–0.6 m and that 70–90% of the total root system was distributed laterally 1.0–1.5 m from the trunk. The root distribution in the soil depends on soil fertility, texture and the availability of soil moisture.

The grey-coloured coconut trunk is unbranched and fibrous, with leaf scars that remain as the older leaves detach. The terminal leaves that make up the crown protect the single terminal bud, which, if damaged, results in the death of the palm. The growth rate of the palm decreases with age, but the trunk can reach a height higher than 20 m in tall coconut varieties and 10 m in dwarf coconut varieties (Fremond *et al.*, 1975). The leaves are pinnate and are arranged in five-merous spirals. They take several years to mature and grow to approximately 6 m long, with 200–300 leaflets from 0.90 to 1.30 m in length. Under favourable environmental conditions, Green Dwarf coconut varieties can produce up to 18 leaves year^{-1} at intervals of 21–24 days (Sobral and Nogueira, 2008). The leaves remain functional for up to 38 months, forming a canopy with 25–30 leaves. Each leaf has a floral primordium in its axil, which turns into a fruiting inflorescence, depending on nutritional conditions and climate (Castro *et al.*, 2009).

The coconut is a monoecious plant, with unisexual flowers in the same inflorescence. It has an axillary paniculate inflorescence protected by large bracts, collectively called a spathe, which lasts around 1 year and is the period required for complete differentiation of the flowers. Dwarf coconut varieties emit their first spathe at around 2.5 years of age, while tall varieties remain vegetative for up to 6.5 years. When the spathe completes its development, in about 3–4 months, it opens, releasing the inflorescence. The number of female flowers is strongly influenced by the plant's nutritional and water status. Usually, around 70% of female flowers are fertilized in the first 2 weeks after the emergence of the inflorescence. The percentage fertilization of female flowers depends on genetic, climatic, pathological and nutritional factors (Fremond *et al.*, 1975).

The coconut fruit consists of a smooth outer layer called the epidermis or exocarp, which surrounds a thick and fibrous middle layer, the mesocarp, and a harder inner layer, the endocarp. The fruit takes on average 12 months to ripen, regardless of the variety, reaching the maximum concentration of sugars (5–6°Bx, or 5–6 g of sucrose in 100 g of solution) in the liquid endosperm from month 5 to month 6. Thereafter, there is a decrease in this sugar value due to biochemical processes of transformation of carbohydrates into fats, and at the same time losing the desired flavour characteristics of the liquid endosperm, or coconut water. To this end, Green Dwarf coconut varieties are more suitable for greater fruit production and are of a more appropriate size for fresh fruit consumption. It is recommended that the green fruit are harvested between months 6 and 7 of development. At this stage, an individual fruit weighs on average 1.5 kg and contains about 300 mL of water (Aragão *et al.*, 1999). However, as the fruits are harvested prior to physiological maturity, palms of this kind tend to produce more fruit in comparison with palms that are allowed to produce mature fruit. Under this form of management, production alternates between highs and lows in consecutive years, a behaviour explained mainly by physiological exhaustion of the palm because of the early harvest of unripe fruit in large quantities.

2.3 Physiological and Environmental Constraints

2.3.1 Abiotic constraints

Salt stress

The wide occurrence of coconut palms in coastal regions around the world leads many to believe that coconut is a halophyte (a native species of saline environments). However, coconut palms only tend to occupy regions of the coastal plateau that have non-saline soils with good natural drainage. Palms that grow closer to the beach, and are intermittently influenced by high tides, often have reduced leaf growth and a decreased rate of CO_2 assimilation (Santos *et al.*, 2020). It is generally accepted that the larger populations of coconut palms in coastal areas are due to the higher annual rainfall, which meets most of the crop's water demand compared with the dryer inland regions. Contrary to this thinking, evidence obtained from young and adult plants cultivated in salt-affected soils (Pereira *et al.*, 2017; Silva *et al.*, 2017, 2018), and under brackish water irrigation (Marinho *et al.*, 2006; Ferreira Neto *et al.*, 2007; Lima *et al.*, 2017) suggests that coconut is a salt-tolerant glycophyte. In addition, studies undertaken in coastal regions, especially in Asia, demonstrated that after the occurrence of tsunamis, the most salt-sensitive fruit trees (e.g. mango, *Mangifera indica* L.; guava, *Psidium guajava* L.; and star fruit, *Averrhoa carambola* L.) died or were damaged, while tolerant plants (coconut and sea hibiscus, *Hibiscus tiliaceus* L.) showed no salt damage (Lee *et al.*, 2008; Marohn *et al.*, 2012).

The studies highlighted above demonstrated that irrigation water salinity of up to 5.0 dS m^{-1} or soil salinity of up to 6.7 dS m^{-1} does not cause significant effects on the growth and production of coconut (Ferreira Neto *et al.*, 2002: Marinho *et al.*, 2006; Ferreira Neto *et al.*, 2007; Silva *et al.* 2016; Lima *et al.*, 2017). Hebbar *et al.* (2021) observed that a nutrient solution containing 10% seawater, with an electrical conductivity of 8.32 dS m^{-1}, did not reduce palm height, leaf area, collar girth or biomass accumulation of coconut seedlings, even with a 19% reduction in the rate of net photosynthesis. According to Marinho *et al.* (2005), coconut germination percentage is not affected by an irrigation water salinity of up to 20 dS m^{-1}, although there was a delay in the germination process. However, seedling establishment is a stage more sensitive to salt stress, with significant reductions in biomass production at salinities greater than 5 dS m^{-1}. Furthermore, where only brackish water is available, sowing should start 2–3 months before the rainy season to make full use of rainwater in the period of greatest sensitivity to salt. Seasonal rainfall can also contribute to reducing the residual effects of salts accumulated in the soil during irrigation in the dry season (Fig. 2.1), as demonstrated by Marinho *et al.* (2005).

The number of female flowers per inflorescence is one of the most important factors that define the most productive palms and varieties. Under favourable nutritional conditions and soil moisture levels, the number of female flowers in adult palms can exceed 16 flowers per inflorescence. Under salt stress conditions, the palms produce more female flowers (Fig. 2.2), but they are infertile. This symptom occurs 8 months after the beginning of salt stress (Ferreira Neto *et al.*, 2007). Saline irrigation water (>10 dS m^{-1}) will also reduce the weight and volume of coconut water and negatively impacts fruit appearance (Fig. 2.3). This reduction in water volume results in higher levels of total soluble solids (°Bx) and electrical conductivity of coconut water (Ferreira Neto *et al.*, 2002).

When the irrigation water contains a high salt concentration, most of the sodium ions (Na^+) are retained in the roots, stem and coconut fruit rind (exocarp and mesocarp), while leaves have a greater accumulation of potassium ions (K^+) and a higher K^+:Na^+ ratio (Hebbar *et al.*, 2021), an important mechanism for salt tolerance in coconut (Yang *et al.*, 2021). The accumulation of organic solutes is another response commonly observed in coconut palms under salt stress and can contribute to osmotic adjustment. Lima *et al.* (2017) found high levels of organic solutes (carbohydrates and soluble amino-nitrogen) and Na^+ in roots of coconut seedlings irrigated with brackish water (electrical conductivity up to 19.3 dS m^{-1}). The high

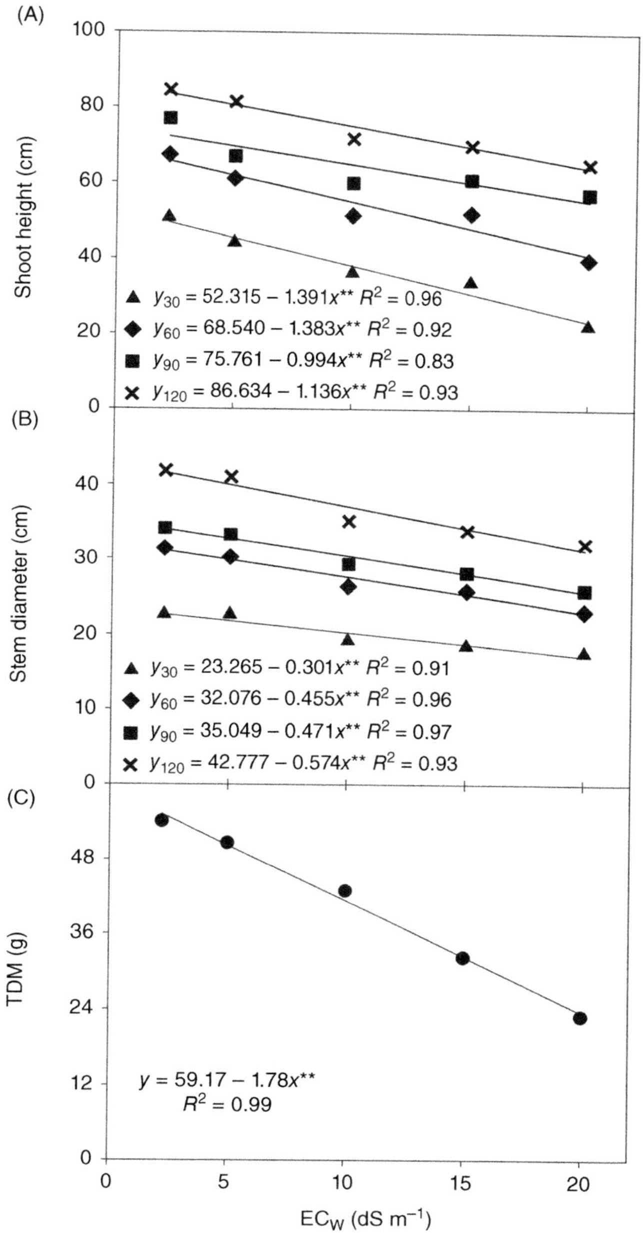

Fig. 2.1. (A) Shoot height, (B) stem diameter at 30, 60, 90 and 120 days after transplanting (DAT), and (C) total dry mass (TDM) at 120 DAT for a dwarf coconut variety cultivated in a sandy loam soil, as a function of the residual effect of irrigation water salinity (EC_w) applied during germination and seedling establishment. From Marinho *et al.*, 2005, CC-BY.

concentration of organic solutes and the retention of Na^+ in the roots allows the maintenance of a low Na^+:K^+ ratio in the leaves (Lima *et al.*, 2017; Hebbar *et al.*, 2021).

The physiological and biochemical responses of palms vary significantly with the intensity of salt stress, and some responses may represent salt tolerance mechanisms, while others may be

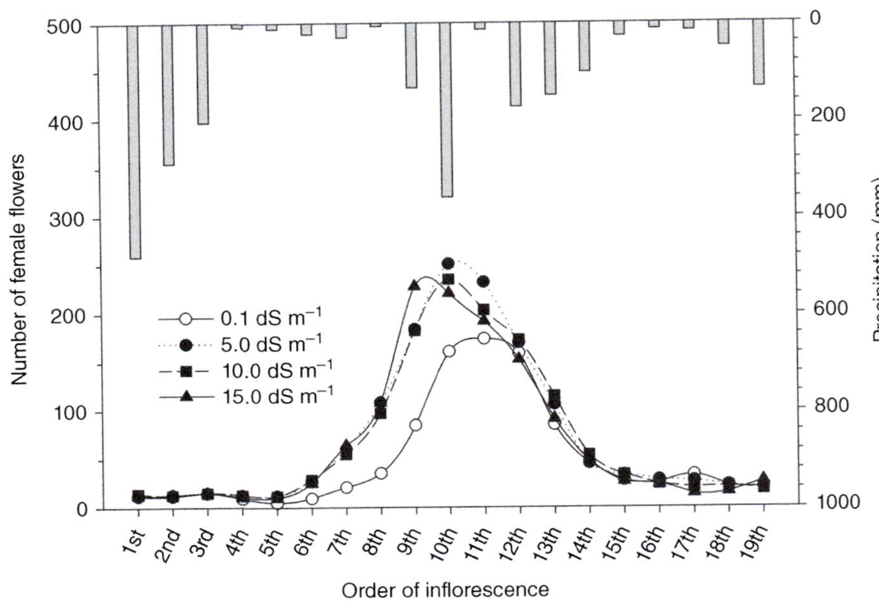

Fig. 2.2. Number of female flowers in different inflorescences of a dwarf coconut variety as a function of irrigation water salinity, grown at the Experimental Farm of Parnamirim in the State of Rio Grande do Norte, Brazil. Adapted from Ferreira Neto *et al.*, 2011, CC-BY.

Fig. 2.3. Coconut fruit harvested 11 months after the initiation of irrigation with water with electrical conductivities of 0.1, 5.0, 10.0 or 15.0 dS m^{-1}. From Ferreira Neto *et al.*, 2002, CC-BY.

induced under severe stress levels, contributing to palm survival. In many studies, an increase in antioxidant protection (activity of enzymes and antioxidant substances) is correlated with salt tolerance, guaranteeing the removal of reactive oxygen species and protection of the primary metabolism of the plant cells. However, a study with young coconut palms identified a negative correlation between superoxide dismutase (EC number 1.15.1.1) and polyphenol oxidase activity (EC number 1.14.18.1) and biomass production, with antioxidant protection being increased only at very high levels of salinity, which might have already severely affected palm growth (Hebbar *et al.*, 2021).

Waterlogging and related constraints

Despite being somewhat salt tolerant, coconut is sensitive to waterlogging, which alters the root environment by depleting O_2, accumulating CO_2 and reducing the iron (Fe) and manganese (Mn) content (Kozlowski, 1997; Hopmans *et al.*, 2021). In non-halophytes, waterlogging inhibits the formation and expansion of leaves and internodes, causes premature leaf abscission and senescence, inhibits photosynthesis and carbohydrate transport, and reduces root growth and nutrient absorption, negatively impacting productivity (Kozlowski, 1997; Benassi *et al.*, 2013).

In coconut plantations, the problem of excess water has been observed in isolation during the rainy season (Prasada Rao, 2016), or associated with salt stress, especially in areas with inefficient drainage of drylands (Kozlowski, 1997; Sousa *et al.*, 2011) or flooding with seawater by high tides (Santos *et al.*, 2020), or during the occurrence of tsunamis (Marohn *et al.*, 2012). This problem is also relevant when the subsoil has low permeability, or when the water table is shallow, which culminates in multiple stresses, including salinity and excess water.

A study carried out in Brazil evaluated the ecophysiological responses of tall coconut varieties under irrigated, rainfed, Foreiro (palms close to the beach, but not influenced by tides) and Preamar (plants close to the beach and influenced by high tides) conditions (Santos *et al.*, 2020). The results indicated that abiotic stress occurred in all palms, except for those that were irrigated. Salt stress and waterlogging affected plants in the Preamar area, while those under rainfed conditions were impacted by water deficit and nutritional stress. The transient salinity and waterlogging stresses observed in the Preamar area resulted in lower nitrogen (N) and K^+ concentrations in leaves, and a significant increase in the Na^+:K^+ ratio. The plants in the Preamar area also had fewer leaflets per leaf, low values of CO_2 assimilation and stomatal conductance, and a high internal concentration of CO_2. These results indicate that seawater causes severe stress that occurs intermittently throughout the year and hinders carbon (C) assimilation, which limits the capacity of fruit production (Santos *et al.*, 2020). This problem may intensify in the future due to climatic change events that result in elevated sea levels or more regular and sustained seawater inundation events (Renaud *et al.*, 2015; Kheir *et al.*, 2019; Corwin, 2021).

Salinity and waterlogging problems can also occur simultaneously in irrigated areas, notably under inadequate drainage conditions, as has been observed in the Brazilian semi-arid regions (Sousa *et al.*, 2011). Under these conditions, the seedlings of Green Dwarf coconut varieties show a reduction in their photosynthetic rate and biomass accumulation (Fig. 2.4). Photosynthesis inhibition was linked to stomatal and non-stomatal causes and depended on the interaction and severity of waterlogging and salinity. Therefore, coconut seedlings may be used in revegetation programmes in salt-affected areas, provided these areas are not exposed to frequent waterlogging cycles (Medeiros *et al.*, 2018).

Water deficit

Among the main environmental factors impacting coconut production, water deficit is one of the most frequently found around the world (Gomes and Prado, 2007; Gomes *et al.*, 2010). Tall coconut varieties have good survival rates when kept under water-deficit conditions, but their productivity is affected, and it takes 1–2 years to recover from a severe water-shortage event (Nair, 1989). If the stem of the coconut palm narrows due to water deficit, the plant's production potential may no longer reach its original level. Conversely, Green Dwarf coconut varieties, having a high rate of transpiration, will consume a large volume of water. Thus, under limited edaphoclimatic conditions, these palms will show increased effects of water stress compared with tall varieties (IRHO-CIRAD, 1992).

The importance of water availability can be seen when observing the mean fruit yield in the south-east regions of Brazil, characterized by having a good average rainfall compared with the north-east region of Brazil, which has high evapotranspiration together with irregular rainfall distribution, causing constant seasonal water deficits. In the former case, the mean yield in 2020 was 15,503 fruit ha^{-1}, while in the latter it was 7943 fruit ha^{-1} (IBGE, 2021).

Water deficit reduces the germination percentage and causes a lag in seedling establishment. When subjected to a water deficit, the

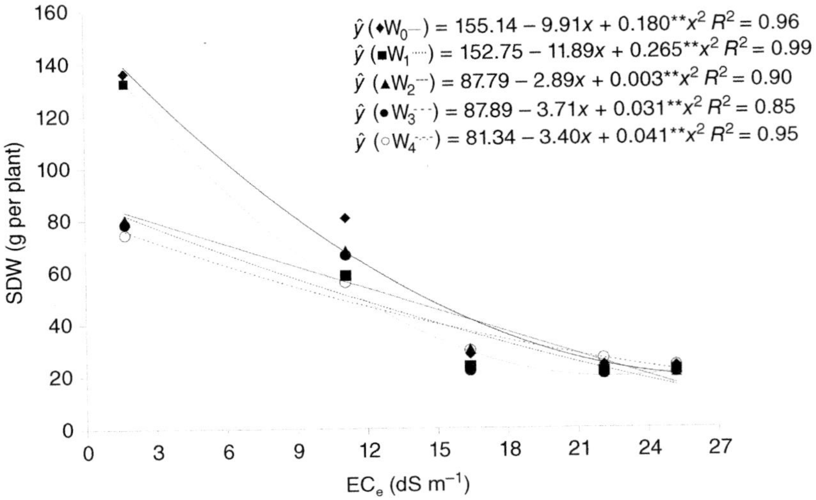

Fig. 2.4. Shoot dry weight (SDW) of a Green Dwarf coconut variety in response to soil salinity (EC_e) and waterlogging cycles. W_0, W_1, W_2, W_3 and W_4 represent zero, one, two, three and four waterlogging cycles, imposed at 30, 60, 90 and 120 days after transplanting. Each cycle lasted 4 days, simulating the waterlogging caused by a typical rainfall event in the north-east region of Brazil. From Medeiros *et al.*, 2018, CC-BY.

coconut palm presents a reduction in the number and size of leaves, exhibits premature leaf fall, and has a delay in the reproductive stage and a decrease in the number of female flowers produced per inflorescence, with a resultant reduction in fruit production (Miranda and Gomes, 2006). As the development of the inflorescence starts 3–4 months before the spathe opens, a water deficit in this period can cause the inflorescence to fail and/or abort, consequently affecting fruit production 7–8 months later (Passos, 2018). A water deficit can delay the start of production for 8–24 months after the deficit ends, generating low-quality fruits.

In several coconut genotypes, in the rainy season, even under limiting conditions of photosynthetically active radiation density (<500 μmol m^{-2} s^{-1} for most of the day), stomatal conductance and transpiration values are higher than in the dry season. Under water-deficit conditions during the dry season, all Green Dwarf coconut varieties showed reduced stomatal conductance (Passos *et al.*, 2005). However, this reduction in gas exchange seemed to be more associated with environmental conditions than with soil moisture availability. In experiments with an adult Green Dwarf coconut variety (Fig. 2.5), cultivated in the coastal region of Ceará State, Brazil, the application of different fractions of crop evapotranspiration (ET_c) (50, 75, 100 and 125% of ET_c) did not cause significant changes in the leaf gas exchange (photosynthesis, stomatal conductance and transpiration) of coconut plants, while over the year there was a reduction in these variables in the months with a higher vapour-pressure deficit (Table 2.1; Gomes do Ó, 2017; Araújo, 2019). Water deficits also cause a reduction in fruit production characteristics. The number of fruit produced per plant and the volume of water per fruit are considerably affected by water deficit, while the content of soluble solids in the coconut water is increased (Fig. 2.6) (Gomes do Ó, 2017; Araújo, 2019).

Temperature stress

All physiological processes and biochemical reactions are impacted by temperature, causing activation or inhibition. The coconut palm has a C3 photosynthetic metabolism (plants in which the initial product of the assimilation of CO_2 through photosynthesis is 3-phosphoglycerate, which contains three C atoms) and has a high ratio of photosynthetic tissues/whole shoot biomass, especially when the production of leaves is not inhibited by low or high temperatures, water

Fig. 2.5. General view of an experiment with a Green Dwarf coconut variety cultivated in Paraipaba (3°26′S; 39°09′W, altitude 39 m), Brazil, and under the application of different fractions of crop evapotranspiration. Photo courtesy of Dr Marlos, Brazil.

Table 2.1. Analysis of variance of leaf gas exchange of Green Dwarf coconut palms subjected to different irrigation levels (50, 75, 100 and 125% of crop evapotranspiration) applied in different months. Adapted from Araújo, 2019, with permission.

		Mean squares			
Year/source of variation	Df	*A*	g_s	*E*	C_i
2017					
Water depth	3	4.00^{NS}	0.00^{NS}	0.94*	234.37^{NS}
Month	4	35.73**	0.02**	0.73 NS	4557.80**
Water depth × month	12	0.2^{NS}	0.00^{NS}	0.47^{NS}	989.99^{NS}
Residue	76	2.61	0.00	0.34	631.95
CV (%)	–	18.22	23.45	17.73	11.62
2018					
Water depth	3	2.89^{NS}	0.00^{NS}	0.76^{NS}	388.84^{NS}
Month	6	52.12**	0.02**	34.60**	4051.01**
Water depth × month	18	1.47^{NS}	0.0^{NS}	0.3^{NS}	269.08^{NS}
Residue	108	215.59	0.00	0.39	413.13
CV (%)	–	13.98	19.49	14.08	8.94

Df, degrees of freedom; *A*, net photosynthesis (μmol m^{-2} s^{-1}); g_s, stomatal conductance (mol m^{-2} s^{-1}); *E*, leaf transpiration (mmol m^{-2} s^{-1}); C_i, internal concentration of CO^2 (μmol mol^{-1}); ^{NS}, not significant; *, significant at p<0.05; **, significant at p<0.01 by F-test.

deficit, salinity or nutritional deficiency, among other stresses. Coconut develops best between the Tropics of Capricorn and Cancer, where temperatures are closer to the optimal value for physiological processes (annual mean temperature of 27–29°C), especially in the coastal areas, where humid to subhumid climates also predominate (Rajagopal *et al*., 2006; Naresh Kumar and Aggarwal, 2013), although temperatures above 40°C, from April to July in the tropical regions of India, decreased the functional leaf area index, the rate of C assimilation and fruit yield

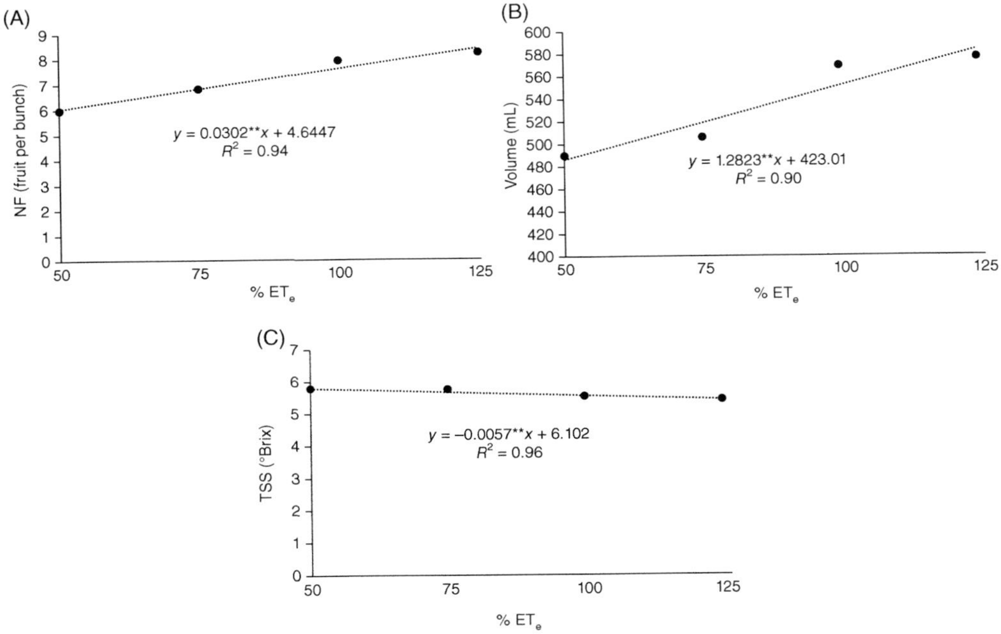

Fig. 2.6. (A) Number of fruit per bunch, (B) volume of water in the fruit, and (C) total soluble solids in coconut water in a Dwarf green coconut variety subjected to different irrigation intensities (50, 75, 100, and 125% ET_e). NF, number of fruit; TSS, total soluble solids. From Araújo, 2019, with permission.

(Naresh Kumar and Aggarwal, 2013). Photosynthesis can also be impacted by the high magnitude of the leaf-to-air vapour pressure difference under high leaf temperature, a key factor influencing the stomatal conductance of coconut (Passos *et al.*, 2005; Hebbar *et al.*, 2013).

The reproductive phase of the coconut palm is extremely sensitive to high temperatures, manifested by button (female flowers) shedding and poor fruit set. Hebbar *et al.* (2020) evaluated the progamic phase (from pollination to fertilization) of the variety West Coast Tall, in Kerala, India, and the effect of high temperatures under both *in vivo* and *in vitro* conditions. High temperatures (>33°C) significantly reduced pollen-tube growth through the pistil, suggesting its inability to reach the ovule in time to undertake fertilization. High temperatures also advanced nectar secretion and stigma receptivity, probably rendering it unappealing to insect pollinators. Thus, both poor pollination and the inability of the pollen tube to reach the ovule in time to undertake fertilization could be the cause of poor fruit set in the West Coast Tall variety under high temperatures (Hebbar *et al.*, 2020).

Low temperature is one of the main environmental factors that can limit coconut productivity. Minimum temperatures above 10°C can trigger flowering, but minimum temperatures below 10°C for 1 month or more will cause fruit fall. Frequent but short periods of temperatures below 15°C have been observed to result in abnormalities of fruit, such as bicarpellate fruit, and a lack of pollination under north Indian conditions (Hebbar *et al.*, 2013). A study performed in Bangladesh showed that temperatures below 16.8°C, with a mean precipitation of 1170 mm and K^+ at 0.11 mEq 100 g^{-1} in the soil severely damaged the palms (Rashid *et al.*, 2015). Findings indicate that younger palms are more adaptable to cold than older palms. Moreover, the availability of water and K^+ in the rhizosphere also helps to reduce the adverse effects of low temperatures on coconut palms. Coconut palms can show cold injury at temperatures of 0°C or below, with desiccated foliage as the primary symptom (Rashid *et al.*, 2015).

A study conducted in China evaluating the physiological responses and cold tolerances (temperatures of 25°C, 15°C, 10°C and 5°C) of ten varieties of tall and dwarf coconut varieties revealed that low temperatures led to increased membrane damage, higher malondialdehyde content and soluble and non-soluble protein content, and increased activities of the defence enzymes superoxide dismutase, peroxidase, catalase (EC 1.11.1.6) and ascorbate peroxidase (EC 1.11.1.11) (Sun *et al.*, 2021). In this study, three coconut genetic resources (two tall and one dwarf coconut variety) were identified as genotypes with strong cold resistance.

Generally, low temperatures reduce fruit production and cause abnormal fruit development in coconut crops grown in Hainan Province, China. Proteomic analysis of two coconut varieties in response to cold stress identified certain proteins involved in the biosynthesis of secondary metabolites, photosynthesis and respiration (Yang *et al.*, 2020). The increased abundance of stress-responsive and general-function proteins under cold stress suggests that the Hainan Tall variety is cold tolerant, compared with an aromatic coconut variety, through enhanced scavenging capacity of reactive oxygen species during oxidative stress.

Nutritional constraints

Knowledge of soil fertility is fundamental for planning and using amendments. In the case of coconut, the desired value for base saturation of the soil corresponds to 60%. For sandy soils, the application of limestone at 2000–3000 kg ha^{-1} is generally used to correct acidity, increase nutrient availability, and improve physical or chemical soil traits. The balance between cations in the soil is important for the availability of nutrients for plants. For K^+, calcium (Ca^{2+}) and magnesium (Mg^{2+}), the ratio in the soil should be between 1:9:3 and 1:25:5 (Malavolta *et al.*, 1997). The ideal N content, in leaf number 14, is 18–20 g N kg^{-1} for tall varieties and 18–22 g kg^{-1} for hybrid varieties (Saldanha *et al.*, 2017). Below this value, N nutrition is compromised, which can, according to Manciot *et al.* (1980), lead the palms to show deficiency symptoms.

Irrigated Green Dwarf coconut varieties have favoured development, with better use of added N, stimulating vegetative growth and reaching an average of 22 leaves per palm annually (Ferreira Neto *et al.*, 2011). An adult Green Dwarf coconut palm, when cultivated under conditions of nutritional deficiency, presents a reduction in the number of leaves, caused by a reduction in the rate of leaf formation (Ferreira Neto *et al.*, 2007, 2011). Therefore, an adequate supply of N is fundamental for the generation of a larger photosynthetically active area, capable of sustaining high fruit production.

K^+ can act as a growth regulator, inhibiting excessive growth when N availability is high. When these factors interact, they induce an increase in the palm crown diameter. As reported by Ferreira Neto *et al.* (2011), a revised palm spacing for irrigated cultivation of the Green Dwarf variety should be considered, especially when being optimally fertilized. Excess K^+ can positively or negatively affect the absorption of other cations by the plants, as they compete directly for the same site on the carrier. Mg^{2+} deficiency can be induced by excessive fertilization with K^+, increasing the K^+:Mg^{2+} ratio. However, in K^+-deficient soils, its addition can cause greater Mg^{2+} absorption (Lins *et al.*, 2021), which is accompanied by an increase in dry matter.

Under fertigation, high nutrient requirements can influence the critical levels of nutrients in leaves. Holanda *et al.* (2007) observed that in fertigated Green Dwarf coconut, the critical levels of K^+ in leaf 14 are 9–10 g kg^{-1}. This is greater than the critical level proposed by Magat (2005) of 6–8 g kg^{-1} for leaf 14 of the Green Dwarf variety without irrigation. However, even in areas where the natural content of K^+ in the soil is high (>0.30 $cmol_c$ kg^{-1}), it can be compromised by exhaustive harvests (Sobral and Nogueira, 2008). The action of K^+ fertilization on production is highly significant, as well as the damage caused to plants by K^+ deficiency during the first year of planting (IRHO, 1989).

Experience has shown antagonism between K^+ and Ca^{2+}, K^+ and Mg^{2+}, and K^+ and Na^+, but the K^+–Mg^{2+} relationship is more divergent than the others (Ferreira Neto *et al.*, 2007, 2014). Coomans (1977) noted that Mg^{2+} deficiency was induced in hybrid coconut palms by K^+ application. However, Mg^{2+} fertilization did not affect K^+ levels in the leaf. In orchards with low productivity, high values of K^+ in the leaves can be detected, predicting accumulation of the nutrient, due to low productivity, as well as when other factors cause fruiting losses (Sobral *et al.*, 2007; Santos *et al.*, 2020).

A Na^+ content of 2.2–2.9 g kg^{-1} is associated with maximum production (Ferreira Neto *et al.*, 2007), and is achieved using NaCl or sometimes naturally from sea spray (Santos *et al.*, 2020). Na^+ concentration in coconut leaves tends to decline with increasing K^+ fertilization (Ferreira Neto *et al.*, 2007), but the Na^+:K^+ ratio becomes very high under severe salt stress (Santos *et al.*, 2020). At adequate levels, Na^+ stimulates the growth of some plant species, contributing to the processes of cell expansion and osmotic balance, replacing K^+ in some specific functions (Marschner, 2012).

The importance of chloride (Cl^-) as a macronutrient in coconut was demonstrated by von Uexkull (1992), who found that the application of KCl increased albumin weight from 117 to 216 g kg^{-1} and the Cl^- content from 0.40 to 2.33 g kg^{-1} in leaf 14, while the K^+ content in the same leaf remained practically constant. Cl^- plays a role in plant photosynthesis and osmotic regulation, as well as promoting resistance to drought and fungal diseases (Ollagnier *et al.*, 1983; Marschner, 2012). Von Uexkull (1992) suggested that the critical level of Cl^- in leaves is around 2.5 g kg^{-1} and that anything lower predisposes the crop to water stress (drought) with symptoms of brittle leaves.

2.3.2 Biotic stress

In general, biotic stress in plants is due to attack by viruses and living organisms such as bacteria, fungi, oomycetes, nematodes, insects, arachnids, weeds and herbivores, which all affect plant growth, development and productivity (Gull *et al.*, 2019). Biotic stresses such as diseases and pests are serious threats that affect the sustainability of coconut production.

Lethal yellowing disease (LYD), caused by a phytoplasma, has been reported to infect coconut plantations in many regions in America, Africa, Asia and Oceania (Gurr *et al.*, 2016), and has impacts on coconut palms both physiologically and biochemically (Oropeza *et al.*, 1995). The noon stomatal conductance of infected coconut palms was found to decrease in association with this disease (Eskafi *et al.*, 1986; Cardena *et al.*, 1991), which consequently led to greater abnormal stomatal closure (León *et al.*, 1996). Phloem and xylem malfunction has also been demonstrated in affected palms. The phloem exudation rate was found to decrease in the inflorescence sap and trunk section of severely affected palms (Eden-Green and Waters, 1982). The malfunction of phloem also led to an increase in the leaf carbohydrate concentration, as the sugar transport to sink tissues was inhibited (Maust *et al.*, 2003), whereas water transport of diseased palms was greatly reduced, as shown by a low uptake rate (10%) of ^{32}P and a 44% water flow rate in the petioles (Eskafi *et al.*, 1986). The infected palms also had a lower photosynthetic rate (reduced from 9.00 to 1.75 μmol m^{-2} s^{-1}) and reduced protein (reduced to half at stage 4 in the lower portion of the crown), carotenoid and chlorophyll content (León *et al.*, 1996). This was supported by a decrease in the number of active photosystem II reaction centres (Maust *et al.*, 2003). Leaf senescence in these palms could be due to plant growth regulator imbalances as the abscisic acid (increased 2.5-fold) and ethylene concentration increased (León *et al.*, 1996). Additionally, the endogenous zeatin-type (Z-type) and isopentenyladenine-type (iP-type) cytokinin contents in the infected leaves were greatly reduced at stage 2-inflorescence necrosis (11–382-fold lower), and became pronounced at stage 6 (leaf chlorosis and senescence), as several cytokinins (zeatin-9-glucoside, zeatin riboside, isopentenyladenine and isopentenyladenine-9-glucoside) were not detected (Aguilar *et al.*, 2009). There were also biochemical changes observed in the roots of infected palms, whereby the respiration rates decreased by 76–79% and protein concentrations were reduced by 50–66% for lateral and subtending roots (Islas-Flores *et al.*, 1999).

Leaf scorch decline, another disease of the coconut industry of unknown cause, can reduce coconut yields within 2–3 years and subsequently kill the palms over the following 5–10 years. The photosynthetic and transpiration rate, functional leaf area, leaf chlorophyll content, water potential, zinc (Zn^{2+}) content and inflorescence sap volume are reduced as the disease develops. Stomatal diffusive resistance as well as abscisic acid concentrations in the xylem sap increase in diseased palms (Ranasinghe, 2005).

Pests pose another serious issue for coconut production. For example, palms with red palm weevil (*Rhynchophorus ferrugineus* Olivier) often die unless it is detected and treated early (Faleiro, 2006). An infestation of red palm weevil has been found to affect the biochemistry of the coconut varieties Matag and Pandan in terms of their oxidative stress responses. The infested tissues (in the lower trunk) were found to have a lower lipid peroxidation level and hydrogen peroxide (H_2O_2) production for both varieties, whereas the percentage of ion leakage (lower and upper trunk tissues) was significantly higher for the infested Matag variety (Yusuf *et al.*, 2017).

The coconut rhinoceros beetle (*Oryctes rhinoceros* L.) is another prime pest that affects coconut growth during its adult stage. Slow physiological growth is observed, which eventually results in secondary fungal infections (Nangai and Martin, 2020). Infestation with coconut rhinoceros beetle was found to influence leaflet nutrient, metal and lignin content. The N and B contents were significantly increased by 15% and 30%, respectively, in infested leaves, but the compounds that were not resorbed (C, Ca, Fe, Mn and Zn) were reduced in the senesced leaves. The infested leaflet was also found to have a lower content of metals such as nickel, chromium, cobalt and lead. The C:N and lignin:N values also decreased, suggesting that the herbivory attack by the beetle on coconut may restrict green leaf nutrition and increase the decomposition speed of senesced leaves (Marler, 2018).

In the plant defence mechanism, feeding by the red palm mite (*Raoiella indica* Hirst) significantly influenced the oxidative enzyme activities of coconut. A resistant variety (Jamaican Tall) was found to show higher oxidative enzyme activities measured by the polyphenol oxidase:peroxidase ratio. The total protein content of the Jamaican Tall variety was found to increase after red palm mite infestation for 72 h (Carlos *et al.*, 2016).

The black-headed caterpillar (*Opisina arenosella* Walker) is another major herbivorous pest in coconut plantations, affecting 1.6 million palms in Karnataka, India (Shivashankar *et al.*, 2000). A biochemical change in volatile organic compounds was observed in affected plants, which differed among the varieties studied and among infested and non-infested palms (Shameer *et al.*, 2017).

Another crucial pest affecting the coconut palm is the coconut mite (*Aceria guerreronis* Keifer), which can reduce yield by up to 30% through physiological and/or biochemical changes. For instance, the moisture and chlorophyll content of the perianth tissue infested by the coconut mite was found to decrease to 15% and 24%, respectively, whereas the total sugars and free amino acid levels increased to 12% and 84%, respectively. Mite attack has also been known to increase the oxidase enzyme activities of defensive compounds such as phenols, peroxidase and indole-3-acetic acid by 55%, 39% and 4%, respectively. The nutrient content in the infested tissues was also impacted, whereby N content increased to 19% and P, Ca, Cu, Mg and Fe contents decreased to 60%, 67%, 19%, 80% and 52%, respectively (Vishnupriya *et al.*, 2017).

In coconut plantations, wide spacing between each palm is needed due to its canopy structure. This can facilitate weed growth, as sufficient sunlight can reach the plantation floor to promote seed germination and growth. These weeds compete with palms for nutrients, soil moisture and light at the seedling stage (Senarathne *et al.*, 2003). For instance, weed competition was found to reduce the dry matter yield of coconut leaves, stems and roots by 61%, 46% and 44%, respectively (Remison and Mgbeze, 1987). Additionally, some weeds were found to act as hosts for the phytoplasma that causes coconut LYD. For instance, plant hoppers (*Fulgoroidea* spp.) and leaf hoppers (*Cicadellidae* spp.) have been found to be vectors for coconut LYD in Nigeria, and these insect vectors were hosted by weed species including Guinea grass (*Panicum maximum* Jacq.), common wild sorghum (*Sorghum arundinaceum* (Desv.) Stapf.), gamba grass (*Andropogon gayanus* Kunth.), African wild rice (*Oryza barthii* A. Chev.) and umbrella sedge (*Cyperus difformis* L.) (Eziashi *et al.*, 2013).

The American palm cixiid (*Haplaxius crudus* (Van Duzee)) is a planthopper that transmits LYD in coconut by acting as a vector for the phytoplasma '*Candidatus* Phytoplasma palmae' (group 16SrIV-A). The principal weed hosts for the nymph of this planthopper are gamba grass, para grass (*Brachiaria mutica* (Forssk.) Stapf.) and dune fingergrass (*Eustachys petraea* (Sw.) Desv.). Recently identified weed hosts for the nymph of the American palm cixiid include signal grass (*Brachiaria decumbens* Stapf.), East African couchgrass (*Digitaria abyssinica* (Hochst. ex A. Rich) Stapf.) and Alabama swamp flatsedge (*Cyperus ligularis* L.) (Ramos Hernández *et al.*, 2018).

Therefore, weeds can be problematic to coconut plantations due to competition and may increase the risk of many diseases by acting as a reservoir for plant pathogens. Weed management in coconut plantations is thus crucial to ensure optimal growth and yield under complex multitrophic interactions of the coconut pathosystem.

In summary, biotic stresses can have a serious impact on the physiology of the coconut palm and can be intensified under climate change conditions. For instance, an increase in temperature is expected to facilitate the distribution and infestation levels of arthropods, pathogens and weeds. In general, pest presence will increase and provide greater pressure in most agroecosystems under climate change (Gullino *et al.*, 2022).

2.4 Conclusion

Different environments and cultivation systems impose a series of abiotic (water deficit, salinity, waterlogging, high temperature, cold and nutritional deficiency) and biotic (diseases, pests and weeds) constraints that impact the physiology of the coconut palm, with the tall type considered more tolerant of these adverse environmental factors compared with hybrid and dwarf types. These stress factors can reduce leaf growth and limit the photosynthetic capacity of the plants, reducing fruit growth and yield, in both irrigated and rainfed systems. Global climate change scenarios may increase farmers' economic losses, with a potential negative impact on the sustainability of coconut cultivation in tropical zones, by reducing total rainfall or by increasing the soil salinity in coastal regions. Ecophysiology studies indicate that the search for coconut genotypes tolerant to different environmental constraints is important to deal with future challenges. The need to improve water- and fertilizer-use efficiency is also of great importance as coconut is a permanent crop with a high demand for these resources.

2.5 References

Aguilar, M.L., Espadas, F., Maust, B. and Sáenz, L. (2009) Endogenous cytokinin content in coconut palms affected by lethal yellowing. *Journal of Plant Pathology* 91, 141–146.

Aragão, W.M., Tupinambá, E.A., Ângelo, P.C.S. and Ribeiro, F.E. (1999) Seleção de cultivares de coqueiro para diferentes ecossistemas do Brasil. In: Queiroz, M.A., Goedert, C.O. and Ramos, S.R.R. (eds) *Recursos Geneticos e Melhoramento de Plantas para o Nordeste Brasileiro*. Embrapa Semi-Árido/ Embrapa Recursos Genéticos e Biotecnologia, Brasília, pp. 709–735.

Araújo, B.A. (2019) Respostas fisiológicas do coqueiro anão verde sob diferentes lâminas de irrigação. Masters' Dissertation – Universidade Federal do Ceará, Centro de Ciências Agrárias, Programa de Pós-Graduação em Engenharia Agrícola, Fortaleza, p. 65.

Azevedo, P.V., Sousa, I.F., Silva, B.B. and Silva, V.D.P.R. (2006) Water-use efficiency of dwarf-green coconut (*Cocos nucifera* L.) orchards in northeast Brazil. *Agricultural Water Management* 84, 259–264.

Benassi, A.C., Fanton, C.J. and Santana, E.N. (2013) O cultivo do Coqueiro-anão-verde: Tecnologias de Produção. Incaper. Documents, 227. Vitória, ES, Brazil.

Cardena, R., Villanueva, M.A., Santamarfa, J.M. and Oropeza, C.M. (1991) Presence in Yucatan of mycoplasma-like organisms in *Cocos nucifera* palms showing lethal yellowing disease symptoms. *Canadian Journal of Plant Pathology* 13, 135–138.

Castro, C.P., Passos, E.E.M. and Aragão, W.M. (2009) Fenologia de cultivares de coqueiro-anão nos tabuleiros costeiros de sergipe. *Revista Brasileira de Fruticultura* 31, 13–19.

Child, R. (1974) *Coconuts*. Longmans, London.

Cintra, F.L.D., Leal, M.L.S. and Passos, E.E.M. (1992) Distribuição do sistema radicular de coqueiros anões. *Oléagineux* 47, 225–234.

Coomans, P. (1977) Premiers résultats expérimentaux sur la fertilisation des cocotiers hybrides em Côte-d'Ivoire. *Oléagineux* 32, 155–166.

Corwin, D.L. (2021) Climate change impacts on soil salinity in agricultural areas. *European Journal of Soil Science* 72, 842–862.

Eden-Green, S.J. and Waters, H. (1982) Collection and properties of phloem sap from healthy and lethal yellowing-diseased coconut palms in Jamaica. *Phytopathology* 72, 667–672.

Eskafi, F.M., Basham, H.G. and McCoy, R.E. (1986) Decreased water transport in lethal yellowing-diseased coconut palms. *Tropical Agriculture* 62, 225–228.

Eziashi, E., Omamor, I., Aisueni, N., Aisagbonhi, C., Airede, C. *et al.* (2013) Potential weed species as alternate hosts of insect vectors of the lethal yellowing disease (LYD) of coconut palms (*Cocos nucifera* L.) in Nigeria. *British Journal of Applied Science and Technology* 3, 123–130.

Faleiro, J.R. (2006) A review of the issues and management of the red palm weevil *Rhynchophorus ferrugineus* (Coleoptera: Rhynchophoridae) in coconut and date palm during the last one hundred years. *International Journal of Tropical Insect Science* 26, 135–154.

Ferreira Neto, M., Gheyi, H.R., Holanda, J.S., Medeiros, J.F. and Fernandes, P.D. (2002) Qualidade do fruto verde de coqueiro em função da irrigação com água salina. *Revista Brasileira de Engenharia Agrícola e Ambiental* 6, 69–75.

Ferreira Neto, M., Gheyi, H.R., Fernandes, P.D., Holanda, J.S.D. and Blanco, F.F. (2007) Emissão foliar, relações iônicas e produção do coqueiro irrigado com água salina. *Ciência Rural* 37, 1675–1681.

Ferreira Neto, M., Holanda, J.S.D., Dias, N.S., Gheyi, H.R. and Folegatti, M.V. (2011) Crescimento e produção de coqueiro anão verde fertigado com nitrogênio e potássio. *Revista Brasileira de Engenharia Agrícola e Ambiental* 15, 657–664.

Ferreira Neto, M., Holanda, J.S., Gheyi, H.R., Folegatti, M.V. and Dias, N.S. (2014) Atributos químicos do solo e estado nutricional de coqueiro-anão fertirrigado com nitrogênio e potássio. *Revista Caatinga* 27, 30–40.

Foale, M. and Harries, H. (2009) Farm and forestry production and marketing profile for coconut (*Cocos nucifera*). In: Elevitch, C.R. (ed.) *Specialty Crops for Pacific Island Agroforestry*. Permanent Agriculture Resources, Holualoa, Hawaii.

Fremond, Y., Zller, R. and Lamonte, M.N. (1975) *El Cocotero: Tecnicas Agrícolas y Producciones Tropicales*. Editorial Blume, Barcelona, Spain.

Gomes, F.P. and Prado, C.H. (2007) Ecophysiology of coconut palm under water stress. *Brazilian Journal of Plant Physiology* 19, 377–391.

Gomes, F.P., Oliva, M.A., Mielke, M.S., Almeida, A.A.F. and Aquino, L.A. (2010) Osmotic adjustment, proline accumulation and cell membrane stability in leaves of *Cocos nucifera* submitted to drought stress. *Scientia Horticulturae* 126, 379–384.

Gomes do Ó, L.M. (2017) Fisiologia e produção de plantas de coqueiro anão sob diferentes níveis de irrigação. Masters dissertation, Universidade Federal do Ceará, Fortaleza, Ceará, Brazil.

Gull, A., Lone, A. and Wani, N. (2019) Biotic and abiotic stresses in plants. In: de Oliveira, A. (ed.) *Abiotic and Biotic Stress in Plants*. Intech Open, London. Available at: www.intechopen.com/chapters/66714 (accessed 11 August 2023).

Gullino, M.L., Albajes, R., Al-Jboory, I., Angelotti, F., Chakraborty, S. *et al.* (2022) Climate change and pathways used by pests as challenges to plant health in agriculture and forestry. *Sustainability* 14: 12421.

Gunn, B.F., Baudouin, L. and Olsen, K.M. (2011) Independent origins of cultivated coconut (*Cocos nucifera* L.) in the Old World tropics. *PLoS One* 6: e21143.

Gurr, G.M., Johnson, A.C., Ash, G.J., Wilson, B.A.L., Ero, M.M. *et al.* (2016) Coconut lethal yellowing diseases: a phytoplasma threat to palms of global economic and social significance. *Frontiers in Plant Science* 7: 1521.

Hebbar, K.B., Balasimha, D. and Thomas, G.V. (2013) Plantation crops response to climate change: coconut perspective. In: Singh, H., Rao, N. and Shivashankar, K. (eds) *Climate-resilient Horticulture: Adaptation and Mitigation Strategies*. Springer, New Delhi.

Hebbar, K.B., Neethu, P., Sukumar, P.A., Sujithra, M., Santhosh, A. *et al.* (2020) Understanding physiology and impacts of high temperature stress on the progamic phase of coconut (*Cocos nucifera* L.). *Plants* 9: 1651.

Hebbar, K.B., Santhosh, A., Sukumar, A.P., Neethu, P., Ramesh, S.V. *et al.* (2021) Effect of sea water substitution on growth, physiological and biochemical processes of coconut (*Cocos nucifera* L.) seedlings – a hydroponic study. *Scientia Horticulturae* 280: 109935.

Holanda, J.S., Ferreira Neto, M., da Silva, R.A., Chagas, M.C.M., Sobral, L.F. *et al.* (2007) *Technologies for the Intensive Production of Green Dwarf Coconut*. Research Bulletin 34. EMPARN, Natal, Brazil.

Hopmans, J.W., Qureshi, A.S., Kisekka, I., Munns, R., Grattan, S.R. *et al.* (2021) Critical knowledge gaps and research priorities in global soil salinity. *Advances in Agronomy* 169, 1–191.

IBGE (2021) Produção agrícola municipal. Brazilian Institute of Geography and Statistics (IBGE), Rio de Janeiro, Brazil. Available at: https://sidra.ibge.gov.br/pesquisa/pam/tabelas (accessed 5 October 2021).

IRHO (1989) Rapport d'activité. *Oléagineux* 44, 1–22.

IRHO-CIRAD (1992) Coconut: water supply and drought tolerance. *Oléagineux* 47, 334–337.

Islas-Flores, I., Santamaria, J.M., Cordova, I. and Oropeza, C. (1999) Biochemical changes in roots of coconut palms (*Cocos nucifera* L.) affected by lethal yellowing. *Journal of Plant Physiology* 155, 48–53.

Kheir, A.M., El Baroudy, A., Aiad, M.A., Zoghdan, M.G., El-Aziz, M.A.A. *et al.* (2019) Impacts of rising temperature, carbon dioxide concentration and sea level on wheat production in North Nile delta. *Science of the Total Environment* 651, 3161–3173.

Kozlowski, T.T. (1997) Responses of woody plants to flooding and salinity. *Tree Physiology* 17: 490.

Kushwah, B.L., Nelliat, E.V., Morkave, V.T. and Sunny, A.F. (1973) Rooting pattern of coconuts (*Cocos nucifera* L.). *Indian Coconut of Agronomy* 18, 71–74.

Lee, K.J., Yoo, S., Park, S.W., Lee, S.J., Suriyapananont, V. *et al.* (2008) Effects of 2004-tsunami on the changes from 2005 to 2006 in salinity of soil profiles and groundwater wells and on mortality of garden fruit trees in Maldives. *Forest Science and Technology* 4, 14–27.

León, R., Santamaría, J.M., Alpizar, L., Escamilla, J.A. and Oropeza, C. (1996) Physiological and biochemical changes in shoots of coconut palms affected by lethal yellowing. *New Phytologist* 134, 227–234.

Lima, B.L.C., Lacerda, C.F., Ferreira Neto, M., Ferreira, J.F.S., Bezerra, A.M.E. *et al.* (2017) Physiological and ionic changes in dwarf coconut seedlings irrigated with saline water. *Revista Brasileira de Engenharia Agrícola e Ambiental* 21, 122–127.

Lins, P.M.P., Viegas, I.D.J.M. and Ferreira, E.V.D.O. (2021) Nutrition and production of coconut palm cultivated with mineral fertilization in the state of Pará. *Revista Brasileira de Fruticultura* 43: e113.

Magat, S.S. (2005) Coconut. In: *World Fertilizer Use Manual*. International Fertilizer Industry Association. Paris.

Malavolta, E., Vitti, G.C. and Oliveira, S.A. (1997) *Avaliação do Estado Nutricional das Plantas: Princípios e Aplicações*. POTAFOS, Piracicaba.

Manciot, R., Ollagnier, M. and Ochs, R. (1980) Nutricion mineral y fertilizacion del cocotero en la mundo. *Oléagineux* 35, 13–27.

Marinho, F.J.L., Ferreira Neto, M., Gheyi, H.R., Fernandes, P.D. and Viana, S.B.A. (2005) Uso de água salina na irrigação do coqueiro (*Cocus nucifera* L.). *Revista Brasileira de Engenharia Agrícola e Ambiental* 9, 359–364.

Marinho, F.J.L., Gheyi, H.R., Fernandes, P.D., Holanda, J.S.D. and Ferreira Neto, M. (2006) Cultivo de coco 'Anão Verde' irrigado com águas salinas. *Pesquisa Agropecuária Brasileira* 41, 1277–1284.

Marler, T.E. (2018) Coconut leaf age and coconut rhinoceros beetle herbivory influence leaflet nutrients, metals, and lignin. *Horticulturae* 4: 9.

Marohn, C., Distel, A., Dercon, G., Wahyunto, T.R., Noordwijk, M.V. *et al.* (2012) Impacts of soil and groundwater salinization on tree crop performance in post-tsunami Aceh Barat, Indonesia. *Natural Hazards and Earth System Sciences* 12, 2879–2891.

Marschner, H. (2012) *Marschner's Mineral Nutrition of Higher Plants*. Academic Press, London.

Maust, B.E., Espadas, F., Talavera, C., Aguilar, M., Santamaría, J.M. *et al.* (2003) Changes in carbohydrate metabolism in coconut palms infected with the lethal yellowing phytoplasma. *Phytopathology* 93, 976–981.

Medeiros, W.J.F., Oliveira, F.Í.F., Lacerda, C.F., Sousa, C.H.C., Cavalcante, L.F. *et al.* (2018) Isolated and combined effects of soil salinity and waterlogging in seedlings of 'Green Dwarf' coconut. *Semina: Ciências Agrárias* 39, 1459–1468.

Miranda, F.R. and Gomes, A.R.M. (2006) *Manejo da Irrigação do Coqueiro Anão*. Embrapa Agroindústria Tropical, Circular Técnica (INFOTECA-E).

Nair, R.R. (1989) Summer irrigation requirement of the coconut palm. *Indian Coconut Journal* 19, 3–7.

Nangai, L. and Martin, B. (2020) Habitational study on RPW of coconut palm tree. *Test Engineering and Management* 81, 3564–3568.

Naresh Kumar, S. and Aggarwal, P.K. (2013) Climate change and coconut plantations in India: impacts and potential adaptation gains. *Agricultural Systems* 117, 45–54.

Ollagnier, M., Ochs, R., Pomier, M. and Taffin, G. (1983) Acción del cloro sobre el cocotero híbrido PB-121 em Costa do Marfim y Indonesia. Desarrollo, tolerancia a la sequia. *Oléagineux* 38, 309–321.

Oropeza, C., Alpizar, L., Islas-Flores, I., Escamilla, A. and Santamaria, J. (1995) Physiology and biochemistry of lethal yellowing-affected *Cocos nucifera* L. palms. In: Oropeza, C., Howard, F.W. and Ashburner, G.R. (eds) *Lethal Yellowing: Research and Practical Aspects*. Springer, Netherlands, pp. 65–77.

Passos, C.D., Passos, E.E.M. and Prado, C.H.B.A. (2005) Comportamento sazonal do potencial hídrico e das trocas gasosas de quatro variedades de coqueiro-anão. *Revista Brasileira de Fruticultura* 27, 248–254.

Passos, E.E.M. (2018) Morfologia. In: Ferreira, J.M.S., Warwick, D.R.N. and Siqueira, L.A. (eds) *A Cultura do Coqueiro no Brasil*. EMBRAPA-CPATC, Aracaju, Brazil, pp. 81–90.
Pereira, T.A., Souto, L.S., Sá, F.V.S., Dutra Filho, J.A., Souza, T.M.A. *et al.* (2017) Growth and production of dwarf coconut in saline-sodic soil under doses of potassium sulfate. *Revista Brasileira de Engenharia Agrícola e Ambiental* 21, 454–458.
Prasada Rao, G.S.L.H.V. (2016) Weather extremes and plantation crops in the humid tropics. *MAUSAM* 67, 251–258.
Rajagopal, V., Parthasarathy, V.A., Naresh Kumar, S., Reddy, D.V.S. and Iyer, R. (2006) Coconut. In: Parthasarathy, V.A., Chattopadyay, P.K. and Bose, T.K. (eds) *Plantation Crops*, Vol. 2. Naya Udyog, Kolkata, India, pp. 1–178.
Ramos Hernández, E., Magaña Alejandro, M.A., Ortiz García, C.F., Oropeza Salín, C., Lesher Gordillo, J.M. *et al.* (2018) The coconut pathosystem: weed hosts of nymphs of the American palm Cixiid *Haplaxius crudus* (Hemiptera: Fulgoroidea). *Journal of Natural History* 52, 255–268.
Ranasinghe, C.S. (2005) Changes in the physiological performance of leaf scorch decline (LSD) affected coconut (*Cocos nucifera*) palms. *Experimental Agriculture* 41, 255–265.
Rashid, M.M., Hasan, M.S., Rahman, M.H., Ahmmed, A.N.F. and Khalequzzaman, K.M. (2015) Physiological and structural abnormalities of coconut plants (*Cocos nucifera* L.) at different stages of growth and plantation sites under low temperature condition in Bangladesh. *The Agriculturists* 13, 113–119.
Remison, S.U. and Mgbeze, G. (1987) Competitive effects of weeds and beneficial effects of mulching on coconut seedlings. *Cocos* 5, 19–28.
Renaud, F.G., Le, T.T.H., Lindener, C., Guong, V. and Sebesvari, Z. (2015) Resilience and shifts in agro-ecosystems facing increasing sea-level rise and salinity intrusion in Ben Tre Province, Mekong Delta. *Climatic Change* 133, 69–84.
Ribeiro, F.E., Baudouin, L., Lebrun, P., Chaves, L.J., Brondani, C. *et al.* (2010) Population structures of Brazilian tall coconut (*Cocos nucifera* L.) by microsatellite markers. *Genetics and Molecular Biology* 33, 696–702.
Saldanha, E.C.M., Silva Junior, M.L., Lins, P.M.P., Farias, S.C.C. and Wadt, P.G.S. (2017) Nutritional diagnosis in hybrid coconut cultivated in northeastern Brazil through Diagnosis and Recommendation Integrated System (DRIS). *Revista Brasileira de Fruticultura* 39: e728.
Santos, M.M.S., Lacerda, C.F., Neves, A.L.R., Sousa, C.H.C., Ribeiro, A.A. *et al.* (2020) Ecophysiology of the tall coconut growing under different coastal areas of northeastern Brazil. *Agricultural Water Management* 232: 106047.
Senarathne, S.H.S., Samarajeewa, A.D. and Perera, K.C.P. (2003) Comparison of different weed management systems and their effects on yield of coconut plantations in Sri Lanka. *Weed Biology and Management* 3, 158–161.
Shameer, K.S., McCormick, A.C., Subaharan, K. and Nasser, M. (2017) Volatile organic compounds in healthy and *Opisina arenosella* Walker (Lepidoptera: Oecophoridae) infested leaves of coconut palms. *ENTOMON* 42, 121–132.
Shivashankar, T., Annadurai, R.S., Srinivas, M., Preethi, G., Sharada, T.B. *et al.* (2000) Control of coconut black-headed caterpillar (*Opisina arenosella* Walker) by systemic application of 'Soluneem' – a new water-soluble neem insecticide formulation. *Current Science* 78, 176–179.
Silva, A.R.A., Bezerra, F.M., Lacerda, C.F., Miranda, R.S., Marques, E.C. *et al.* (2016) Organic solutes in coconut palm seedlings under water and salt stresses. *Revista Brasileira de Engenharia Agrícola e Ambiental* 20, 1002–1007.
Silva, A.R.A., Bezerra, F.M.L., Lacerda, C.F., Sousa, C.H.C. and Bezerra, M.A. (2017) Physiological responses of dwarf coconut plants under water deficit in salt-affected soils. *Revista Caatinga* 30, 447–457.
Silva, A.R.A., Bezerra, F.M.L., Lacerda, C.F., Miranda, R.S. and Marques, E.C. (2018) Ion accumulation in young plants of the 'green dwarf' coconut under water and salt stress. *Revista Ciência Agronômica* 49, 249–258.
Sobral, L.F. and Nogueira, L.C. (2008) Influence of N and K applied by fertirrigation on soil properties, critical leaf levels and yield of dwarf coconut. *Revista Brasileira de Ciência do Solo* 32, 1675–1682.
Sobral, L.F., Freitas, J.A.D. de, Holanda, J.S. de, Fontes, H.R., Cuenca, M.A.J. *et al.* (2007) 5. Coconut – Green Dwarf variety. In: Crisóstomo, L.A. and Naumov, A. (eds) *Fertilizing for High Yield and Quality – Tropical Fruits of Brazil*. Bulletin No. 18. International Potash Institute, Horgen, Switzerland, pp. 89–102.
Sousa, C.H.C., Silva, F.L.B., Lacerda, C.F., Costa, R.N.T. and Gheyi, H.R. (2011) Instalação de um sistema de drenagem subterrânea em um solo salino-sódico cultivado com coco em Pentecoste – Ceará. *Revista Brasileira de Agricultura Irrigada* 5, 16–23.

Sun, C., Zhang, R., Yuan, Z., Cao, H. and Martin, J.J.J. (2021) Physiology response and resistance evaluation of twenty coconut germplasm resources under low temperature stress. *Horticulturae* 7: 234.

Vásquez, C., Dávila, M., Méndez, N., Jiménez, M.A., Sandoval, M.F. and Alcalá, F.J. (2016) Oxidative enzymes in coconut cultivars in response to *Raoiella indica* feeding. *African Journal of Biotechnology* 15, 1755–1762.

Vishnupriya, R., Umapathy, G., Venugopal, S. and Manivannan, V. (2017) Phytochemical interaction between coconut, *Cocos nucifera* L., and perianth mite, *Aceria guerreronis* Keifer. *Journal of Entomology and Zoology Studies* 5, 791–794.

von Uexkull, H.R. (1992) Drought and chlorine deficiency – a deadly combination for hybrid coconuts. *Better Crops International* 8, 8–9.

Yang, Y., Saand, M.A., Abdelaal, W.B., Zhang, J., Wu, Y. *et al.* (2020) iTRAQ-based comparative proteomic analysis of two coconut varieties reveals aromatic coconut cold-sensitive in response to low temperature. *Journal of Proteomics* 220: 103766.

Yang, Y., Bocs, S., Fan, H., Armero, A., Baudouin, L. *et al.* (2021) Coconut genome assembly enables evolutionary analysis of palms and highlights signaling pathways involved in salt tolerance. *Communications Biology* 4: 105.

Yusuf, N., Nassir Azmi, W. and Ahamad Zakeri, H. (2017) Early oxidative stress responses in coconut (*Cocos nucifera*) stem of matag and pandan cultivars due to red palm weevil (*Rhynchophorus ferrugineus* Olivier) infestation. *Malaysian Applied Biology* 46, 187–193.

3 Management Practices for Coconut Production

Ravi Bhat*, Surekha Rajkumar, Neenu Satyaseelan and Periasamy Subramanian

Division of Crop Production, Indian Council of Agricultural Research (ICAR)-Central Plantation Crops Research Institute (CPCRI), Kasaragod, Kerala, India

Abstract

Coconut (*Cocos nucifera* L.) is an important plantation crop sustaining the livelihood of small and marginal farming families. It is grown mainly for its edible oil; however, a number of other products with industrial applications and nutritional and medicinal value are also produced. In recent times, a marginal decline in coconut production has been observed due to a decline in soil fertility, changing climate, an increased number of senile palms, and the emergence of new pests and diseases. Coconut growers are also exposed to economic risk and uncertainties due to market price fluctuations. Hence, the coconut sector requires innovative, multi-pronged strategies to address these challenges and to accelerate industry growth. This chapter provides a glimpse of the scientific aspects of coconut production technologies that can enhance farm income by improving fruit yield and reducing the adverse impacts of biotic and abiotic stress.

3.1 Introduction

Coconut (*Cocos nucifera* L.) is a perennial plantation crop that supports the livelihood of millions of families across the world and is intrinsically linked to their socio-cultural life. Coconut is botanically a monocot with a ploidy of $2n = 32$, in the family Arecaceae and order Arecales. Coconut is believed to have originated in the region of the Malayan Peninsula and Archipelago, New Guinea, and the Bismarck Archipelago, and from there it spread to other tropical regions (Ignacio and Miguel, 2021). The impermeable exocarp and the mesocarp's dense fibre layer of both the immature (green) and mature (brown and dry) fruit make it exceptionally buoyant, and this could be a reason for its widespread and rapid movement to many remote locations (Harries, 2012). Thus, the crop has spread naturally to regions of Madagascar and eastern Africa, as well as southern India and Sri Lanka, where it has a documented history dating back 2000–3000 years.

In more than 91 countries across the tropical regions of Asia, America and East Africa, coconut is grown on an area of 11.81 million ha. The Philippines and Indonesia have the highest area under coconut cultivation, followed by India, in terms of both fruit production and area coverage. In productivity, Myanmar ranks first followed by Vietnam, Malaysia and India (Table 3.1). In the years 2005–2020, coconut fruit production, productivity and acreage increased globally. India makes up about 20% and 25% of the

* Email: ravi.bhat@icar.gov.in

 The Coconut: Botany, Production and Uses (eds S. Adkins *et al.*)
DOI: 10.1079/9781789249736.0003

world's production area and fruit production, respectively (Table 3.1; FAOSTAT, 2021). In India, coconut cultivation is practised in 17 states and three territories under diverse soil and climatic conditions (Table 3.2).

Coconut offers food, drink, oil, fibre, timber, fuel and medicines, and hence is popularly referred to as the 'Tree of Life', 'tree of a thousand uses', 'Tree of Abundance' or 'Tree of Heaven' (Chan and Elevitch, 2006; Angeles *et al.*, 2018). Owing to its intrinsically rich macro- and micronutrient profile, it is also known as the 'wonder fruit' for human health and nutrition (Mat *et al.*, 2022).

Every part of the coconut palm can be utilized; for instance, the trunk and leaves are used

Table 3.1. Ranking of coconut area planted, production per annum and productivity of different countries worldwide. Adapted from FAOSTAT, 2021.

Country	Area (× 1,000 ha)	Production (million t)	Productivity (fruit ha^{-1})
Philippines	3,651,873 (1)	14.77 (2)	40,431 (11)
Indonesia	2,800,000 (2)	17.13 (1)	61,174 (8)
India	2,151,000 (3)	14.68 (3)	68,257 (4)
Tanzania	657,658 (4)	0.43 (12)	6,517 (15)
Sri Lanka	503,452 (5)	2.47 (4)	49,037 (10)
Mexico	204,133 (6)	1.29 (7)	63,094 (7)
Papua New Guinea	188,841 (7)	1.19 (8)	63,165 (6)
Brazil	186,950 (8)	2.33 (5)	12,683 (13)
Vietnam	158,959 (9)	1.68 (6)	105,502 (2)
Thailand	124,374 (10)	0.81 (9)	64,807 (5)
Vanuatu	89,817 (11)	0.30 (14)	33,683 (12)
Malaysia	76,776 (12)	0.54 (10)	69,892 (3)
Ghana	75,195 (13)	0.40 (13)	53,714 (9)
Solomon Islands	66,796 (14)	0.08 (15)	12,043 (14)
Myanmar	48,902 (15)	0.53 (11)	108,550 (1)
Total	**10,984,726**	**58.63**	

Table 3.2. Estimates of area, production of fruit per annum and productivity of coconut in 2019–2020 in India. From Horticulture Division, Department of Agriculture and Cooperation, Ministry of Agriculture and Farmers Welfare, Government of India, with permission.

State/Union Territory	Area (× 1,000 ha)	Production (million fruit)	Productivity (fruit ha^{-1})
Andhra Pradesh	111.38	1,555.82	13,969
Arunachal Pradesh	0.07	0.30	4,605
Assam	20.75	159.87	7,704
Bihar	12.16	78.39	6,444
Chhattisgarh	1.58	0.96	611
Gujarat	27.40	264.88	9,667
Karnataka	624.03	4,300.69	6,892
Kerala	760.78	6,980.30	9,175
Maharashtra	29.95	523.66	17,485
Mizoram	0.03	0.15	4,350
Nagaland	1.06	9.02	8,482
Odisha	51.71	354.57	6,857
Others	58.37	292.80	5,017
Tamil Nadu	437.57	5,373.21	12,280
Telangana	0.53	6.49	12,330
Tripura	4.61	18.43	3,996
West Bengal	31.30	389.16	12,433
Total	**2,173.28**	**20,308.70**	**9,345**

in handicrafts and as building materials for fences and houses (Prades *et al.*, 2012). The coconut husk is processed into carpets, geotextiles, rope and compressed wood. It is also used as a nutritional fibre source, as a sterile inert support medium for plant growth, and as seat padding in trains and vehicles (Rencoret *et al.*, 2013). Activated charcoal can be made from the dense endocarp (shell). The liquid endosperm or water has a high mineral, sugar and vitamin content and is consumed fresh or as a sports beverage (Hidalgo, 2017). The mature solid endosperm or copra is used for cooking oil extraction and in different edible products (Patil *et al.*, 2017). Coconut oil cake, a by-product of the oil extraction process, is a rich source of vegetable protein used in livestock feed. The unopened floral spathe is tapped for sap, or neera, to make palm nectar and toddy, which are used for the preparation of palm sugar, jaggery and vinegar. Development of enhanced processing techniques and the use of integrated processing facilities have ensured that a variety of processed end products can be made from these basic raw materials.

Medicinal preparations from coconut include virgin coconut oil (Sharma, 1996; Winarsi *et al.*, 2008), and more recently, interest in coconut products has grown due to its potential antiviral properties and its value in the treatment of coronavirus disease 2019 (COVID-19; Dayrit and Newport, 2020).

At present, coconut cultivation is witnessing a decline in the traditional coconut-growing areas due to a variety of factors including pest and disease outbreaks, prevalence of senile palms, land conversion to other high-value crops, outdated production techniques, lack of crop management, and cultivation by small and marginal resource-poor coconut farmers. The low income of the farmers and the meagre profit of the traditional coconut industries can cause irregular and inadequate supply of coconut raw materials. Sole cropping of coconut, ineffective supply chains, product price volatility, limited product diversification, and the low-quality and non-utilization of certain by-products are some of the reasons for the decline in coconut production. Increased coconut productivity could be achieved by adopting more effective crop-management practices.

3.2 Management Practices for Increased Production

3.2.1 Climate requirements

Coconut is a tropical crop that thrives well in humid environments. It grows well within a latitude of 20° on either side of the equator and up to an altitude of 600–900 m above mean sea level (Persley, 1992). The required temperature range is between 20°C and 32°C, with a mean annual temperature of 27°C. A low variation between daytime and night-time temperatures is best for coconut growth and development. At extreme temperatures, inflorescence drying occurs, causing decreased fruit production. The relative humidity should be around 80–85% (Debmandal and Mandal, 2010), as it influences pollination and fertilization (Krishnakumar, 2017). High humidity throughout the life of the palm also inhibits transpiration, limits nutrient uptake, and increases the risk of incidence of pests and diseases (Magat, 1999a). A well-distributed annual rainfall of 1300–2300 mm is more crucial than the overall volume of rainfall for coconut, and there should not be a dry spell longer than 3 months (Rajagopal *et al.*, 1996). Healthy palm growth requires nearly 2000 h of sunshine year^{-1} (Ohler, 1999).

3.2.2 Site selection for coconut cultivation

Sandy loam, coastal sandy, laterite and alluvial soils with good drainage and a lack of a hard substratum within 2 m of the surface, which allows unrestricted root growth and aeration, are most suitable for growing coconut (Murray, 1977). The depth of soil is an important physical property as it provides improved physical stability and strong anchorage against wind gusts (Ganarajah, 1953; Kushwah *et al.*, 1973). Coconut palms thrive under a wide range of pH values (5.0–8.0), but the ideal pH range is 6.4–7.0 (Manciot *et al.*, 1979). Coconut can resist salt concentrations of up to 0.6% as they are semi-halophytic, and it is safe to irrigate with up to 10% seawater (electrical conductivity of 8.32 dS m^{-1}) (Hebbar *et al.*, 2021). However, using seawater exclusively for coconut is harmful to its growth (Sankaranarayanan *et al.*, 1958). The soil should be fertile, containing all the major and trace nutrient elements and at least 1% organic matter (Manciot *et al.*, 1979) but devoid of toxic elements (Okeri *et al.*, 2007).

3.2.3 Varietal selection

Coconut offers a huge variability from tall to dwarf varieties, large- to small-sized fruit, susceptibility and resistance to drought and salinity, and high to low oil and fibre content, among other characteristics. There are two primary types of the species: tall (*Cocos nucifera* var. *typica*) and dwarf (*Cocos nucifera* var. *nana*) (Satyabalan, 1997). Improved varieties are available that are suitable for various agroecological regions, production of products such as copra and tender coconut, biotic and abiotic stress tolerance, or having ornamental value.

3.3 Optimum Spacing and Planting Geometry

Land preparation for coconut planting should consider the topography, soil type and environmental factors. In areas with a high-water table or in low-lying areas, planting should be done on a mound, whereas on slopes and steep terrain, coconut should be planted on contour terraces. Young coconut seedlings are transplanted due to the palm's ability to produce adventitious roots (Tomlinson, 2006), and the ability of the palm's trunk to store a large quantity of water (Holbrook and Sinclair, 1992) and sugar (Mialet-Serra *et al.*, 2005). However, it is imperative to use vigorously growing seedlings that are 7–8 months old and that have been raised from fully developed, mature fruit. An optimum planting density is necessary to harvest the maximum sunshine during their active growth phase. It is advised to space dwarf palms approximately 7–9 m apart, hybrid palms at 8–9 m and tall palms at 9–10 m. Under the square system of planting, 100–200 palms ha^{-1} can be planted, and under the triangle system of planting, 115–236 palms ha^{-1} can be established (Table 3.3). The hedge system of planting utilizes spacing of 6.5 × 9.5 m (Magat, 1999a).

Deep and wide pits of 1.5 × 1.5 × 1.2 m (length × breath × depth) are recommended for laterite soils with a rocky substratum, 1.0 m^3 pits are recommended for loamy soils with a low water table, and for sandy soils, pits should be 0.75 m^3. Before planting, the pits are filled with two layers of coconut husks to ensure soil moisture conservation. A mixture of topsoil and farmyard manure is placed over the top to cover 60% of the pit. Similarly, to loosen laterite soils, 2 kg of common salt is also applied (Krishnakumar, 2017). When planting, the seedling is placed in the centre of the pit (Fig. 3.1), so that its collar region is above the soil surface, which avoids soil entering the leaf axils. The soil around the seedlings must then be tightly pressed. The optimum depth for the fruit body of the seedling is 5–8 cm below the filled soil level in the pit.

Table 3.3. Square and triangular systems for planting coconut seedlings and the planting density created. From Kerala Agricultural University, 2011, with permission.

	Population density (palms ha^{-1})	
Spacing (m)	Square method	Triangular method
7.6	177	198
8.0	156	180
8.5	138	160
9.0	134	143
10.0	100	115

Fig. 3.1. Planting a mature coconut seedling at the centre of a pit (1 m^3) filled with coconut husks (with concave surface facing up) in two layers at the bottom, together with a mixture of topsoil and farmyard manure. Photo courtesy of R. Bhat.

To conserve rainwater, a little depression in the soil surface should be made towards the crown base (Santos *et al.*, 1996).

3.4 Care of Juvenile Palms

The palms should be given proper care during the early years of their growth (the first 5 years). To safeguard the seedlings from strong wind, staking should be provided. To protect from sun scorching, there should be adequate shade with plaited coconut leaves or other suitable shading materials (Fig. 3.2). During summer, to maintain adequate soil moisture, irrigation needs to be provided at the rate of 45 L of water once every 4 days, along with proper drainage (Krishnakumar, 2017). If a drip system of irrigation is employed, 10 L of water day^{-1} is recommended. Periodic weeding should be done, and during heavy rains, soil covering the seedling's collar region should be removed. Gradually, the basin region around the palm is widened, manure is applied and the basin is topped up with soil. By doing this, a basin radius of 1.8 m will be formed by year 4 of planting. Insect and disease infection of the palms should be monitored regularly and corrective measures implemented.

3.5 Fertilization for Palm Nutrition

Coconut produces one leaf and one inflorescence every month. Fertilizing at the right time and in the right amount can greatly enhance fruit production. For young palms, the nutrients required are, in the order of importance, nitrogen > phosphorus > potassium > magnesium (N > P > K > Mg), but for the adult bearing palms, K > Mg > N > P is the order of nutrient priority (Tennakoon, 2003).

To promote vegetative growth, early flowering and bearing and to achieve a higher yield, regular fertilization from the first year of planting is essential. A soil test is advised once every 3 years to determine the type and rate of chemical fertilizers. After 3 months of planting, one-tenth of the recommended chemical fertilizer for an adult palm should be provided. In the second year of planting, one-third of the recommended dose of fertilizer for an adult palm is recommended and applied in two stages. Twice the quantity of the second-year dose may be applied

Fig. 3.2. A recently planted young coconut seedling, protected and shaded by a dry coconut leaf barrier. Photo courtesy of R. Bhat.

in the third year, and from the fourth year onwards, the full recommended dose of fertilizer for an adult palm should be applied (Table 3.4). The general recommended dose of fertilizer per palm year^{-1} is 500:320:1200 g of N, phosphorus pentoxide and potassium oxide (Subramanian *et al.*, 2005).

Fertilizer is ideally applied in two dosages: one-third of the recommended dose during pre-monsoon season and the remaining two-thirds during the post-monsoon season in the circular basin (now 1.8 m in radius and 20 cm deep). It should be followed by a spread of 50 kg of green leaf manure, compost or farmyard manure. If irrigation is possible, it is preferable to apply fertilizer in more split intervals, such as March, June, September and December. Fertilizer recommendations for different soil types are given in Table 3.5.

For acidic and light sandy soils, along with recommended manures, the addition of dolomite or limestone at a rate of 1 kg per palm year^{-1}, or magnesium sulfate at a rate of 0.5 kg per palm year^{-1} should be done. Lime or dolomite should not be applied along with chemical fertilizers and should be applied to the soil and forked in before commencement of the monsoon. Based on soil test value, micronutrients should be applied carefully. It is feasible to apply water-soluble fertilizers such as urea, phosphoric acid (commercial grade), diammonium phosphate and potassium chloride by drip irrigation (fertigation) in six equal split doses (Subramanian *et al.*, 2012).

To improve the soil quality and nutrient recycling, organic manures such as compost, farmyard manure, vermicompost, coir pith compost or green leaf manures are recommended. It is recommended to apply 10 kg of organic manure per palm as compost, vermicompost or farmyard manure, 6 months after planting in the main field. In year 2, this can be increased to 20 kg per palm, in year 3 to 30 kg per palm, in year 4 to 40 kg per palm, and in year 5 onwards to 50 kg per palm.

Coconut leaves, spathe, bunch waste and husk and biomass of other intercropped species available from the coconut garden can be recycled *in situ* utilizing earthworms (*Eudrilus* spp.) to efficiently break it down to form a granular and nutrient rich vermicompost (1.8% N, 0.21% P, 0.16% K and 17.84% organic matter) to ensure

Table 3.4. Fertilizer recommendation for coconut (g per palm) for the first 4 years after planting and the season in which to apply.

	May–June			September–October		
Year of planting	N	P_2O_5	K_2O	N	P_2O_5	K_2O
Year 1		Planting		110	200	225
Year 2	110	200	225	240	400	450
Year 3	240	400	450	480	800	900
Year 4 onwards	370	600	670	715	1000	1330

K_2O, potassium oxide; N, nitrogen; P_2O_5, phosphorus pentoxide.

Table 3.5. Fertilizer recommendations for mature coconut palms growing in different agronomic and soil conditions.

	Quantity (g per palm year^{-1})		
Type of management	N	P_2O_5	K_2O
Blanket recommendation			
For maintained palms			
Moderately maintained palms	340	170	680
Well-maintained palms	500	320	1200
For reclaimed clay-rich soils	250	350	900
For red loam soils	680	230	900
High yielding and hybrid palms			
Palms maintained under irrigation	1000	500	2000
Palms maintained under rainfed conditions	500	320	1200

K_2O, potassium oxide; N, nitrogen; P_2O_5, phosphorus pentoxide.

slow nutrient release throughout the year. In the coconut palm basin areas, legume cover crops such as wild groundnut (*Calopogonium mucunoides* Desv.) or giant sensitive plant (*Mimosa invisa* Martius ex Colla) can be grown, and after 140 days, about 20.5 and 12.8 kg of the above-ground wet biomass can be incorporated into the basin area, and 134.8 g and 70.0 g of N per basin, respectively, is added (Thomas *et al.*, 2001). In the Philippines, multi-nutrient fertilizers at the rate of 2–3 kg per palm $year^{-1}$, containing a minimum of 15% Cl^-, 0.2% B and 5% S, is applied as a blanket recommendation to optimize coconut yields and achieve higher economic yields. In addition to this, 1–2 kg sodium chloride is recommended for inland regions and other El Nino-vulnerable areas (Magat, 2003).

3.6 Optimizing Water Use

Moisture stress is one of the most detrimental environmental factors for coconut growth and development as it influences a variety of biochemical and physiological processes in palms (Repellin *et al.*, 1997; Lambers *et al.*, 1998; Rajagopal and Kasturi Bai, 1999; Prado *et al.*, 2001; Passos *et al.*, 2005; Gomes *et al.*, 2008; Ramesh *et al.*, 2020).

Palms raised under rainfed conditions will experience dry spells during summer, causing leaf drop, bending and breaking of leaves, button shedding and fall of immature fruits. They may take 3–4 years to recover once soil moisture conservation practices are adopted. Flooding, basin irrigation, sprinklers, spray gun irrigation and drip irrigation are the most widely used irrigation systems in coconut gardens. In the flood system, considerable amounts of water get wasted and this can result in poor soil aeration, mainly in heavy clay soils, causing water stagnation, which in turn leads to spread of soil-borne pathogens. Under a basin irrigation system, irrigation channels are made between the two rows of coconut, and to connect each basin, subchannels are prepared and water is supplied to the basin via these channels. Here, water loss occurs due to evaporation, seepage and deep percolation. Sprinkler or spray gun irrigation can be adopted in coconut-based intercropping systems where the whole surface needs to be irrigated (Fig. 3.3). Drip irrigation is the most efficient microirrigation technique and ensures higher

Fig. 3.3. A sprinkler irrigation system applying water to the intercrop in a coconut mixed farming system. Photo courtesy of R. Bhat.

water-use efficiency while saving labour and energy. For installation of drip irrigation, four pits of size 30 cm^3 diameter can be opened at 1 m distances from the palm trunk. In each pit, a plastic pipe of approximately 16 mm diameter can be installed with drippers placed so that water will penetrate 30 cm below the soil surface.

The water requirement of the cropping system is determined by evaporation (E_o) and the crop coefficient (K_c), and is expressed as the rate of evapotranspiration (ET) in mm day^{-1} or mm per period. Coconut is extravagant in its water requirement, with the volume depending on the age of the palm, its height and leaf area, the type of soil and other environmental conditions (Nogueira *et al.*, 1998). In young palms (up to 3 years old), the water requirement is lowest. About 1591 and 1533 mm of water $year^{-1}$ is required to irrigate to 80–100% and 60–100%, respectively (Shanthamallaiah *et al.*, 1978). After the juvenile phase, the growth and reproductive stages occur simultaneously, and the palms require more readily available moisture throughout their life cycle (Murray, 1977). An adult palm requires 600–800 l of water once every 4–7 days. Copeland (1906) suggested that the daily water requirement of a coconut palm is 25–45 l per palm. Azevedo *et al.* (2006), in north-east Brazil, applied 50, 100 or 150 l per palm day^{-1} (equivalent to 1.0, 2.0 and 3.0 mm of rainfall day^{-1}, respectively) to 2–6-year-old dwarf palms growing on a sandy soil, and found no yield differences among the treatments, but application of irrigation water in excess of these rates increased the volume of coconut water by 16% per fruit. Arachchi (1998) developed a drip irrigation system for coconut palms growing on a sandy soil in Sri Lanka and recommended 30 l of water h^{-1} for 2.5 h and from each of four drippers. It was also noticed that coconut husk or coir dust mulching in a 1.75 m radius around the coconut trunk reduced evaporation from the soil surface, and increased soil water retention and water availability to the palms.

In India, it has been reported that drip irrigation daily at 66% of the pan evaporation rate is adequate (~32 l per palm day^{-1}) during December to May (summer months) with 2–4 l h^{-1} per dripper, saving 34% water. In sandy soils, six drippers are required, whereas for other soil types, four would suffice (Dhanapal *et al.*, 2007). A drip irrigation system design is shown in Fig. 3.4.

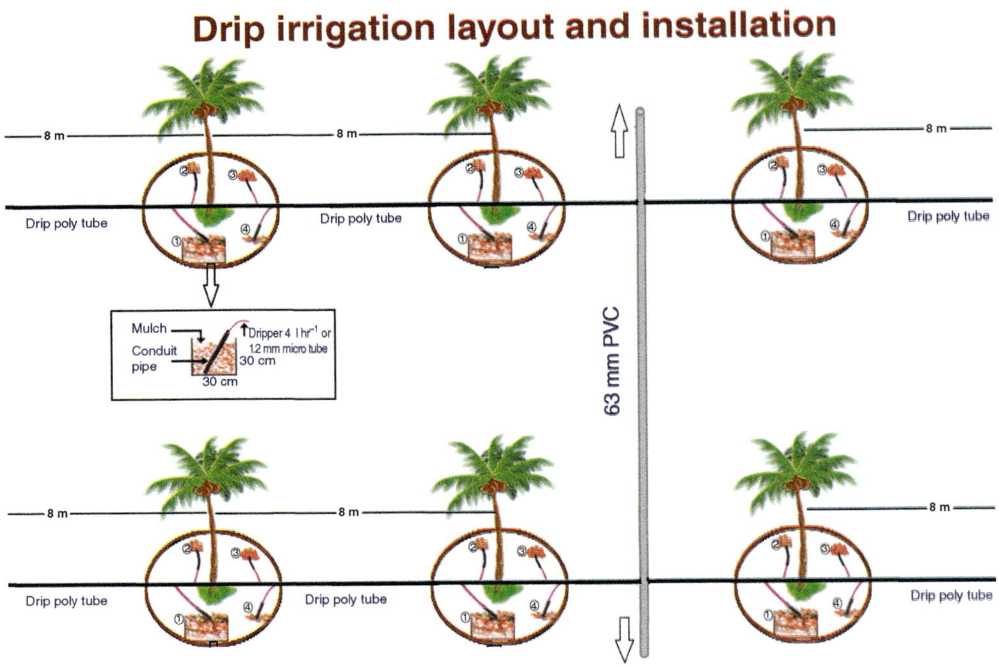

Fig. 3.4. A drip irrigation system layout in coconut showing adoption of four drippers 1 m away from the palm basin. From Subramanian *et al.*, 2018, with permission.

3.7 Needs-based Soil and Moisture Conservation Measures

Mulching with easily available materials such as coconut leaves and husk applied in two to three layers and coir pith at a thickness of 10 cm is a viable way to conserve soil moisture in coconut gardens. It not only reduces the soil temperature and evaporation rate but also creates a conducive environment for good root growth and growth of desirable soil flora and fauna. The appropriate period for mulching is before the end of the monsoon when the topsoil still has some moisture remaining. Coconut husk burial in trenches (50 cm wide × 50 cm deep and a convenient length), dug within coconut interspaces in layers, with the bottom layer facing up and top layer facing down, is another method to conserve moisture. When the land has a mild slope of 15–20%, a half-moon shaped barrier (bund) can be built in the coconut basin to prevent water runoff and to improve water penetration into the soil. Pineapple (*Ananas comosus* (L.) Merr.) grown in interspaces protects the bunds and provides stability. When the land is characteristically slopy, the trenches of 50 × 50 cm and of desired length could be prepared between a couple of rows of coconut palms. Catch pits (1.5 m long × 0.5 m wide × 0.5 m deep), empty or filled with coconut husks, are another land configuration tool advised for soil moisture conservation.

3.8 Crop Diversification

Crop diversification and intensive cultivation is suggested in coconut gardens to enhance palm productivity as well as system productivity. This may be achieved by adopting intercropping or mixed-cropping practices, or by the introduction of other enterprises such as dairy cattle or poultry into the system. The components of a successful production system could be: (i) intercropping (short duration/seasonal crops, or long-duration crops) and multistorey or mixed cropping; (ii) livestock rearing (e.g. poultry, cattle, goat, swine, game birds); (iii) a nursery to produce planting materials such as fruit crops, horticultural or ornamental plants, and other high-value crops; and (iv) pisciculture.

The phyllotaxy of the coconut palm ensures adequate sunlight for the intercrops. Thus, the coconut-based cropping system can be designed to utilize the vertical and horizontal space dimensions between coconut palms to maximize crop biomass productivity by effectively utilizing solar energy and nutrients in the different layers of the soil (Magat, 1990). Coconut-based mixed-cropping systems positively influence the microclimatic conditions by modifying air and soil temperatures, vapour pressure deficits and soil moisture content (Ranasinghe *et al.*, 2014).

The lifespan of a coconut palm is categorized into three phases depending on the growth of its canopy and the amount of light penetration through the canopy (Table 3.6):

1. From planting to full canopy development (0–8 years). In this phase, a major portion of the sunlight is transmitted to the soil because the canopy is not fully developed, and as the coconut palm canopy development progresses, the percentage utilization of sunlight progressively increases and suitable intercrops can be introduced into the inter-row spaces.
2. Juvenile phase (8–25 years). In this phase, the canopy is fully developed, and there is moderate to poor light penetration through the canopy; hence, only shade-loving intercrops can be introduced.
3. Mature palms (>25 years). In this phase, there will be a reduction in crown size and the trunk of a coconut tree reaches such a height that it will not obstruct the penetration of light to the interspaces, so this phase is the ideal time to raise annual or perennial intercrops as well as a multistorey cropping system.

A high-density multi-species cropping system (HDMSCS), practised in a coconut garden, enables farmers to meet their diverse requirements for food, fuel, fodder and timber, as well as their economic well-being. The HDMSCS envisaged in India during the 1980s, followed by further refinement, is a versatile, economically practical, environmentally sustainable system with an underlying theme of effectual use of natural resources. Various models have been developed to suit the different agroecological zones and requirements of households in other countries. The component crops can be selected depending on the nutritional preferences and economic requirements of the family involved in farming.

Table 3.6. Suitable intercrop species for different growth phases of coconut. Adapted from Magat, 2003a.

Phase	Age of coconut crop	Sunlight availability	Suitable intercrops
1	Planting to 8 years	High to moderate	Field crops: maize, upland rice, beans, cowpea, mung bean, groundnut Fruit crops: citrus, papaya, watermelon, pineapple, banana Vegetables: sweet potato, tomato, cabbage, aubergine, sweet pepper, hot pepper, okra
2	8–25 years	Moderate to low	Multiple crops: black pepper, cacao, coffee, tomato, vanilla, ginger, langsat, rambutan, durian, mangosteen, gmelina
3	26–60 years	High	Field crops: maize, upland rice, groundnut, mung bean, cowpea, beans Vegetables: tomato, aubergine, cabbage, sweet pepper, hot pepper, okra, ginger, sweet potato, taro, cassava, ube Beverage crops: coffee, cacao Fruit crops: langsat, rambutan, durian, mangosteen, citrus (pomelo, calamansi) Others: gmelina, ramie, abaca

An integrated farming system comprises a wide range of resource-saving practices to attain economic benefits with high and sustainable farm outputs, but unlike intensive farming, it has minimal or no negative impact on the environment (IFAD, 2005). This farming system has many benefits, such as increased farm income and maintaining soil fertility and productivity through organic waste recycling. It yields nutritious food, helps in protecting the environment, enables effective recycling of waste, reduces the production cost, and offers stable income and employment.

The social benefits of coconut-based mixed farming systems include a nutritional stock of fresh coconuts and products, and other crops such as cereals (providing quality carbohydrates, proteins, fat and oils), root crops (a source of carbohydrates and minerals), legumes (serving as a source of quality vegetarian protein, fat and oils), fruit crops, leafy and fruit vegetables, spices (providing vitamins, minerals and dietary fibres), coffee (*Coffea* spp.) and cocoa (*Theobroma cacao* L.; good and natural source of antioxidants and phenolics), fibre crops (serve as raw material for textile, paper and packaging industries), and wood and timber (source of housing materials) (Maheswarappa *et al.*, 2017).

The ecological benefits include stable soil and land for intensive and sustainable agricultural production. These benefits are mainly attributed to the efficient utilization of the land resources through the optimization of moisture use and retention capacities, maximization of solar energy capture, minimal soil erosion and nutrient losses, and enhancement of soil fertility build-up through the generation of more biomass (Maheswarappa *et al.*, 2017). The functions and services of a coconut-based mixed farming system are given in Table 3.7.

Soil fertility can be improved by farmers tethering two cattle each night to a palm where a shallow trench has been cut into the soil to capture dung and urine. This method has been referred to as 'compost made in the foot of the palm' (Schrader, 1950). In some coconut lands, cattle are kept in a shed during the night. The straw on the floor that becomes mixed with dung and urine can be raked and collected in a pit outside the shed. The straw in the pit can be mixed with manure and sprinkled with water twice a week to promote decomposition. This has been referred to as the 'pen manure' approach (Schrader, 1950). In this way 10 t of pen manure could be prepared per head of cattle annually, which could be used to fertilize 200 coconut palms.

In Sri Lanka, pasture plants such as legumes like stylosanthes (*Stylosanthes* spp.), palisade grass (*Urochloa brizantha* (Hochst. ex A. Rich.) R. Webster), para grass (*Urochloa milliformis* (Forssk.) Stapf.), green panic grass (*Panicum maximum* Jacq.), knotroot bristlegrass (*Setaria geniculata* (Willd.) P. Beauv.), Egyptian crowfoot grass (*Dactyloctenium aegyptium* (L.) Willd), summer grass (*Alloteropsis cimicina* (L.) Stapf), false couch

Table 3.7. A broad list of functions and services provided by coconut-based mixed-cropping systems. From Maheswarappa *et al.*, 2017, with permission.

Ecosystem function	Ecosystem services	Ecosystem services in coconut-based mixed-cropping systems
Modification of atmospheric conditions	Gas regulation	Carbon sequestration potential
Retention of ground water	Supply of water	Enhanced subsoil water infiltration and retention of moisture
Preservation of soil	Control of erosion, sediment retention, and nutrient losses	Soil loss due to runoff is prevented; further silt and clay transported is stored
Soil genesis	Formation of soils	Aids the process of rock weathering and helps in accumulation of organic matter
Soil water conservation and suppression of weeds	Weed control, reduction of soil evaporation and nutrient recycling	Coconut fronds, husk and legumes grown between coconut help largely in the suppression of weed growth and conservation of soil moisture
Stabilization of ecosystem integrity in response to environmental stress	Disturbance regulation	Regulation of farm incomes safeguarding against typhoon damages

grass (*Digitaria longiflora* (Retz.) Pers.), crown grass (*Paspalum* spp.), jungle rice grass (*Echinochloa colonum* (L.) Link), roadside flatsedge (*Cyperus zollingeri* Steud.), finger grass (*Digitaria marginata* (Retz.) Hook. F. ex Ridl.), bahia grass (*Paspalum commersonii* Lam.) and Indian muraina grass (*Ischaemum ciliare* Retz.) and various legume species such as touch-me-not plant (*Mimosa pudica* L.), centro or butterfly pea (*Centrosema pubescens* Benth.), tropical kudzu (*Pueraria phaseoloides* (Roxb.) Benth.) and three-flower ticktrefoil (*Desmodium triflorum* (L.) DC) have all been successfully cultivated under coconut (Ferdinandez, 1973).

3.9 Harvesting for Different Purposes

Coconut palms produce fruit throughout the year, and it takes 12 months for their complete development and maturity from pollination to harvest. It is advisable to harvest coconuts upon attaining full maturity for culinary purposes and for making copra. If harvested fruits include those at the colour break stage, they should be segregated and stored for about 2–4 weeks to reach the maturity required for maximum copra and oil yield (Punchihewa and Arancon, 1999). According to the International Coconut Community, the harvesting interval should be 45 days to obtain high copra and oil recovery. However, for tender coconut production, 7–8-month-old fruit are required, so more frequent harvesting is necessary. Eleven-month-old coconut fruit have the best quality fibre and hence, for coir production, the fruit are harvested while still green. Therefore, based on the intended use, the number of harvests per year may range from four to ten.

Various methods for harvesting have been adopted in different countries, and sometimes within different regions of a country. The climber can climb the palm with or without a climbing device. To make climbing easy, some palms have grooves carved into their trunks. Although it is risky, it is still the most widely used method. Using palm-climbing devices reduce the risk when climbing tall palms. In India, three types of harvesting are recommended: (i) the manual climbing method; (ii) using a power-operated ladder; and (iii) using a climbing gadget (Thamban and Leena, 2013). Plucking with long bamboo poles is also practised in some parts of India. In the pole method of harvesting, long bamboo poles, with a harvesting scythe attached at the end of the pole, is used. This method is faster and more efficient as well as less risky compared with climbing methods.

In Sri Lanka, the methods of harvesting are: (i) using a long bamboo pole with an attached knife; (ii) manual climbing; (iii) using a bamboo ladder, or (iv) using a climbing gadget (Thampan, 1994). In Malaysia, Indonesia and Thailand, trained monkeys (northern and southern pig-tailed macaques, *Macaca leonina* Blyth. 1863 and *M. nemestrina* L.) are sometimes employed (La Rue, 1919; Azis *et al.*, 1980; Sponsel *et al.*, 2002). The monkeys are trained to pluck the mature coconuts. The monkey will twist each fruit in the bunch using both hands and a foot. One of the disadvantages of this method is that the bunch stalk will remain in the tree until it falls naturally.

Another method is connecting the crowns of palms by ropeways, which enable the climber to reach nearby palms without climbing down and up each palm. These ropeways are constructed in such a manner that the climber can walk along these ropeways after reaching the crown of one tree. This method is commonly used when the entire plantation is used for toddy tapping. Another method followed in parts of Africa and the Pacific Islands is to allow the mature fruit to fall naturally to the ground (Punchihewa and Arancon, 1999). However, this causes the palm to produce less fruit and there will be a considerable reduction in annual yield from each palm. For tender coconuts, the bunches should be harvested carefully by tying with a rope and lowering to the ground. In some places, water drains are dug out in between coconut rows and this can provide a cushion effect for fruits falling to the ground.

Coconut farming communities have been encountering a scarcity of workers skilled in the harvesting of fruit so there is a genuine need to develop efficient coconut-harvesting robots or other technologies to support revival of the coconut industry. Several groups such as Igus® Robot Control from Germany, Prakasan Thattari from the southern Indian State of Kerala (Abraham *et al.*, 2014) and CocoBot from Amrita Vishwa Vidyapeetham University, Kerala (Megalingam *et al.*, 2013) are working on these types of technologies. In Indonesia, an end-to-end autonomous coconut-harvesting robot system, involving mechanical climbing and harvesting, has been developed (Wibowo *et al.*, 2016). Although these systems have the potential to revolutionize harvesting, none of these technologies has yet been widely adopted by the coconut industry.

3.10 Yield

The yield of coconut palm varies depending on rainfall, soil type, seasonal conditions, variety, age of the palm, fertilizer use, intercropping practices, occurrence of pests and diseases, and other factors. The average yield of a coconut palm is around 80–100 fruit per palm year^{-1}. In India, well-maintained coconut gardens on loamy soils, under rainfed conditions, yield about 60–100 fruit per palm year^{-1}, whereas in alluvial soils the yield would be much higher. In soils with poor nutrient status or gardens that are neglected, the yield can be drastically low (≤20 fruit per palm year^{-1}). Generally, a dwarf variety produces 70–80 fruit per palm year^{-1}, and a tall variety produces 80–100 fruit per palm year^{-1} under good management practices. For hybrids, the yield may vary from 100 to 130 fruit per palm year^{-1}.

3.11 Conclusion

At present, coconut production is witnessing a decline in the traditional coconut-growing areas due to a variety of factors. The farmers involved gain only a low income and the traditional coconut industries are achieving a meagre profit. Increased coconut productivity through widespread adoption of sound crop-management practices, planting of improved varieties, the use of proper planting methods with adequate nutrient and water applied, and using a coconut-based mixed-cropping/farming system model may all enhance the income and livelihoods of farmers and help revive the industry.

3.12 References

Abraham, A., Girish, M., Vitala, H.R. and Praveen, M.P. (2014) Design of harvesting mechanism for advanced remote-controlled coconut harvesting robot (A.R.C.H-1). *Indian Journal of Science and Technology* 7, 1465–1470.

Angeles, J.G.C., Lado, J.P., Pascual, E.D., Cueto, C.A., Laurena, A.C. *et al.* (2018) Towards the understanding of important coconut endosperm phenotypes: is there an epigenetic control? *Agronomy* 8: 225.

Arachchi, L.V. (1998) Preliminary requirements to design a suitable drip irrigation system for coconut (*Cocos nucifera* L.) in gravelly soils. *Agricultural Water Management* 38, 169–180.

Azevedo, P.V., Sousa, I.F., Silva, B.B. and Silva, V.P.R. (2006) Water use efficiency of dwarf-green coconut (*Cocos nucifera* L.) orchards in northeast Brazil. *Agricultural Water Management* 84, 259–264.

Azis, H., Liyanage, D.V. and Davis, T.A. (1980) Coconut harvesting practices in Indonesia: role of the pig-tailed macaque monkey. *Pemberitaan Lembaga Penelitian Tanaman Industri* 36, 58–73.

Chan, E. and Elevitch, C.R. (2006) Cocos nucifera *(Coconut).*In: Elevitch, C.R. (ed.) *Species Profiles for Pacific Island Agroforestry*. Permanent Agriculture Resources, Holualoa, Hawaii.

Copeland, E.H. (1906) On the water relations of the coconut palm (*Cocos nucifera*). *Journal of Science* 1, 6–57.

Dayrit, F.M. and Newport, M.T. (2020) The potential of coconut oil and its derivatives as effective and safe antiviral agents against the novel coronavirus. *Indian Coconut Journal* 62, 21–23.

DebMandal, M. and Mandal, S. (2010) Coconut (*Cocos nucifera* L.: Arecaceae): in health promotion and disease prevention. *Asian Pacific Journal of Tropical Medicine* 4, 241–247.

Dhanapal, R., Upadhyay, A.K., Subramanian, P., Sairam, C.V. and Maheswarappa, H.P. (2007) Influence of drip irrigation on growth and yield of coconut (*Cocos nucifera* L.) in laterite soil. *Journal of Plantation Crops* 32, 26–30.

FAOSTAT (2021) Statistical database. Food and Agricultural Organisation of the United Nations, Rome. Available at: www.fao.org/faostat/ (accessed 14 August 2023).

Ferdinandez, D.E.F. (1973) Utilization of coconut lands for pasture development. *Ceylon Coconut Planters Review* 7, 14–19.

Ganarajah, T. (1953) Planting coconuts on old rubber lands. *Ceylon, Coconut Quarterly* 4, 17–18.

Gomes, F.P., Oliva, M.A., Mielke, M.S., de Almeida, A.A.F., Leite, H.G. *et al.* (2008) Photosynthetic limitations in leaves of young Brazilian Green Dwarf coconut (*Cocos nucifera* L. 'nana') palm under well-watered conditions or recovering from drought stress. *Environmental and Experimental Botany* 62, 195–204.

Harries, H.C. (2012) Germination rate is the significant characteristic determining coconut palm diversity. *AoB Plants* 2012: pls045.

Hebbar, K.B., Santhosh, A., Sukumar, A.P., Neethu, P., Ramesh, S.V. and Selvamani, V. (2021) Effect of sea water substitution on growth, physiological and biochemical processes of coconut (*Cocos nucifera* L.) seedlings – a hydroponic study. *Scientia Horticulturae* 280: 109935.

Hidalgo, H.A. (2017) Market potential of pasteurized coconut water in the Philippine beverage industry. *International Journal on Advanced Science Engineering Information Technology* 7, 898–903.

Holbrook, N.M. and Sinclair, T.R. (1992) Water balance in the arborescent palm, *Sabal palmetto*. I. Stem structure, tissue water release properties and leaf epidermal conductance. *Plant Cell & Environment* 15, 393–399.

IFAD (International Fund for Agricultural Development) (2005) *Integrated Crop-Livestock Farming System*. Burkina Faso, Rome.

Ignacio, I.F. and Miguel, T.S. (2021) Research opportunities on the coconut (*Cocos nucifera* L.) using new technologies. *South African Journal of Botany* 141, 414–420.

Kerala Agricultural University (2011) *Package of Practices Recommendations: Crops*, 14th edn. Kerala Agricultural University, Thrissur, India.

Krishnakumar, V. (2017) Establishing coconut garden. *Indian Horticulture* 62, 23–28.

Kushwah, B.L., Nelliat, E.V., Markose, V.T. and Sunny, A.F. (1973) Rooting pattern of coconut (*Cocos nucifera* L.). *Indian Journal of Agronomy* 18, 71–74.

Lambers, H., Chapin, F.S. and Pons, T.L. (1998) Role in ecosystem and global processes. In: *Plant Physiological Ecology.* Springer, New York, pp. 495–517.

La Rue, C.D. (1919) Monkeys as coconut pickers. *Science* 50, 187.

Magat, S.S. (1990) Growing conditions and growth habits of coconut in relation to coconut-based farming systems. In: de Silva, S. (ed.) *Coconut Based Farming Systems. Proceedings of the XXVII COCOTECH Meeting,* 25–29 June, 1990, Manila, the Philippines, pp. 17–40.

Magat, S.S. (1999a) Coconut production and productivity: current status and research, development and extension (RD&E) needs in agriculture modernization. In: *National Coconut Research, Development and Extension (RD&E) Network Forum,* February 1999. Agricultural Research and Development Branch, Philippine Coconut Authority, Quezon City. Philippines.

Magat, S.S. (1999b) *Handbook of Fertilizer Recommendations for Coconut and Selected Intercrops.* Asian and Pacific Coconut Community, Jakarta.

Magat, S.S. (2003) Achieving coconut supply reliability through research-based crop nutrition management of coconut farms in the Philippines. *CORD* 19, 34–34.

Maheswarappa, H.P., Subramanian, P. and Chowdappa, P. (2017) Integrated coconut farming system for stable income. *Indian Horticulture* 62, 23–33.

Manciot, R., Ollagnier, M. and Ochs, R. (1979) Mineral nutrition and fertilization of the coconut around the world. Part I. *Oleagineux* 34, 499–515.

Mat, K., Abdul Kari, Z., Rusli, N.D., Che Harun, H., Wei, L.S. *et al.* (2022) Coconut palm: food, feed, and nutraceutical properties. *Animals (Basel)* 12: 2107.

Megalingam, R.K., Venumadhav, R., Pavan, A., Mahadevan, A. and Charly, T. (2013) Kinect based wireless robotic coconut tree climber. In: *3rd International Conference on Advancements in Electronics and Power Engineering, 2013*, Jan 8–9, Kuala Lumpur, Malaysia. ISBN: 978-93-82242-21-5.

Mialet-Serra, I., Clement, A., Sonderegger, N., Roupsard, O., Jourdan, C. *et al.* (2005) Assimilate storage in vegetative organs of coconut (*Cocos nucifera*). *Experimental Agriculture* 41, 161–174.

Murray, D.B. (1977) Coconut palm. In: Alvin, P.T. and Kozlowski, T.T. (eds) *Ecophysiology of Tropical Crops*. Academic Press, New York, pp. 24–27.

Nogueira, L.C., Nogueira, L.R.Q. and Miranda, F.R. (1998) Irrigaçao do coqueiro. In: Ferreira, J.M.S., Warwick, D.R.N. and Siqueira, L.A. (eds) *A Cultura do Coqueiro no Brasil.* EMBRAPA, Brasília, pp. 159–187.

Ohler, J.G. (1999) Climate and soils. In: Ohler, J.G. (ed.) *Modern Coconut Management: Palm Cultivation and Products*. Food and Agriculture Organization of the United Nations, Intermediate Technology Publications, London, pp. 35–53.

Okeri, H.A., Alonge, P.O. and Udoh, J.J. (2007) Assessment of nutritional status of soil supporting coconut (*Cocus nucifera*) cultivation in some localities of Edo State of Nigeria. *African Journal of Biotechnology* 6, 258–262.

Passos, C.D., Passos, E.E.M. and Prado, C.H.B.A. (2005) Comportamento sazonal do potencial hídrico e das trocas gasosas de quatro variedades de coqueiroanão. *Revista Brasileira de Fruticultura* 27, 248–254.

Patil, U., Benjakul, S., Prodpran, T., Senphan, T. and Cheetangdee, N. (2017) A comparative study of the physicochemical properties and emulsion stability of coconut milk at different maturity stages. *Italian Journal of Food Science* 29, 145–157.

Persley, G.J. (1992) *Replanting the Tree of Life: Towards an International Agenda for Coconut Palm Research*. CAB International, Wallingford, UK.

Prades, A., Dornier, M., Diop, N. and Pain, J.P. (2012) Coconut water uses, composition and properties: a review. *Fruits* 67, 87–107.

Prado, C.H.B.A., Passos, E.E.M. and de Moraes, J.A.P.V. (2001) Photosynthesis and water relations of six tall genotypes of *Cocos nucifera* in wet and dry seasons. *South African Journal of Botany* 67, 169–176.

Punchihewa, P.G. and Arancon, R.N. (1999) *Coconut: Post-harvest Operations*. Asian and Pacific Coconut Community, Jakarta.

Rajagopal, V. and Kasturi Bai, K.V. (1999) Water relations and screening for drought tolerance. In: *Advances in Plant Physiology and Biochemistry of Coconut Palm*. Asian and Pacific Coconut Community, Jakarta, pp. 56–70.

Rajagopal, V., Jacob, M., Mathew, J. and Shivashankar, S. (1996) Impact of dry spells on the ontogeny of coconut fruits and its relation to yield. *Plant Research and Development* 3, 251–255.

Ramesh, S.V., Hebbar, K.B., Rajesh, M.K., Sukumar, P.A., Gangaraj, K.P. *et al.* (2020) Transcriptome analysis of *Cocos nucifera* L. seedlings having contrasting water-use efficiency (WUE) under water-deficit stress: molecular insights and genetic markers for drought tolerance. *Biology and Life Sciences Forum* 4: 73.

Ranasinghe, C.S., Premasiri, R.D.N. and Pradeep A.P.C. (2014) Vulnerability of coconut-based agroforestry systems in Sri Lanka to climate change. In: Wachira, M.A., Rabar, B. and Magaju, C. (eds) *Abstracts of the 3rd World Congress on Agroforestry – Trees for Life: Accelerating the Impact of Agroforestry* 10–14 February, New Delhi, India, p. 27.

Rencoret, J., Ralph, J., Marques, G., Gutierrez, A., Martínez, A.T. *et al.* (2013) Structural characterization of lignin isolated from coconut (*Cocos nucifera*) coir fibers. *Journal of Agricultural Food Chemistry* 61, 2434–2445.

Repellin, A., Pham, A.T., Tashakorie, A., Sahsah, Y., Daniel, C. *et al.* (1997) Leaf membrane lipids and drought tolerance in young coconut palms (*Cocos nucifera* L.). *European Journal of Agronomy* 6, 25–33.

Sankaranarayanan, M.P., Verghese, E.J. and Menon, K.P.V. (1958) A note on the tolerance of salinity by coconut palm. *Indian Coconut Journal* 11, 133–139.

Santos, G.A., Batugal, P.A., Othman, A., Baudouin, L. and Labouisse, J.P. (1996) *Manual on Standardized Research Techniques in Coconut Breeding*. International Plant Genetic Resources Institute – Asia, the Pacific and Oceania, Singapore.

Satyabalan, K. (1997) *Coconut Varieties and Cultivars: Their Classification*. Asian and Pacific Coconut Community, Jakarta.

Schrader, R.S. (1950) Grass under coconuts. *Ceylon Coconut Quarterly* 1, 17–20.

Shanthamallaiah, N.R., Murugendra Kumar, C. and Krishna Manohar, R. (1978) Effect of coir dust mulching on water requirement and on the performance of coconut seedlings in the nursery. In: *Proceedings of PLACROSYM-1*. Indian Society for Plantation Crops, Kasaragod, pp. 261–266.

Sharma, P.V. (1996) *Classical Uses of Medicinal Plants*. Chaukhamba Vishwabharati, Varanasi, India.

Sponsel, L.E., Ruttanadakul, N. and Natadecha-Sponsel, P. (2002) Monkey business? The conservation implications of macaque ethnoprimatology in southern Thailand. In: Fuentes, A. and Wolfe, L.D. (eds) *Primates Face to Face: The Conservation Implications of Human–Nonhuman Primate Interconnections*. Cambridge University Press, Cambridge, pp. 288–309.

Subramanian, P., Dhanapal, R., Mathew, A.C., Palaniswami, C., Upadhyaya, A.K. *et al.* (2012) Effect of fertilizer application through micro-irrigation technique on nutrient availability and coconut productivity. *Journal of Plantation Crops* 40, 168–173.

Subramanian, P., Dhanapal, R., Sanil, P., Palaniswami, C., Sairam, C.V. *et al.* (2005) Glyricidia (Glyricidia sepium) as green manure in improving soil fertility and productivity of coconut under coastal littoral sandy soil. *Journal of Plantation Crops* 33 (3), 179–183.

Subramanian, P., Thamban, C., Josephrajkumar, A., Vinayaka, H., Hebbar, K.B. *et al.* (2018) *Coconut*. ICAR-Central Plantation Crops Research Institute, Kasaragod, India.

Tennakoon, A. (2003) *Soil Fertility Status and Fertilizer Recommendation for Coconut in Sri Lanka*. Soil and Plant Nutrition Division Coconut Research Institute, Lunuwila, Sri Lanka.

Thamban, C. and Leena, S. (2013) Climbing coconut trees… past and present. *Indian Coconut Journal* 56, 7–9.

Thampan, P.K. (1994) Plucking coconut: a dangerous and exhausting activity. *Cocoinfo International* 1, 24–27.

Thomas, G.V., Biddappa, C.C. and Prabhu, S.R. (2001) Evaluation of N_2-fixing cover legumes as green manures for N substitution in coconut (*Cocos nucifera* Linn.) palm. *Tropical Agriculture* 78, 13–18.

Tomlinson, P.B. (2006) The uniqueness of palms. *Botanical Journal of the Linnean Society* 151, 5–14.

Wibowo, T.S., Sulistijono, I.A. and Risnumawan, A. (2016) End-to-end coconut harvesting robot. In: *2016 International Electronics Symposium (IES)*. Institute of Electrical and Electronics Engineers, New York, pp. 444–449.

Winarsi, H., Hernayanti and Purwanto, A. (2008) Virgin coconut oil (VCO) enriched with Zn as immune-stimulator for vaginal candidiasis patient. *HAYATI Journal of Biosciences* 15, 135–139.

4 Coconut Production under a Changing Climate

Kukkehalli Balachandra Hebbar[1]*, Shunmugiah Veluchamy Ramesh[1], Sundaravelpandian Kalaipandian[2] and Steve W. Adkins[2]

[1]Division of Physiology, Biochemistry, and Post-Harvest Technology, Indian Council of Agricultural Research (ICAR), Central Plantation Crops Research Institute ICAR-CPCRI, Kasaragod, Kerala, India; [2]School of Agriculture and Food Sustainability, The University of Queensland, Gatton, Queensland, Australia

Abstract

The coconut palm (*Cocos nucifera* L.) experiences numerous environmental stresses throughout its lifetime. Climate change-induced global warming and its influence on current and future climate scenarios may pose additional constraints on productivity. Among the various concerns, high-temperature stress, moisture-deficit stress and sea-level rise are three major impediments to the future cultivation of coconut. Research efforts have begun to help understand how these changes affect the physiological, biochemical and anatomical aspects of coconut development and reproduction. Climate-smart cultivation of perennial crops such as coconut warrants the adoption of adaptive approaches that have been formed on sound scientific principles. This chapter considers the multifarious effects of climate change on coconut growth and development, the potential consequences of climate change-induced stresses, location of suitable production regions, research advancements to date that might help counteract the negative aspects of climate change, and the adaptive strategies to be adopted to negate these effects.

4.1 Introduction

Coconut (*Cocos nucifera* L.) is appropriately called 'Kalpavriksha' or the 'Tree of Life' as every part of the palm is useful. The palm has provided food, fibre, fuel, medicine, shelter and timber for at least 500,000 years (Foale, 2003). Considering the current area under coconut cultivation, which is over 12 million ha in over 90 countries (FAOSTAT, 2020), and its role in life and culture, it must be one of the most important crop plants sustaining human life. Coconut is known to thrive best between the latitudes of 23°N and 23°S of the equator. Beyond these latitudes, temperature becomes a major limiting factor to its growth. Coconut cultivation requires an evenly distributed annual rainfall (1300–2300 mm), an abundance of sunlight (250–350 Wm^{-2}), large numbers of sunshine hours (120 h month^{-1}), a warm average annual temperature (27°C), and a high ambient relative humidity (RH; >60%) (Murray, 1977). Coconut cultivation in many regions of the world is already frequently exposed to drought stress resulting from low soil moisture availability and low atmospheric RH (caused by high temperature and low RH), and such conditions are substantially impacting productivity.

Global climate change is now a serious threat to the world community. A report by the Intergovernmental Panel on Climate Change

* Email: hebbar.kb@icar.gov.in

 The Coconut: Botany, Production and Uses (eds S. Adkins *et al.*)
DOI: 10.1079/9781789249736.0004

and the third part of the Sixth Assessment Report states that if emissions of greenhouse gases continue to rise unabated, the plan to address climate change will not be adequate to address the high temperature-induced climate emergency (IPCC, 2022). In this context, decision making regarding the future cultivation of perennial plants, such as coconut, needs to be based on sound scientific principles involving system efficiency and economic feasibility. This will require multiple approaches. This chapter therefore discusses the holistic assessment of the consequences of climate change for the growth and development of the coconut palm, its potential effect on future production regions, especially in South Asia, and strategies for developing resilience to climate change-induced stress.

4.2 Impact of Climate Change Variables on Growth

Climate variables (i.e. temperature, rainfall, occurrence of soil salinity or alkalinity, and elevated atmospheric CO_2 concentration) all have serious bearings on the growth and development of coconut palms. These climate variables play an enormous role in defining the growth behaviour, phenology, physiology and ecological relevance of this crop, and ultimately affect fruit yield and quality. The positive impact of elevated atmospheric CO_2 concentration in enhancing the palm's biomass, as observed in many C3 crops (those that utilize the C3 carbon-fixation pathway as the sole mechanism to fix CO_2), can be offset by the plant's response to the steep rise in atmospheric temperature. For instance, growth of coconut within northern Indian conditions is luxuriant under normal conditions; however, the high temperatures (>40°C, under low RH conditions) that are now becoming more common during the summer season hamper pollination, fertilization and fruit set. Although it is difficult to efficiently forecast the impacts of various climate change variables on the growth and development of coconut palms because of its perennial nature, studies have begun to investigate the potential effects of severe climatic stresses on coconut growth and development.

The concurrence of water-deficit stress due to dry spells with the important fruit developmental stages, (i.e. from ovary to button development) will lead to severe yield losses (Rajagopal *et al.*, 1996). Tall coconut varieties, being cross-pollinated, are more affected by increases in atmospheric temperature, and their yield will be severely affected if high temperatures coincide with their peak flowering time. All coconut varieties are susceptible to external climatological influences from floral primordium initiation to the late development stages of fruit development, a period of around 44 months. Hence, any dry spells during this period will be detrimental to fruit production (Rajagopal *et al.*, 1996). Given this 44-month long reproductive phase of the palm (pre-fertilization of ~32 months and a post-fertilization period of 12 months), the effect of an extreme weather event can occur over a prolonged period. It has been shown that a severe water-deficit stress event during this time of development can result in yield reductions that can be detectable for the next 4 years (Naresh Kumar *et al.*, 2007).

All stages of the reproductive process in coconut are very sensitive to high-temperature stress conditions. Although the most sensitive stages are during the inflorescence development (i.e. the time of pollen and ovule formation, pollination and fruit formation), the prevailing climatic conditions during the next 3 months post-anthesis are also important and can affect the rate of fruit set and economic yield.

Hence, drought and/or high-temperature stress during any of these sensitive developmental stages can negatively affect fruit set and eventually yield. The germination of coconut pollen on the stigmatic surface and pollen-tube growth through the style are processes extremely sensitive to high-temperature conditions causing severe reductions in fertilization and fruit set. Similarly, conditions of water shortage will further affect the growth of the palm as more than 60% of the cropped area in India, for example, is under rainfed cultivation. The ultimate production potential of coconut is therefore dependent on the rate of fruit set, and this is highly susceptible to the weather conditions prevailing at that time (Ranasinghe *et al.*, 2015). In Sri Lanka, rainfall during the January–February period is desirable; however, high intense rainfall during the months of May–August will adversely affect the yield of coconut (Peiris and Peries, 1993). In addition, the amount of rainfall, the average temperature and the RH prevailing during the previous year have been shown to affect the yield of palms in the following year (Peiris and Thattil, 1998). Statistical model relationships developed for three distinct climate zones of Sri Lanka (dry, wet and intermediate zones) have revealed a close relationship between the frequency of extreme climatic conditions and the productivity of palms. More precisely, increased precipitation in the dry and intermediate zones, and high temperatures in the dry zone during the initial 4 months following inflorescence opening, will all adversely affect productivity (Pathmeswaran *et al.*, 2018). Thus, various climatic events could severely affect the fruit production potential, but this depends on the climatic zone where the palms are growing (Pathmeswaran *et al.*, 2018). In India, the effect of seasonal climatic variation on water-use efficiency of coconut seedlings showed there to be high water use in summer, even though the biomass of the seedlings remained unchanged. In addition, the relatively unresponsiveness of leaf stomata to high vapour-pressure deficits in dwarf compared with tall varieties was ascribed to their poor biomass gain (Hebbar *et al.*, 2022).

4.2.1 Drought stress

Drought affects almost all of the coconut-production regions of the world as the crop is generally grown in rainfed conditions. Besides the occurrence of drought years, coconut also encounters frequent dry spells during the summer production months. In India, analysis of the long-term rainfall records indicates a declining trend in precipitation and therefore a greater possibility of more dry spells in the future (Krishnakumar *et al.*, 2008). The anatomical, biochemical, physiological and molecular repercussions of moisture-deficit stress have been documented for coconut (Rajagopal and Kasturi Bai, 2002; Gomes *et al.*, 2007; Hebbar *et al.*, 2013a; Ramesh *et al.*, 2020). Among the many adaptive traits to drought stress, the roots of tall coconut varieties have a better ability to extract moisture from deep soil layers than dwarf varieties. Tall varieties also have naturally thicker epicuticular wax layers on their leaves, tracheids with greater scalariform thickenings, and substomatal cavities on their leaves to aid in water-loss prevention, compared with dwarf varieties. The severe decline in photosynthetic efficiency in both the tall and dwarf varieties during drought-stress conditions could be attributed to the limited CO_2 diffusion and to other non-stomatal factors including drought-induced damage to the photosynthetic apparatus (Kasturi Bai *et al.*, 2006; Gomes *et al.*, 2007; Hebbar *et al.*, 2013b). Reductions in photosynthesis under drought-stress conditions have been shown to be the result of the partial or complete closure of the stomata and to the subsequent decrease in the flow of CO_2 into the mesophyll cells. The closure of the leaf stomata under drought-stress conditions is associated with the production of the plant growth regulator abscisic acid, which transfers a stress signal to the whole plant, with one response being the closure of the stomata (Sunoj *et al.*, 2017). Tall varieties of coconut have better water-use efficiency owing to their lower leaf stomatal density, and their higher epicuticular wax thickness. In addition, a further study has reported that varietal variations occur with respect to water-use efficiency (Prado *et al.*, 2001).

4.2.2 High-temperature stress

High-temperature stress, if it occurs at any one of the reproductive stages of palm development, can cause a severe reduction in fruit yield. Such stresses enhance the processes of photorespiration and dark respiration, thereby bringing down the overall photosynthetic efficiency of the palm. Coconut, being a perennial palm, can flower at any time of the year, with pollen grains exhibiting a germination range from 23% to 40% (Ranasinghe *et al.*, 2010). However, high-temperature stress reduces pollen viability and therefore germination. Studies examining coconut pollen viability during the hot summer months have been used to help identify varieties that possess high-temperature stress tolerance (Ranasinghe *et al.*, 2015; Hebbar *et al.*, 2018).

The response of pollen to environmental temperature fluctuations provides further indications of a variety's tolerance to heat stress. Analysis of *in vitro* pollen germination of diverse coconut varieties (tall, dwarf and hybrid) showed wide variability with an average pollen germination of 48% (Hebbar *et al.*, 2018). The mean cardinal temperatures for coconut pollen grain germination were determined to be 42.3°C (maximum, T_{max}), 26.5°C (optimum, T_{opt}) and 12.9°C (minimum, T_{min}). Thus, the T_{opt} determined for coconut was much lower than that determined for many warm-season field crops (Hebbar *et al.*, 2018). Coconut varieties, such as Chowghat Orange Dwarf (COD), Laccadive Ordinary Tall (LCT), West Coast Tall (WCT) and various hybrids all exhibited a better tolerance to high-temperature stress than the sensitive Malayan Yellow Dwarf (MYD) (Fig. 4.1; Hebbar *et al.*, 2018). Breeding approaches to introduce high-temperature stress tolerance into popular varieties is a possible way of sustaining yield production under climate change. Categorization of

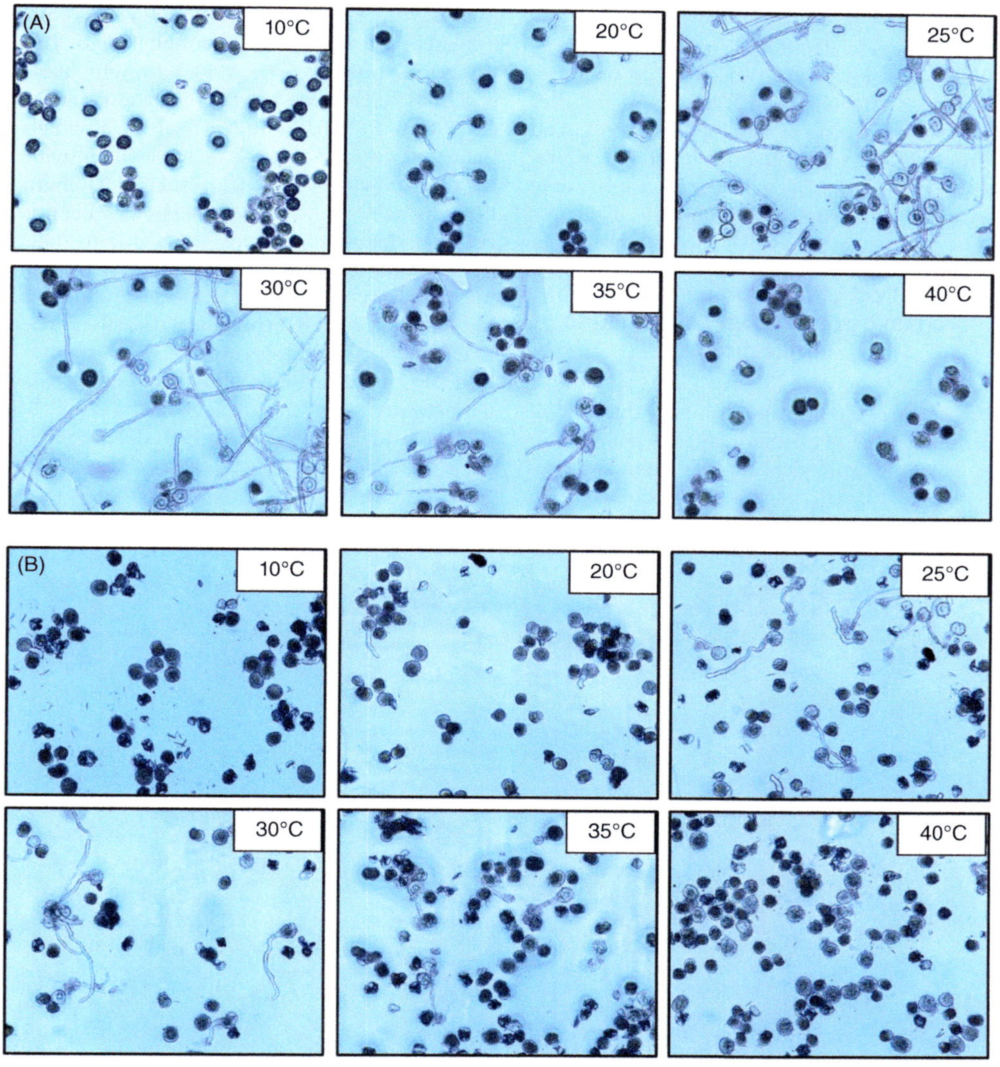

Fig. 4.1. *In vitro* pollen germination and pollen-tube growth of (A) the West Coast Tall variety and (B) the Malayan Yellow Dwarf variety. The images show the effect of temperature on pollen germination of a tolerant (A) and susceptible (B) variety. High levels of pollen germination and pollen-tube growth can be seen in (A) compared with (B) at the higher incubation temperatures. From Hebbar *et al.*, 2018, with permission.

coconut varieties for their tolerance to high-temperature stress showed that varieties possessing a high T_{opt} and a low T_{min} (i.e. COD, LCT, WCT and various hybrids) had a broader temperature adaptability, whereas varieties having a relatively high T_{opt} but also a relatively high T_{min} (i.e. Cochin China Tall, Chowghat Green Dwarf, Cameroon Red Dwarf, Gangabondam Green Dwarf and Philippines Ordinary Tall) had only a moderate tolerance to high-temperature stress. However, varieties that showed poor pollen germination (i.e. MYD), a low T_{max} and a high T_{min} exhibited the lowest tolerance to high-temperature stress (Hebbar *et al.*, 2018). In another study, the screening of varieties for their pollen germination identified hybrids involving several dwarfs × Sri Lanka Tall to be the most tolerant to high-temperature stress (Ranasinghe *et al.*, 2018).

4.2.3 Effect of high temperature on the progamic phase (from pollination to fertilization)

The temperature prevailing during the time of coconut fruit set is an important determinant for the eventual fruit yield, as high temperatures during this period can cause inflorescence and fruit shedding (Ranasinghe *et al.*, 2015). Thus, poor fruit production is observed when the period of inflorescence opening occurs from January to March, and from April to May in Sri Lanka and India, respectively, due to the high daytime temperatures (>33°C) at those times (Ranasinghe *et al.*, 2015). Under such high-temperature conditions, the activity of the enzymes ribulose 1,5-bisphosphate carboxylase/oxygenase (Rubisco; EC 4.1.1.39) and Rubisco activase are inhibited, which leads to a decrease in photosynthesis and a higher photorespiration rate (Sunoj *et al.*, 2017). The effect of high-temperature stress (T_{max} >33°C) during the progamic phase of flowering revealed a significant reduction in pollen-tube growth through the pistil and a low fertilization rate (Fig. 4.2; Hebbar *et al.*, 2020). In addition, high-temperature stress can advance nectar secretion and stigmatic receptivity, rendering the flower unsuited for insect pollination. Nevertheless, high RH conditions can offset these ill effects of high-temperature stress, implying that the climate change phenomenon of high-temperature stress in coastal regions could be mitigated due to the relatively high RH present in these regions (Hebbar *et al.*, 2020).

Fig. 4.2. *In vivo* experimental set-up for the induction of a high-temperature treatment on the inflorescence of the coconut variety West Coast Tall. The inflorescence was enclosed in a transparent chamber fitted with a microprocessor-controlled heater (A) and a mist generator (B) to maintain the temperature and humidity, respectively, inside the chamber. From Hebbar *et al.*, 2020, CC BY 4.0.

4.2.4 Elevated CO_2: stress and benefits

A global meta-analysis involving over 600 observations on the responses of various woody plant species to elevated atmospheric CO_2 concentrations has shown this effect to be beneficial for growth (Mndela *et al.*, 2022). Such treatments are thought to increase the plants' photosynthetic rate and water-use efficiency, which in turn stimulates their growth and development. Another study reported that in most C3 plant species, such as coconut, elevated atmospheric CO_2 concentrations increases leaf thickness, a trait useful under high-temperature stress (Muralikrishna *et al.*, 2013).

Elevated atmospheric CO_2 concentration treatments applied to coconut seedlings have been shown to reduce their stomatal aperture and guard cell area on the leaf surface, as well as increasing the thickness and the cuticular layer of their leaves (Muralikrishna *et al.*, 2013). The authors surmised that the production of such leaves indicates an increase in CO_2 fixation. Even at such reduced stomatal density, photosynthetic rates under elevated CO_2 concentrations were found to be increased (Naresh *et al.*, 2007; Muralikrishna *et al.*, 2009). A controlled field facility utilizing a series of open-top chambers has been used to grow coconut seedlings under elevated CO_2 concentrations of either 550 or 700 ppm. The seedlings exhibited an 8% or 25% increase in biomass, respectively. In addition, the plant water-use efficiency also increased under both elevated CO_2 conditions (Hebbar *et al.*, 2013b). Nevertheless, this enhancement of biomass production due to CO_2 elevation was reduced to 8% under drought stress. Disappointingly, the growth effect resulting from the elevated CO_2 condition also created conditions for spindle leaf disease to take hold.

Carbon sequestration is the fixation of atmospheric CO_2 into biomass or the soil carbon pool, which reduces the CO_2 in the atmosphere and eventually reduces the impact of climate change. Coconut is one of the more important plantation crops that can fix large amounts of atmospheric CO_2. On average, a 10-year-old coconut palm can fix approximately 18–28 kg of CO_2 per palm year^{-1} (Boomiraj *et al.*, 2020). Another 2-year study reported that coconut intercropping could fix 3.01 t ha^{-1}, compared with 2.31 t ha^{-1} under a monocropping system (Kumar and Maheswarappa, 2019). These results indicate that coconut plantations could potentially be used not only for sustainable agriculture production but also to reduce the impacts of global climate change.

4.2.5 Low-temperature stress

Environmental temperatures below 13°C adversely affect coconut flowering, certain leaf anatomical features and the palm's fruiting characteristics (e.g. producing wrinkled fruit and low yields) (Cao *et al.*, 2014). An iTRAQ (isobaric tags for relative and absolute quantitation)-based comparative proteomic analysis determined the molecular response of two Chinese coconut varieties subjected to low-temperature stress. Abundant production of various photosynthesis-related proteins was seen in the cold-tolerant Hainan Tall variety compared with a cold-sensitive variety. This abundant production of photosynthesis-related proteins in the Hainan Tall variety compensated for the stress-induced damage seen in its thylakoid membranes (Yang *et al.*, 2020). Analysis of 20 coconut varieties subjected to cold stress (ranging from 25°C to 4°C) determined various physiological and biochemical responses, and from these responses the study identified certain varieties (e.g. C2, C7 and C10) with enhanced resistance to cold-temperature stress (Sun *et al.*, 2021).

4.2.6 Seawater substitution

Sea-level rises attributed to climate change-induced global warming is occurring in many places around the globe. Various simulation models (e.g. the Representative Concentration Pathway 4.5 model) have predicted that global mean sea levels are set to increase by 0.3–0.6 m by the turn of this century (Church *et al.*, 2013). Although coconut is a moderately salt-tolerant species, its growth response to seawater is unknown. In a study by Hebbar *et al.* (2021) to help devise an appropriate management strategy for plantations in coastal regions that may encounter future sea-level rises, coconut seedlings were grown under salt solution hydroponics. The seedlings could tolerate saltwater equivalent to 8.32 dS m^{-1} electrical conductivity without any negative effects on their growth and physiological development. The underlying physiological studies suggested that the absorbed Na^+ ions were differentially compartmentalized in root and shoot tissues, thereby maintaining a normal leaf K^+:Na^+ ratio (Hebbar *et al.*, 2021). In addition, coconut leaves have a natural ability to tolerate Cl^- and this would also play a considerable role in coconut's adaptation to sea-level rises (Hebbar *et al.*, 2021). Another earlier study reported that green coconut seedlings subjected to a combined drought and salinity stress were severely affected in their growth, as shown by certain physiological measurements on leaf gaseous exchange, their relative chlorophyll index and chlorophyll fluorescence (da Silva *et al.*, 2017). This study also demonstrated that the impact of drought stress was more noticeable than that of salinity stress, with a response generally involving stomatal activity regulation (da Silva *et al.*, 2017). Exposure of coconut seedlings to 0%, 10%, 25%, 50%, 75% and 100% seawater substitution (equivalent to 2.17, 8.32, 16.32, 30.03, 42.14 and 53.69 dS m^{-1}, respectively) significantly changed their physiological processes, with the aerial biomass being reduced by 47% in the 8.32 dS m^{-1} treatment. Further studies are now needed to narrow down the level of tolerance to seawater substitution and to explore the response of tall varieties to seawater substitution (Fig. 4.3; Hebbar *et al.*, 2021).

4.2.7 Interaction effect of elevated CO_2, high temperature and drought

Coconut seedlings grown in open-topped chambers and subjected to the interaction effects of elevated atmospheric CO_2 concentration and high-temperature, drought and high-nutrient conditions showed that the adverse effects of high temperature (+3°C above ambient) could be alleviated to some extent by elevated atmospheric CO_2 concentration. High-temperature stress alone substantially decreased the biomass production by 10%, whereas high-temperature stress combined with drought stress reduced the biomass by 16%. However, the beneficial effect of elevated atmospheric CO_2 concentration (700 ppm), under conditions of drought and high-temperature stress, resulted in an increase in biomass by 8% (Hebbar *et al.*, 2013b). The effect of elevated CO_2 concentration, under high-temperature stress, was found to reduce the leaf density of stomata and guard cells, while increasing the thickness of the upper leaf cuticle (Muralikrishna *et al.*, 2013). Overall, this study on coconut seedlings indicated that, through an adjustment of stomatal density and leaf cuticle thickness in response to extreme environmental stresses, coconut has a natural ability to adapt to certain aspects of climate change (Muralikrishna *et al.*, 2013).

4.3 Impact on Production Regions

Coconut cultivation is undertaken mainly by resource-poor farmers in South-east Asia, South Asia and the Pacific. The 62 million t of fruit harvested and its products are produced on an area of around 12 million ha. Considering the perennial nature of the crop, assessing the impact that climate change may have on the extent of its cultivation is challenging. Although

Fig. 4.3. Growth of coconut seedlings in hydroponics 3 months after substituting Hoagland solution with seawater (10%, 25%, 50%, 75% or 100%). From Hebbar *et al.*, 2021, with permission.

minimum temperatures of 13°C can allow some flowering to occur, temperatures at or below this will cause a severe reduction in flowering and induce fruit fall. When temperatures increase to over 40°C, there is a decrease in the leaf-area index and dry-matter production, and this results in decreased fruit yield (Naresh Kumar *et al*., 2007). Peiris *et al*. (2008) utilized a multiple linear regression model to correlate coconut production potential in various agroecological regions of Sri Lanka with different regional rainfall patterns. A simulation modelling approach (InfoCrop-COCONUT) was used to predict coconut production in India under a changing climate (Naresh Kumar *et al*., 2008; Naresh Kumar and Aggarwal, 2013). The model identifies regions within India that have additional potential for coconut growth under a changing climate (i.e. western coastal regions, north-eastern states and certain islands) and regions where additional agricultural inputs and management approaches (drip and summer irrigation) would be needed to help offset the ill effects of climate change and to maintain yield. The model also identifies regions where yield would be severely reduced. In another study, MaxEnt model-based predictions have identified climatically suitable coconut-production regions under various future climate scenarios for the years 2050 and 2070, using two Representative Concentration Pathway models (4.5 and 8.5) (Fig. 4.4; Hebbar *et al*., 2022). This analysis reveals a contraction in the size of the suitable climatic area for coconut cultivation, with unfavourable climatic conditions

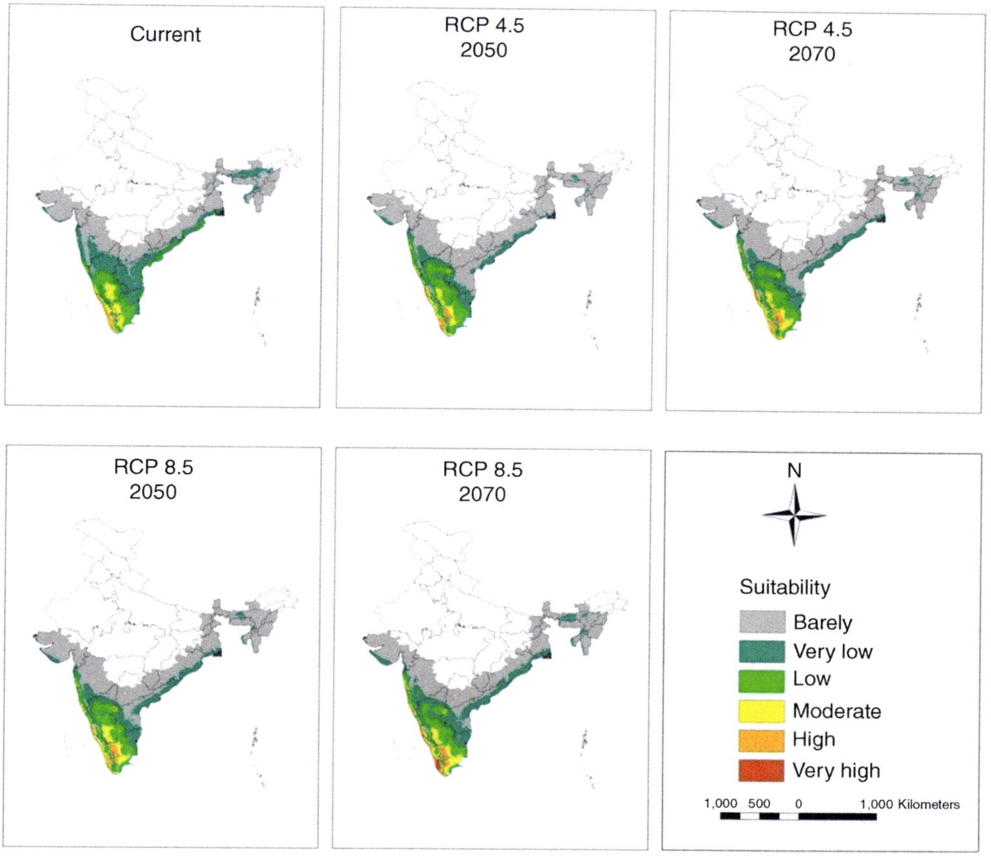

Fig. 4.4. Climatically suitable areas for coconut production in India under the current and future climates predicted for 2050 and 2070 using Representative Concentration Pathway (RCP) models 4.5 and 8.5, as modelled by MaxEnt. From Hebbar *et al.*, 2022, CC BY 4.0.

anticipated in the southern interior and eastern regions of India (Hebbar *et al.*, 2022).

Coconut is an important horticultural crop grown in the Pacific Island countries. In Kiribati, prolonged droughts have severely reduced fruit production and quality, triggered premature fruit drop and accelerated palm senility (McGregor and McGregor, 1999). In Vanuatu, high temperatures coupled with low rainfall have been factors that intensified an outbreak of coconut foliar decay virus (Taylor *et al.*, 2016), mainly by increasing the activity of the planthopper *Myndus taffini* Bonfils, 1983, the vector that transmits the disease. In addition, the coastal regions of Kiribati have witnessed stunted tree growth due to the rising sea levels caused by seawater incursion. Similarly, rising sealevels in Tuvalu have caused yellowing of coconut palms, and coastal erosion has resulted in the loss of many coastal palms (Freeman, 2010).

4.4 Adaptation Strategies

Coconut's perennial nature, coupled with its long breeding cycle, pose a serious challenge to coconut breeders wishing to develop climate-tolerant varieties. The use of screening approaches to select coconut varieties for various tolerance traits (i.e. high-temperature tolerance, improved pollen germination and fruit retention under high-temperature stress (Hebbar *et al.*, 2018), *in situ* drought tolerance (Kasturi Bai *et al.*, 2009), enhanced water-use efficiency (Hebbar *et al.*, 2018; Ramesh *et al.*, 2020) and cold-stress tolerance (Sun *et al.*, 2021) are all strategies being attempted. Such screening approaches have already identified a number of Indian varieties (Chandra Kalpa, Chandra Laksha, Kalpa Dhenu, Kalpatharu, Kalpa Mitra, Kera Keralam and Kera Sankara) that exhibit certain kinds of stress tolerance (Kasturi Bai *et al.*, 2009). Another study undertaken on the morphological and fruit-quality traits of various coconut varieties identified several features (e.g. fruit flavour, fat content and weight, and the number of female flowers produced) that can be used for the selection and breeding of new varieties with improved resilience to climate change (Zhang *et al.*, 2021).

In Sri Lanka, the coconut lands predicted to be lost under climate change have been determined to be approximately 88,000 ha in highly vulnerable regions and 108,000 ha in moderately vulnerable regions. Plantation-level adaptations are thought to be helpful in mitigating the effects of climate change in Sri Lanka, and include: (i) mulching (utilizing coconut husks, fronds, farm-derived green matter and weed trash) to diminish weed growth and water loss from the soil surface; (ii) incorporation of organic matter (especially in areas of sandy or clay soils); (iii) growing cover crops; (iv) developing measures to harvest and utilize rainwater; and (v) supplying irrigation water during summer, (Pathiraja *et al.*, 2015). In addition, the identification of drought-tolerant lines and breeding for drought-tolerant varieties are also thought to be part of the long-term strategy to alleviate the effects of climate change in Sri Lanka.

In India, scientific measures are being used to help offset the harmful effects of climate change, such as: (i) improving soil moisture conservation (e.g. through mulching with coir dust, husk burial, addition of tank silt or using other aspects of organic agriculture); (ii) using drip irrigation, especially in summer; (iii) applying ample fertilizer; and (iv) using and maintaining appropriate palm spacing to reduce shading (Hebbar *et al.*, 2013a).

Genomic approaches (i.e. development of molecular markers and quantitative trait locus mapping, and deciphering the transcriptomic responses of varieties to stresses) are being used to help develop new genetic materials for climate-smart coconut cultivation (Cardena *et al.*, 2003; Shalini *et al.*, 2007; Ramesh *et al.*, 2019, 2020). Identification of genetic markers linked to agronomically essential traits (e.g. early flowering (Herran *et al.*, 2000), yield of fruit (Lebrun *et al.*, 2001), characteristics of fruit (Baudouin *et al.*, 2006), leaf epicuticular wax content (Riedel *et al.*, 2009), and resistance to biotic stresses such as lethal yellowing disease (Cardena *et al.*, 2003) and eriophyid mites (*Aceria guerreronis* Keifer, 1965) (Shalini *et al.*, 2007)) have been undertaken. In addition, the newly acquired whole-genome sequence of coconut (e.g. for Hainan Tall (Xiao *et al.*, 2017), Catigan Green Dwarf (Lantican *et al.*, 2019) and Chowghat Green Dwarf (Rajesh *et al.*, 2020)) can now be used to develop high-quality linkage maps to examine quantitative trait loci of agronomic importance, and this should lead to the development of new, climate-smart coconut varieties.

4.5 Conclusion

Coconut is known to experience multiple environmental stresses during its lifetime, and in addition to these events, climate change will pose many more constraints on productivity. Among the most significant concerns are the occurrence and impacts of high-temperature stress, soil moisture-deficit stress and the rise in sea level that will reduce the land on which coconut can be grown. Research has been under way to better understand the effects of these stresses on coconut growth and development. Work is proceeding on various climate-smart adaptive cultivation strategies, which are being based on sound scientific principles. New tolerant varieties are being developed, and work is under way to identify land areas that will become suitable for further coconut production under climate change.

4.6 References

Baudouin, L., Lebrun, P., Konan, J.L., Ritter, E., Berger, A. *et al.* (2006) QTL analysis of fruit components in the progeny of a Rennell Island Tall coconut (*Cocos nucifera* L.) individual. *Theoretical and Applied Genetics* 112, 258–268.

Boomiraj, K., Jagadeeswaran, R., Karthik, S., Poornima, R., Jothimani, S. and Sudhagar, R.J. (2020) Assessing the carbon sequestration potential of coconut plantation in Vellore district of Tamil Nadu, India. *International Journal of Environment and Climate Change* 10, 618–624.

Cao, H., Huang, H., Lei, X., Zhang, D. and Zhang, R. (2014) Difference of the leaf anatomical structure of coconut varieties under low temperature treatments. *Chinese Journal of Tropical Crops* 35, 2420–2425.

Cardena, R., Ashburner, G.R. and Oropeza, C. (2003) Identification of RAPDs associated with resistance to lethal yellowing of the coconut (*Cocos nucifera* L.) palm. *Scientia Horticulturae* 98, 257–263.

Church, J.A., Clark, P.U., Cazenave, A., Gregory, J.M., Jevrejeva, S. *et al.* (2013) Sea level change. In: Stocker, T.F., Qin, D., Plattner, G.K. *et al.* (eds) *The Physical Science Basis. Contribution of Working Group I to the Fifth Assessment Report of the Intergovernmental Panel on Climate Change*. Cambridge University Press, Cambridge, New York, pp. 1137–1216.

da Silva, A.R.A., Bezerra, F.M.L., de Lacerda, C.F., de Sousa, C.H.C. and Bezerra, M.A. (2017) Physiological responses of dwarf coconut plants under water deficit in salt-affected soils. *Caatinga* 30, 447–457.

FAOSTAT (2020) Countries by commodity. Food and Agricultural Organisation of the United Nations, Rome. Available at: www.fao.org/faostat/en/#rankings/countries_by_commodity (accessed 14 August 2023).

Foale, M.A. (2003) *The Coconut Odyssey: The Bounteous Possibilities of the Tree of Life*. Australian Centre for International Agricultural Research (ACIAR), Canberra.

Freeman, A. (2010) Impact of climate change on food security and biosecurity of crop production systems in small Pacific nations. Final report. Asia-Pacific Network for Global Change Research (APN), Kobe, Japan. Available at: www.nab.vu/impact-climate-change-food-security-and-biosecurity-crop-production-system (accessed 14 August 2023).

Gomes, F.P., Oliva, M.A., Mielke, M.S., Almeida, A.A.F., Leite, H.G. and Aquino, L.A. (2007) Photosynthetic limitations in leaves of young Brazilian Green Dwarf coconut (*Cocos nucifera* L. '*nana*') palm under well-watered conditions or recovering from drought stress. *Environmental and Experimental Botany* 62, 195–204.

Hebbar, K.B., Balasimha, D. and Thomas, G.V. (2013a) Plantation crops response to climate change: coconut perspective. In: Singh, H.C.P., Rao, N.K.S. and Shivashankara, K.S. (eds) *Climate-resilient Horticulture: Adaptation and Mitigation Strategies*. Springer, New Delhi, pp. 177–187.

Hebbar, K.B., Sheena, T.L., Kumari, K.S., Padmanabhan, S., Balasimha, D. *et al.* (2013b) Response of coconut seedlings to elevated CO_2 and high temperature in drought and high nutrient conditions. *Journal of Plantation Crops* 41, 118–122.

Hebbar, K.B., Rose, H.M., Nair, A.R., Kannan, S., Niral, V. *et al.* (2018) Differences in *in vitro* pollen germination and pollen tube growth of coconut (*Cocos nucifera* L.) cultivars in response to high temperature stress. *Environmental and Experimental Botany* 153, 35–44.

Hebbar, K.B., Neethu, P., Sukumar, P.A., Sujithra, M., Santhosh, A. *et al.* (2020) Understanding physiology and impacts of high temperature stress on the progamic phase of coconut (*Cocos nucifera* L.). *Plants (Basel)* 9: 1651.

Hebbar, K.B., Santhosh, A., Sukumar, A.P., Neethu, P., Ramesh, S.V. *et al.* (2021) Effect of sea water substitution on growth, physiological and biochemical processes of coconut (*Cocos nucifera* L.) seedlings – a hydroponic study. *Scientia Horticulturae* 280: 109935.

Hebbar, K.B., Abhin, P.S., Jose, V.S., Neethu, P., Santhosh, A. *et al.* (2022) Predicting the potential suitable climate for coconut (*Cocos nucifera* L.) cultivation in India under climate change scenarios using the MaxEnt model. *Plants (Basel)* 11: 731.

Herran, A., Estioko, L., Becker, D., Rodriguez, M.J.B., Rohde, W. *et al.* (2000) Linkage mapping and QTL analysis in coconut (*Cocos nucifera* L.). *Theoretical and Applied Genetics* 101, 292–300.

IPCC (2022) Climate Change 2022: Mitigation of Climate Change. Working Group III Contribution to the Sixth Assessment Report of the Intergovernmental Panel on Climate Change. IPCC, Geneva. Available at: www.ipcc.ch/report/ar6/wg3/ (accessed 15 August 2023).

Kasturi Bai, K.V., Rajagopal, V. and Naresh Kumar, S. (2006) Chlorophyll fluorescence transients with response to leaf water status in coconut. *Indian Journal of Plant Physiology* 11, 410–414.

Kasturi Bai, K.V., Naresh Kumar, S. and Rajagopal, V. (2009) *Abiotic Stress Tolerance in Coconut*. Central Plantation Crops Research Institute Publications, Kasaragod, India.

Krishnakumar, K.N., Rao, G.S.L.H.V.P. and Gopakumar, C.S. (2008) Climate change at selected locations in the Kerala state, India. *Journal of Agrometeorology* 10, 59–64.

Kumar, K.N. and Maheswarappa, H.P. (2019) Carbon sequestration potential of coconut-based cropping systems under integrated nutrient management practices. *Journal of Plantation Crops* 47, 107–114.

Lantican, D.V., Strickler, S.R., Canama, A.O., Gardoce, R.R., Mueller, L.A. *et al.* (2019) *De novo* genome sequence assembly of dwarf coconut (*Cocos nucifera* L. 'Catigan Green Dwarf') provides insights into genomic variation between coconut types and related palm species. *G3: Genes, Genomes, Genetics* 9, 2377–2393.

Lebrun, P., Baudouin, L., Bourdeix, R., Konan, J.L., Barker, J.H. *et al.* (2001) Construction of a linkage map of the Rennell Island Tall coconut type (*Cocos nucifera* L.) and QTL analysis for yield characters. *Genome* 44, 962–970.

McGregor, A. and McGregor, K.L. (1999) *Disasters and Agriculture in the Pacific Islands*. United Nations Development Programme (UNDP), New York. Available at: www.nab.vu/disaster-and-agriculture-pacific-islands (accessed 14 August 2023).

Mndela, M., Tjelele, J.T., Madakadze, I.C., Mangwane, M., Samuels, I.M. *et al.* (2022) A global meta-analysis of woody plant responses to elevated CO_2: implications on biomass, growth, leaf N content, photosynthesis and water relations. *Ecological Processes* 11, 1–21.

Muralikrishna, K.S., Naresh Kumar, S., John Sunoj, V.S. and Kasturi Bai, K.V. (2009) Elevated carbon dioxide and temperature reduce stomatal density in coconut (*Cocos nucifera* L.). In: *Proceedings of National Workshop on Climate and Development*. Kerala Agricultural University, Thrissur, Kerala.

Muralikrishna, K.S., Naresh Kumar, S. and Sunoj, V.S.J. (2013) Elevated CO_2 and temperature affect leaf anatomical characteristics in coconut (*Cocos nucifera* L.). In: Sundaresan, J., Sreekesh, S., Ramanathan, A., Sonnenschein, L. and Boojh, R. (eds) *Climate Change and Island and Coastal Vulnerability*. Springer, Dordrecht, Netherlands, pp. 141–153.

Murray, D.B. (1977) Coconut palm. In: Alvin, P.T. and Kozlowski, T.T. (eds) *Ecophysiology of Tropical Crops*. Academic Press, New York, pp. 24–27.

Naresh Kumar, S. (2007) *Climate Change Effects on Growth and Productivity of Plantation Crops with Special Reference to Coconut and Black Pepper: Impact, Adaptation and Vulnerability and Mitigation Strategies*. Final Report on ICAR Network Project on Impact of Climate Change on Indian Agriculture. Indian Council of Agricultural Research (ICAR), New Delhi.

Naresh Kumar, S. and Aggarwal, P.K. (2013) Climate change and coconut plantations in India: impacts and potential adaptation gains. *Agricultural Systems* 117, 45–54.

Naresh Kumar, S., Rajagopal, V., Thomas, T.S., Cherian, V.K., Narayanan, M.K.R. *et al.* (2007) Variations in nut yield of coconut (*Cocos nucifera* L.) and dry spell in different agroclimatic zones of India. *Indian Journal of Horticulture* 64, 309–313.

Naresh Kumar, S., Bai, K.V.K., Rajagopal, V. and Aggarwal, P.K. (2008) Simulating coconut growth, development and yield with the InfoCrop-coconut model. *Tree Physiology* 29: 751.

Pathiraja, P.M.E.K., Griffith, G.R., Farquharson, R.J. and Faggian, R. (2015) The Sri Lankan coconut Industry: current status and future prospects in a changing climate. *Australasian Agribusiness Perspectives* 18: 106.

Pathmeswaran, C., Lokupitiya, E., Waidyarathne, K.P. and Lokupitiya, R.S. (2018) Impact of extreme weather events on coconut productivity in three climatic zones of Sri Lanka. *European Journal of Agronomy* 96, 47–53.

Peiris, T.S.G. and Peries, R.R.A. (1993) Effects of bimonthly rainfall on coconut yield in the low country intermediate zone (IL_1) of Sri Lanka. *Cocos* 9, 1–11.

Peiris, T.S.G. and Thattil, R.O. (1998) The study of climate effects on the nut yield of coconut using parsimonious models. *Experimental Agriculture* 34, 189–206.

Peiris, T.S.G., Hansen, J.W. and Zubair, L. (2008) Use of seasonal climate information to predict coconut production in Sri Lanka. *International Journal of Climatology* 28, 103–110.

Prado, C.H.B.A., Passos, E.E.M. and Moraes, J.A.P.V. (2001) Photosynthesis and water relations of six tall genotypes of *Cocos nucifera* in wet and dry seasons. *South African Journal of Botany* 67, 169–176.

Rajagopal, V. and Kasturi Bai, K.V. (2002) Drought tolerance mechanism in coconut. *Bulletin BuroTrop* 17, 21–22.

Rajagopal, V., Jacob, M., Mathew, J. and Shivashankar, S. (1996) Impact of dry spells on the ontogeny of coconut fruits and its relation to yield. *Plantation Research and Development* 3, 251–255.

Rajesh, M.K., Chowdappa, P., Behera, S.K., Kasaragod, S., Gangaraj, K.P. *et al.* (2020) Assembly and annotation of the nuclear and organellar genomes of a dwarf coconut (Chowghat Green Dwarf) possessing enhanced disease resistance. *Journal of Integrative Biology* 24, 726–742.

Ramesh, S.V., Hebbar, K.B., Rajesh, M.K. and Archana, P. (2019) Soil water-deficit differentially modulates the expression of stress associated genes (SAGs) of *Cocos nucifera* L. seedlings with contrasting water-use efficiency (WUE). In: *Book of Abstracts. 23rd Plantation Crops Symposium,* 6-8 March 2019, Chikkamagaluru, India, p. 93.

Ramesh, S.V., Hebbar, K.B., Rajesh, M.K., Sukumar, P.A., Gangaraj, K.P. *et al.* (2020) Transcriptome analysis of *Cocos nucifera* L. seedlings having contrasting water-use efficiency (WUE) under water-deficit stress: molecular insights and genetic markers for drought tolerance. *Biology and Life Sciences Forum* 4: 73.

Ranasinghe, C.S., Waidyarathna, K.P., Pradeep, A.P.C. and Meneripitiya, M.S.K. (2010) Approach to screen coconut varieties for high temperature tolerance by *in-vitro* pollen germination. *Cocos* 19, 1–11.

Ranasinghe, C.S., Silva, L.R.S. and Premasiri, R.D.N. (2015) Major determinants of fruit set and yield fluctuation in coconut (*Cocos nucifera* L.). *Journal of the National Science Foundation of Sri Lanka* 43, 253–264.

Ranasinghe, C.S., Kumarathunge, M.D.P. and Kiriwandeniya, K.G.S. (2018) Genotypic differences in cardinal temperatures for *in vitro* pollen germination and pollen tube growth of coconut hybrids. *Experimental Agriculture* 54, 731–743.

Riedel, M., Riederer, M., Becker, D., Herran, A., Kullaya, A. *et al.* (2009) Cuticular wax composition in *Cocos nucifera* L.: physicochemical analysis of wax components and mapping of their QTLs onto the coconut molecular linkage map. *Tree Genetics and Genomes* 5, 53–69.

Shalini, K.V., Manjunantha, S., Lebrun, P., Berger, A., Baudouin, L. *et al.* (2007) Identification of molecular markers associated with mite resistance in coconut (*Cocos nucifera* L.). *Genome* 50, 35–42.

Sun, C., Zhang, R., Yuan, Z., Cao, H. and Martin, J.J.J. (2021) Physiology response and resistance evaluation of twenty coconut germplasm resources under low temperature stress. *Horticulturae* 7: 234.

Sunoj, V.S.J., Hebbar, K.B. and Prasad, P.V.V. (2017) Phenotyping tools to understand effects of climate change. In: Hebbar, K.B., Naresh Kumar, S. and Chowdappa, P. (eds) *Impact of Climate Change on Plantation Crops*. Daya Publishing House, New Delhi.

Taylor, M., McGregor, A. and Dawson, B. (2016) *Vulnerability of Pacific Island Agriculture and Forestry to Climate Change*. Pacific Community, New Caledonia.

Xiao, Y., Xu, P., Fan, H., Baudouin, L., Xia, W. *et al.* (2017) The genome draft of coconut (*Cocos nucifera* L.). *GigaScience* 6: gix095.

Yang, Y., Saand, M.A., Abdelaal, W.B., Zhang, J., Wu, Y. *et al.* (2020) iTRAQ-based comparative proteomic analysis of two coconut varieties reveals aromatic coconut cold-sensitive in response to low temperature. *Journal of Proteomics* 220: 103766.

Zhang, R., Cao, H., Sun, C. and Martin, J.J.J. (2021) Characterization of morphological and fruit quality traits of coconut (*Cocos nucifera* L.) germplasm. *HortScience* 56, 961–969.

5 Pests and Diseases Affecting Coconut

Carlos F. Ortiz[1], Eder Ramos[2], Carmen Silverio[2], Augusto Peña[3], María Narváez[3] and Carlos Oropeza[3*]

[1]*Colegio de Postgraduados. Cárdenas, Tabasco, Mexico;* [2]*Instituto Nacional de Investigaciones Forestales, Agrícolas y Pecuarias, Campo Experimental Huimanguillo, Mexico;* [3]*Centro de Investigación Científica de Yucatán, Mérida, Yucatán, Mexico*

Abstract

The coconut palm (*Cocos nucifera* L.) is an important crop worldwide, because it is a source of raw materials for value-added products, whose markets have been growing rapidly in recent years. Unfortunately, production is decreasing because most coconut plants around the world are old and affected by devastating pests and diseases. The present chapter reports the main pests and diseases affecting coconut and other palm species around the world, focusing on the risk of dispersal through a single continent and from one continent to another, taking as an example the case of lethal yellowing disease, with the objective of highlighting the consequences that could result if we do not take the appropriate care, revising what has been done in each case, and talking about initiatives at the international level, with an interregional, intercontinental or global approach to reduce risks of dispersal of agents of diseases or pests.

5.1 Introduction

The coconut palm (*Cocos nucifera* L.) in the family Arecaceae is distributed worldwide, cultivated mainly by small producers, most with less than 4 ha of land, for whom coconut is used principally for their subsistence, providing building materials, food security and nutrition. Coconut is so important for different cultures that it is considered a sacred tree in India, the 'Tree of a Thousand Uses' in Malaysia, the 'Tree of Life' in the Philippines and the 'Tree of Abundance' in Indonesia (Earthstoriez, 2018). The coconut palm is also an important source of raw materials for value-added products. In recent years, there has been a rapid growth in various coconut markets, particularly for packed coconut water, virgin oil, milk and sugar (Prades *et al.*, 2016). To keep markets growing, production must also grow. Unfortunately, most coconut palms around the world are now old and affected by devastating pests and diseases, reducing their productivity. To avoid further reductions in production and to stimulate growth in production, we need to deal with the issues caused by pests and diseases and promote the replanting of land with selected coconut palms that are high yielding and, if possible, resistant or tolerant to the most devastating pests and diseases. This chapter reports on the main pests and diseases affecting

* Email: cos@cicy.mx

 The Coconut: Botany, Production and Uses (eds S. Adkins *et al.*)
DOI: 10.1079/9781789249736.0005

coconut and other palm species around the world, focusing on the risk of dispersal within a single continent and from one continent to another, taking as an example the case of lethal yellowing (LY) and similar LY-type diseases (LYDs) with the aim of highlighting the consequences that could result if we do not take the appropriate care and learn from what has been done in each case, and consider initiatives at the international level, with interregional, intercontinental and global approaches to the reduction of risk of dispersal of pests or agents of disease.

5.2 Diseases with Wide Geographical Distribution

5.2.1 Lethal yellowing disease

First reports and symptoms

LY is a disease that affects coconut, other palms and non-palm species. The symptoms for coconut are fruit drop, inflorescence necrosis, leaf yellowing and loss of the crown leaving a bare trunk (Fig. 5.1). Palms showing these symptoms were first reported in the 19th century in several countries in the Caribbean (Eden-Green, 1995).

Causal agent and host plant species

Studies based on electron microscopy were the first to associate LY symptomatology with the presence of phytoplasmas (formerly mycoplasma-like organisms; Plavsic-Banjac *et al.*, 1972). In the Americas, there are now five strains of phytoplasma that are identified as belonging to group 16SrIV and subgroups A, B, D, E and F, and they have been reported in more than 45 plant species (Oropeza *et al.*, 2020). The most devastating strain has been 16SrIV-A, causing several epidemic outbreaks and killing millions of coconut palms in the Caribbean, most of which were of the highly susceptible Atlantic Tall variety. However, some palm and non-palm species, when infected with strain 16SrIV-A, do not show symptoms, or if they do, it is only a few plants within the population, and the population

Fig. 5.1. Lethal yellowing symptoms in coconut showing (A) fruit drop, (B) inflorescence necrosis, (C, D) yellowing progressing from older (C) to younger (D) leaves, and (E) loss of the crown leaving the bare trunk. Photos courtesy of C.F Ortiz.

size does not seem to be affected in the longer term, as with the Florida thatch palm (*Thrinax radiata* Lodd. ex Schult. & Schult.f.; Centro de Investigación Científica de Yucatán, unpublished data). This occurrence, therefore, might represent the existence of permanent phytoplasma reservoirs from which insect vectors can feed.

LYDs

Besides LY, associated with phytoplasma 16SrIV-A affecting coconut, there are similar syndromes associated with other 16SrIV strains in the Americas. For instance, there are reports of epidemic outbreaks caused by the strain 16SrIV-D, affecting *Phoenix* spp. in Mexico and the USA (Palma-Cancino *et al.*, 2020). There are also strains of other phytoplasma groups affecting coconut and other palm or non-palm species in over 20 countries around the world. Together, these cases associated with coconut and 40 other palm species involving eight different phytoplasma groups and subgroups are collectively referred to as LYDs (Oropeza *et al.*, 2020).

Insect vectors and transmission

Phytoplasmas can be transmitted by phloem-feeding insects. In the case of LY, a planthopper species (*Haplaxius crudus* Van Duzee, 1907) was found to be associated with coconut in areas being infected with 16SrIV phytoplasmas, as evidenced in PCR-based studies in Florida (Mou *et al.*, 2020), and Mexico (Narváez *et al.*, 2018). In addition, transmission of 16SrIV phytoplasmas by *H. crudus* to palms, including coconut, has also been reported in Florida (Howard, 1987) and Mexico (Dzido *et al.*, 2020; Narváez *et al.*, 2022). There are another nine insect species in the suborder Auchenorrhyncha that have been found to carry 16SrIV phytoplasmas, but their ability to transmit them to plants is yet to be determined (Tropicsafe, 2022).

Integral management

In Jamaica, LY is managed using a package of measures that include surveillance, elimination of the symptomatic plants, replanting with LY-resistant varieties, weeding, control of the vectors and keeping plantations healthy (Oropeza *et al.*, 2020). The most important measure is replanting with LY-resistant varieties. Therefore, it is very important to have, at the country and regional levels, an LY-resistance screening programme, including tests for LYD phytoplasmas in other continents with phytoplasmas different from those in the Americas.

Spread of LY and LYD phytoplasmas and the related risk

Natural spread of 16SrIV phytoplasmas that cause LY (or LYD) involves the spread of the insect vectors. The vectors are unlikely to fly more than a few kilometres; however, LY apparently spread from the Cancún area in Mexico to the Roatan Island in Honduras, more than 400 km (Ashburner *et al.*, 1996). Thus, as suggested by Weintraub *et al.* (2019), people may unintentionally mediate phytoplasma spread over long distances or inadvertently disperse the vectors through international transportation. It has been reported that LY phytoplasmas can occur in zygotic embryos of seed (Cordova *et al.*, 2003), and that these embryos, when germinated *in vitro*, could develop into plantlets harbouring phytoplasmas (Oropeza *et al.*, 2017). Exchange of goods carrying infected planthoppers or infected seed from one continent or subcontinent to another could cause a catastrophe for the industry.

5.2.2 Bud rot

First reports and symptoms

Bud rot (BR) is caused by oomycetes in the genus *Phytophthora* and causes severe damage to palms with agricultural value worldwide, such as coconut and oil palm (*Oleis guineensis* Jacquin) (Dollet *et al.*, 2012), and several other susceptible Arecaceae with ornamental value (Chase and Broschat, 1991). Symptoms of BR observed in coconut are leaf chlorosis, and drying and collapse of the spear leaves, followed by the leaves at the top of the canopy turning yellow and then brown, leaving the stem with only a basal crown of leaves (Fig. 5.2A). Internally, the bud tissue rots, typically due to attack by *Phytophthora* spp. (Fig. 5.2B). As the disease progresses, the central and lower leaves are also affected. After a few months, they dry out and fall, and the palm

Fig. 5.2. Symptoms of bud rot in coconut showing (A) the collapse of the spear leaf (arrow), (B) bud rot symptoms, (C) the top leaves becoming affected, with the central and lower leaves following, (D) the fruit being attacked and falling prematurely and (E) the sporangia structures of the causal agent, *Phytophthora palmivora*. Photos courtesy of J.M. Thévenin (A–C) and C.F. Ortiz (D, E).

dies. The fruit are also attacked and may fall prematurely (Fig. 5.2C, D). Similarly, seedlings will die if infected (Tuckek, 1926; Thévenin *et al.*, 1995). The first cases of BR were reported in the Cayman Islands and Puerto Rico in the 19th century (Tuckek, 1926), and the disease has now been reported in Jamaica, the Dominican Republic, Nicaragua, Honduras, Trinidad and Tobago, Ghana, India, Ceylon, Indonesia, the Philippines, New Guinea, Fiji, Papua New Guinea, the Solomon Islands, Vanuatu and the USA (Quillec and Renard, 1984).

Causal agent

The causal agent of BR was initially reported in India as *Phytophthora palmivora* (Butler) (Tuckek, 1926) and later confirmed by Brasier and Griffin (1979). *P. palmivora* is a polyphagous soil-dwelling oomycete (Fig. 5.2E) that attacks multiple botanical families besides Arecaceae.

Disease management

In countries where the incidence of BR is relatively low, removal of infected plants may be sufficient. However, countries with a higher incidence require integrated management. This includes surveillance for early detection of symptoms (spear wilting), weeding, an efficient drainage system to avoid waterlogging, and avoiding fertilization with excess nitrogen, which favours *Phytophthora* spp. infections (Garcia *et al.*, 1999). In places with high precipitation, the crown should be sprayed with Bordeaux mixture (a solution of copper sulphate and lime) prior to the start of the rainy season, and at least once every 35–40 days. This helps reduce the incidence of BR (Iyer *et al.*, 2018). Application of the fungicide *Trichoderma harzianum* Rifai, 1969 into the leaf axils and inside the heart, prior to the rainy season, also prevents infection (Coconut Development Board, 2022).

5.2.3 Red ring

First report and symptoms

Red ring (RR) is a deadly disease that attacks coconut palms and other Arecaceae. The symptoms vary depending on the variety: the Atlantic Tall and Malayan Dwarf varieties show leaf yellowing progressing from older to younger leaves (Fig. 5.3A), which gradually turn to brown or bronze. The leaves remain attached to the trunk for a while, while the bud and upper crown collapse despite still carrying some green leaves. A cross-section of the stem reveals a red ring (Fig. 5.3B), giving the disease its name (Moscoso-Ramírez *et al.*, 2002; Chinchilla and Escobar, 2007). Symptoms seen in the Pacific Tall variety and its hybrids have the same disease development, but the change in colour of the leaflets is faster, exhibiting distal areas with necrotic edges, yellow regions and green areas simultaneously (Moscoso-Ramírez *et al.*, 2002).

RR was first described in Trinidad and Tobago in 1906, where more than 18,000 coconut palms were removed exhibiting this disease (Griffith, 1987). By 1920, RR was reported in Guyana, Venezuela, Panama, Honduras, Trinidad, St Vincent and Grenada. It was subsequently described as being present in Belize, Tobago, Surinam, French Guiana, Brazil, Colombia, Ecuador, Peru, Mexico, Costa Rica, Nicaragua, Guatemala and El Salvador (Griffith *et al.*, 1990). In Trinidad, it was reported to be present in 35% of all palms (Griffith, 1983).

Causal agent and host plants

The causal agent of RR disease is the nematode *Bursaphelenchus cocophilus* (Cobb) (formerly *Rhadinaphelenchus cocophilus* (Cobb)) (Fig. 5.3C; Martyn, 2021). This pathogen is a thin, migratory endoparasitic nematode with conical ends at the tail and head. Females are approximately 1 mm long. Among the host plants is the oil palm, to which RR causes severe damage of up to 20% in commercial plantations that are over 6 years of age (Sáenz, 2005). Other susceptible Arecaceae species are the black palm (*Astrocaryum standleyanum* L.H. Bailey), *Attalea butyracea* (Mutis ex L.f.) Wess. Boer, peach palm (*Bactris gasipaes* Kunth), bataua (*Jessenia polycarpa* H. Karst. (1857)), date palm (*Phoenix dactylifera* L.) and royal palm (*Roystonea regia* (Kunth) O.F. Cook) (Griffith *et al.*, 1990).

Vector insect and transmission

The adult of the South American palm weevil or black palm weevil (*Rhynchophorus palmarum* L.; Coleoptera: Curculionidae) (Griffith *et al.*, 1990) is the most efficient means of transmitting the nematode *B. cocophillus* to healthy coconut palms (Fig. 5.3D). Females carry the nematode in their excretory apparatus or attached to their abdomen (Sáenz, 2005), and leave the nematodes at

Fig. 5.3. Symptoms of red ring in coconut showing (A) the yellowing progressing from older to younger leaves, (B) a stem cross-section revealing a red ring, (C) the causal agent, the nematode *Bursaphelenchus cocophilus* and (D) its vector, *Rhynchophorus palmarum*. Photos courtesy of C.F. Ortiz.

the base of healthy palms, in the leaf axil or on the stem surface, and in small perforations made by the females while ovipositing (Fenwick, 1968). The nematodes reach the parenchymatous tissues of the stem, roots and basal leaves and there induce the rapid appearance of the foliar symptoms (Griffith, 1983; Mirisola Filho, 2002).

Disease management

Management of RR is carried out by integrating a series of practices for effective reduction of the weevil population, including baited traps containing food attractants, the use of insecticides and using pheromone (rhynchophorol) diffusers to attract the weevil (Chinchilla and Escobar, 2007; Ovando-Cruz *et al.*, 2019; Castañeda *et al.*, 2022). For further information on *R. palmarum*, see Section 5.4.1.

5.2.4 Hartrot

First report and symptoms

Hartrot is a fatal disease for coconut, first reported in Suriname in 1919 (Holmes, 1925). It is now present in many countries including Cuba, Guatemala, Honduras, Costa Rica, Suriname, French Guiana, Guyana, Trinidad and Tobago, Venezuela, Colombia, Ecuador and Brazil (Mirisola Filho, 2002). Symptoms in coconut and oil palm resemble those of LY (see Section 5.2.1), beginning with yellowing of the lower leaves (Fig. 5.4A), followed by the fall of unripe and ripe fruit, inflorescence necrosis and yellowing ascending to the uppermost leaves (Fig. 5.4B), although necrosis of the arrow leaf can occur before the younger leaves turn yellow (Dollet *et al.*, 2012).

Causal agent and host plants

The aetiological agent of hartrot is *Phytomonas staheli* Donovan, 1909, a flagellate protozoan 12–20 µm long, that belongs to the family Trypanosomatidae (McCoy and Martinez-Lopez, 1982). It colonizes the phloem in host palms (Dollet and Lopez, 1978), including coconut, peach palm (*Bactris gasipaes* Kunth), golden cane palm (*Dypsis lutescens* (H. Wendl.) Beentje & J. Dransf.), Bentinck palm (*Bentinckia nicobarica* (Kurz) Becc. Berry ex Roxb.), royal palm (*Roystonea regia* (Kunth) O.F. Cook), maripa palm (*Maximiliana maripa* (Aubl.) Drude) and fishtail palm (*Caryota mitis* Lour.), which are ornamental plants (Dollet *et al.*, 2012). *P. staheli* has also been reported in oil palm, in which damage is of great magnitude and very frequent. In this palm, the disease is known as 'marchitez sorpresiva' or sudden wilt (Dollet and Lopez, 1978).

Vector and transmission

The weevil *Lincus* spp. (Hemiptera: Pentatomidae) has been reported as a vector of *P. staheli*, as it was capable of acquiring *P. staheli* when feeding on diseased plants and transmits it to healthy plants when feeding on them (Louise *et al.*, 1986). Adult insects of *Lincus curvatus* (Campos & Roell) and *Macropygium reticulare* (Fabricius) were found to be capable of transmitting *P. staheli* to oil palm (Asipuela-Haro *et al.*, 2017).

Disease management

As there are no curative strategies for hartrot, preventative measures are applied, including biweekly supervision of plantations for early detection, eradication of diseased plants, and measures to control vector populations, such as keeping cultivated and neighbouring areas free from Poaceae (Garcia *et al.*, 1999).

5.3 Diseases with Restricted Geographical Distribution

5.3.1 Cadang-cadang

First report and symptoms

In the 1930s, a lethal and epidemic disease of coconut was reported from a southern region of Luzon in the Philippines. It was called 'cadang-cadang', from a local dialect meaning 'death' or 'dying' (Hanold and Randles, 1991). It is estimated that more than 40 million coconut palms have died from this disease in the past century (Rodriguez *et al.*, 2017). The symptoms develop through a series of stages (Fig. 5.5). The first stage (which takes about 2–4 years) involves the appearance of yellow spots on leaves and the fruit become smaller. In the second stage (a further 2 years), leaf spots become numerous giving the lower crown a yellowish appearance, the inflorescences become necrotic, fruit production ceases, and frond production and size decline. In the late stage (5 years), the leaf spots merge, the whole crown becomes yellow/bronze in colour, there is a reduction in frond size and number, the leaflets become brittle and the palm dies. Palms usually become infected after reaching the flowering stage. The time from first symptoms

Fig. 5.4. Symptoms of hartrot disease in oil palm, starting with (A) yellowing of the lower leaves, followed by (B) yellowing of the upper leaves. Photos courtesy of M. Elliott.

to death ranges from 8 to 16 years, but generally takes longer in more mature palms.

Causal agent

Cadang-cadang disease is caused by coconut cadang-cadang viroid (CCCVd), which was identified in vascular tissues and mesophyll cells of diseased coconut palms (Vadamalai *et al.*, 2017). CCCVd is composed of two associated ribonucleic acids (ccRNA-1/ccRNA-2) (Randles and Hatta, 1979). The younger leaves in a palm tree contain lower amounts of CCCVd than mature ones (Mohamed and Imperial, 1984). The viroid is transmitted by the fruit, with a low transmission rate, or rarely is transmitted by pollen (Lédo *et al.*, 2019). So far, no other means of dispersal have been reported and this has probably helped to keep the spread of the disease low.

Integrated management

Starting a new crop with certified clean materials is the basis for successful control of cadang-cadang (Wierzbicki and Eiras, 2021). Other measures are: (i) prevention or reduction of the movement of propagation materials from diseased areas; (ii) implementation of national and international phytosanitary certification programmes; (iii) use of effective quarantine and

Fig. 5.5. Symptoms of cadang-cadang in coconut start with (A) the appearance of yellow spots in leaves and (B) fruit becoming smaller; (C) later, frond number and size decline and the crown becomes yellowish. Photographs courtesy of J. Randles.

diagnostic service; (iv) giving priority to symptom identification; (v) elimination of suspicious plants that may act as alternative hosts to the viroid; and (vi) disinfection of machinery and tools (Pedaute and Duran-Vila, 2016).

5.3.2 Tinangaja

First report and symptoms

Tinangaja is a lethal disease of coconut first described in 1917 in Guam (Hodgson *et al.*, 1998). The first symptom is yellow spotting on the leaves (Fig. 5.6A). Next, the production of leaves, flowers and fruit slows, and wrinkling and deformation of the fruit take place (Fig. 5.6B). Finally, there is a general yellowing of the leaves, thinning of the trunks and eventually palm death (Fig. 5.6C). Infected coconut palms stop producing fruit years before dying (Hadidi *et al.*, 2003). From the first symptom to death can take up to 15 years (Vadamalai *et al.*, 2017). Like cadang-cadang, tinangaja mostly affects trees 25–30 years of age (Boccardo *et al.*, 1981).

Causal agent and spread

The disease is associated with coconut tinangaja viroid (CTiVd), which belongs to the genus *Cocadviroid* in the family *Pospiviroidae*. It is related to CCCVd, the viroid that causes the cadang-cadang, but with only 65% DNA homology (Koltunow and Rezaian, 1989). It is not known how the disease spreads, although it is common for only a small group of palms to become affected. The only way known for CTiVd transmission to take place in palms is through mechanical transmission by injecting nucleic acid inoculum under laboratory conditions (Hadidi *et al.*, 2022). Thus, spread of CTiVd may occur during regular plantation management practices. Any natural means of transmission remain to be elucidated.

Fig. 5.6. Symptoms of tinangaja disease in coconut. (A) Leaves develop yellow spots and the fruit are smaller and distorted. (B) Fruit production is reduced, and the leaves develop general yellowing. (C) Palms with advanced symptoms show 'pencil-point' tapering of the trunk. Photos courtesy of G. Wall.

Disease management

Management recommendations according to Vadamalai *et al.* (2017) are: (i) to remove infected palms and replace with healthy ones; (ii) to disinfect cutting/pruning tools after use; and (iii) to use fruit for propagation purposes from disease-free areas. Long-distance germplasm exchange as embryo cultures is not recommended until it has been determined whether embryos can be infected and can carry the disease.

5.3.3 Foliar decay

First report and symptoms

Foliar decay is a serious endemic disease of coconut palms in the islands of Vanuatu, first reported around 1964. Symptoms usually appear 1–4 months after infestation with the causal agent and start with the yellowing of intermediate leaves (Fig. 5.7A, B), followed by lateral necrosis near the base of the leaf petiole. The leaves gradually dry out, turning brown, and the base of the petiole cracks prematurely. Later, yellowing extends to the upper leaves (Fig. 5.7C), inflorescences become affected and the fruit turn brown and fall (Randles *et al.*, 1992).

Causal agent

The disease is caused by coconut foliar decay virus (CFDV), belonging to the family *Alphasatellitidae*. The virus consists of isometric particles made up of 1291-nucleotide, single-stranded, circular DNA molecules (Gronenborn *et al.*, 2018) that express replication-associated proteins (Varsani *et al.*, 2021).

Epidemiology

CFDV inhabits the phloem of palms and is distributed heterogeneously in infected palms, so for detection of infection, several samples must be tested for reliability (Randles *et al.*, 1992). CFDV is transmitted to coconut by the cixiid *Myndus taffini* Bonfils, 1983, whose nymphs develop in sea hibiscus (*Hibiscus tiliaceus* L.; Wefels *et al.*, 2015). The virus and vector are confined to the Vanuatu archipelago.

Fig. 5.7. Symptoms of foliar decay disease coconut showing (A, B) the initial yellowing of intermediate leaves (as shown in young palms) and (C) the yellowing extending to the upper leaves (as shown in a mature palm). Photos courtesy of G. Wall (A) and J. Randles (B,C).

Disease management

This disease affects only introduced varieties and hybrids, whereas the local Vanuatu Tall variety is the only material completely tolerant to CFDV. Thus, the main strategy for CFDV management is through conventional genetic improvement to develop good-yielding tolerant hybrid varieties, such as Vanuatu Tall × Rennell Island Tall, which has a high tolerance to CFDV (Labouisse *et al.*, 2011). In addition, it is important to eliminate weeds within and around the crop area as *H. tiliaceus* hosts the nymphal stages of the vector *M. taffini*.

5.4 Pests with Wide Geographical Distribution

5.4.1 *Rhynchophorus palmarum*

Importance and distribution

The South American palm weevil (*Rhynchophorus palmarum* L.) and the red palm weevil (*Rhynchophorus ferrugineus* (Olivier, 1790); see Section 5.4.2), belonging to the order Coleoptera of the genus *Rhynchophorus*, are two major pests affecting coconut palms. *R. palmarum* is found in several South American countries (Fig. 5.8), with its distribution restricted for many years to the tropical forests of the Neotropics (EPPO Global Database, 2005). However, in the last decade, these weevils have been reported outside this range and in southern states of the USA (Fig. 5.8).

Morphological description

Adults of *R. palmarum* (Figs 5.9 and 5.10A) have a black body 29–44 mm long and 11.5–18.0 mm wide. Adult males have silks in the rostrum, while females do not (Fig. 5.10B, C). The eggs are yellowish brown and cylindrical with rounded ends, with an average size of 0.9 × 2.5 mm. The larvae are legless with a creamy-white to ivory colour, and a length of 44–57 mm and a width of 22–25 mm (Watanapongsiri, 1966).

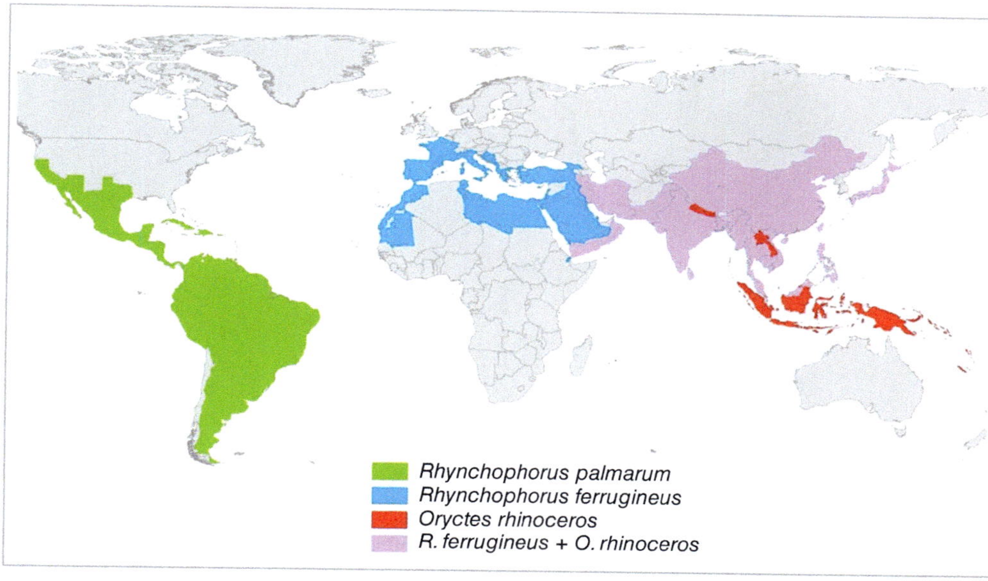

Fig. 5.8. Worldwide distribution of palm weevil pests causing damage to palms. Figure courtesy of D. Palma and A. Peña.

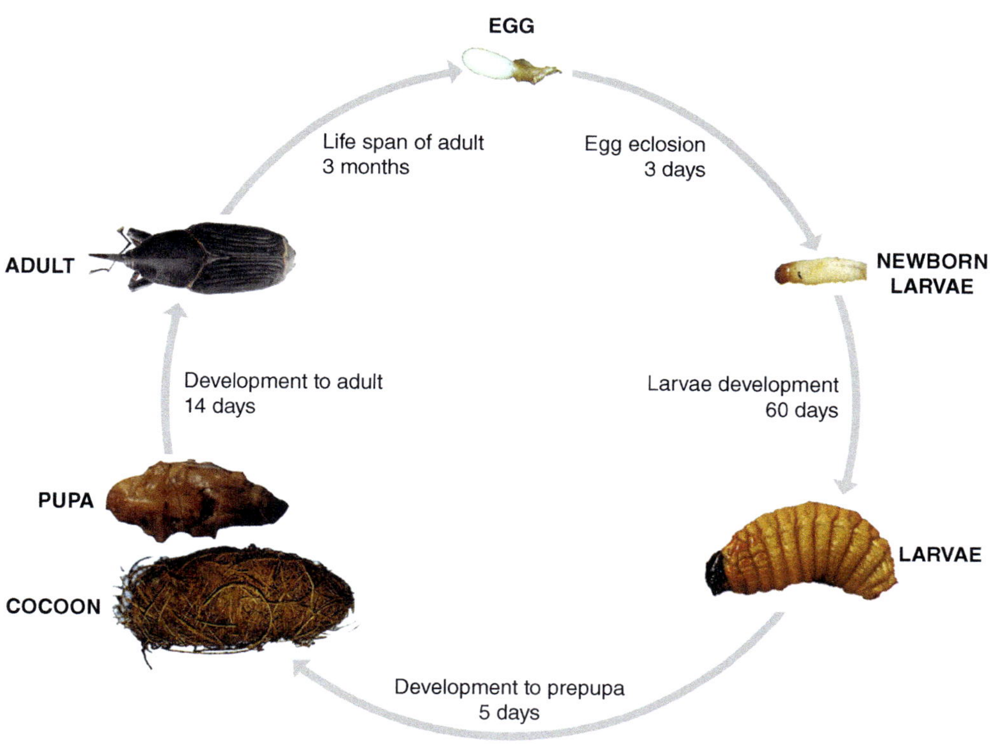

Fig. 5.9. The life cycle of *Rhynchophorus palmarum*. Figure courtesy of A. Peña.

Fig. 5.10. (A) *Rhynchophorus palmarum* adult insect. (B, C) Distinguishing features of males (B) and females (C). (D–F) Symptoms of *R. palmarum* infestation in coconut: progressive yellowing in leaves (D, E), and internal destruction of the trunk by larvae (F). Photos courtesy of A. Peña (A, B, C, D, E, F) and G.A. García-Luna (E).

Biological aspects

The adults disperse by walking or flying and take refuge in the bud, leaf axils and fibrous layers that wrap the stipe of coconut palms (CABI, 2022). Females oviposit inside the palm, and a single female can lay up to 880 eggs, which need 3 days of incubation before hatching (Restrepo *et al.*, 1982). The larvae develop inside the palm, passing through nine stages over a period of 60 days (Vásquez-Ordóñez *et al.*, 2020). They then take a further 5 days to pupate, and 10–14 days to form adults, which can live for up to 2–3 months (Fig. 5.10A; Restrepo *et al.*, 1982). Adults of *R. palmarum* have been shown to affect 35 plant species from 12 different families, but they prefer Arecaceae including coconut and oil palm (CABI, 2020). Damage (Fig. 5.10D, E) is caused by larvae as they feed inside the palm forming galleries that eventually kill the palm (Fig. 5.10F; CABI, 2022). Adults of *R. palmarum* are also a vector of *Bursaphelenchus cocophilus*, which causes RR disease (see Section 5.2.3) that can severely affect coconut palms (Chinchilla and Escobar, 2007).

Phytosanitary management

Palm weevils of both sexes are attracted by an aggregation pheromone (rhynchophorol) produced by males (Jaffé *et al.*, 1993). Based on this, the main control approach is a trap containing a septum impregnated with the pheromone, and possibly a food bait with an insecticide (Camino *et al.*, 2000). For massive trappings, six traps ha^{-1} are recommended (de la Torre *et al.*, 2010). Pheromone efficiency can be increased by using ethyl acetate (Vásquez-Ordóñez *et al.*, 2020). Integral management also includes: (i) the destruction of weevil nests in nearby sites (Chinchilla and Escobar, 2007); (ii) the use of geomatics to aid strategic trap placement (Hurtado and Rincón, 2009); and (iii) employment of biological control agents such as entomopathogenic fungi or parasitoids (Löhr *et al.*, 2019).

5.4.2 *Rhynchophorus ferrugineus*

Economic importance and distribution

The red palm weevil (*R. ferrugineus*) is native to South and South-east Asia, and is currently found in about 50 countries in Africa, Asia, Europe and the Americas (Fig. 5.8; Al-Dobai *et al.*, 2017). It feeds on more than 40 palm species, but in terms of economic importance, coconut, oil palm and date palm are the most important (EPPO, 2008). Losses in palm plantation production in the Mediterranean region alone amount to US$480 million year^{-1} (Al-Dobai *et al.*, 2017).

Morphological description

Red palm weevil adults are 2.0–4.2 cm long and 8–16 mm wide, usually reddish to black, with the legs lighter in colour (Fig. 5.11A). Males have brownish hair on their rostrum, while females do not (Fig. 5.11B, C). The larvae are creamy-white to ivory in colour, have a reddish-brown head, and are about 5 cm long and 2 cm wide (EPPO, 2008).

Biological aspects

R. ferrugineus is an oligophagous insect and the adults prefer to attack palms younger than 20 years of age and complete their life cycle of 3–4 months inside the trunk of the palm (Faleiro, 2006). Females prefer laying their eggs in palms and can oviposit 180–400 eggs during their life. The larvae develop inside the trunk of the palms, from which they feed, completing their development phase in 210 days and their pupal stage in 14–21 days, to form emerging adults (EPPO, 2008). During development, they damage the stem tissues causing death of the palm (Fig. 5.11D).

Phytosanitary management

Management of *R. ferrugineus* is based on detection of early symptoms, the use of baited traps using a pheromone (ferruginol) and trunk injection of certain insecticides (imidacloprid or fenilpirazole) at the initial stages of attack (Kaakeh, 2006). Methods of detection of the larvae include sniffer dogs detecting the volatile compounds produced during larval damage (Suma *et al.*, 2014) and more sophisticated methods based on the use of acoustic sensors (Ashry *et al.*, 2020).

5.4.3 *Oryctes rhinoceros*

Economic importance

The coconut rhinoceros beetle (*Oryctes rhinoceros* L.) is one of the most harmful insects for palm trees found in the Asia-Pacific region (Fig. 5.8). Adults eat the young leaves and cause holes to appear in the tops of the palms, slowing or preventing the development of the plant (Giblin-Davis, 2001). It is estimated that their damage can reduce production by 69%, and can cause death for 25% of young palms that are attacked (Santi *et al.*, 2022). In addition, infestations can occur in conjunction with *R. ferrugineus* (Fig. 5.8), increasing losses further (Manjeri *et al.*, 2014).

Distribution

The origin of *O. rhinoceros* lies between India and Indonesia and has spread to several countries in different regions of the world (Fig. 5.8), including as far away as Hawaii (USA), where it was detected in 2013 (Bedford, 2022).

Morphological description

The adults are 3.0–6.3 cm long and dark brown or black in colour (Fig. 5.12A). Males have

Fig. 5.11. (A) *Rhynchophorus ferrugineus* adult insect. (B, C) Distinguishing features of males (B) and females (C). (D) *Phoenix* palm without damage (left) and damage in an affected palm (right). Photos courtesy of K. Schulz (A), J. Rupérez (B, C) and of A. Torres Ruiz (D; public domain – released by Denghiù, via Wikipedia).

reddish-brown silks in the ventral zone, while females have them in diffuse clusters at the tip of their abdomen. The length of the horn is usually longer in males than in females (Fig. 5.12B–C). The larvae are milky-white in colour with red heads, have three larval stages and grow to 10.0 cm in length (Bedford, 2022).

Biological aspects

O. rhinoceros has about 40 species of host plants, including Arecaceae and non-Arecaceae species, but prefers coconut and oil palm (Bedford, 2022). Damage to the palms is caused by the feeding activity of *O. rhinoceros* (Fig. 5.12D–E), and less so by

Fig. 5.12. (A) *Oryctes rhinoceros* adult insect. (B, C) Distinguishing features of (B) males and (C) females. (D–E) Damage in coconut palms. Photos courtesy of N. Yahya (A), L. Day (B, C) and G. Jackson (D–E).

the larvae feeding on decaying palm materials (Giblin-Davis, 2001), with the females usually ovipositing within the interior tissues of dead palms or decaying plant material (Manjeri *et al.*, 2014).

Phytosanitary management

Males of *O. rhinoceros* produce an aggregation pheromone (4-methyloctanoate ethyl; Hallett *et al.*, 1995) and this is used as an attractant agent in traps for catching the insect. This approach can be used for monitoring using one trap in 2 ha of plantation (Ponnamma *et al.*, 2002), or as a control method using five traps ha^{-1} (Chung, 1997). In addition, there is a biological control approach using *Rhabdovirus oryctes*, a negative-sense RNA virus affecting *O. rhinoceros* (Huger, 2005). The beetle is infected orally, and from it, the virus reaches the larvae causing their death (Bedford, 2014). The virus was released in the late 1960s in various countries, reducing the population of *O. rhinoceros* drastically within 3 years, but recently there has been an increase in the population, possibly due to a decrease in virulence of the biological control isolate (Tanaka *et al.*, 2021).

5.5 Pests with Restricted Geographical Distribution

5.5.1 *Opisina arenosella*

First report and symptoms

The black-headed coconut caterpillar or coconut leaf worm (*Opisina arenosella* Walker; Fig. 5.13A) affects coconut and other palm species (Ramachandran *et al.*, 1979; Cock and Perera, 1987). Reports of *O. arenosella* as a palm pest were rare before 1920. In 1898 it was reported in Sri Lanka and 1909 in India where it was

Fig. 5.13. (A) *Opisina arenosella* adult insect. (B–D) Damage caused in coconut palms. Photos courtesy of K. Moomaung CC BY-SA 3.0 (A), L. Shayma (B), and H. Tsatsia and G. Jackson (C, D).

considered a main defoliator of coconut, spreading from there to other countries in the Asia-Pacific region after 1920 (EPPO Global Database, 2017). It causes damage to coconut palms at all stages of development (Fig. 5.13B–D; Ramachandran *et al.*, 1979). The symptoms of a mild infestation (5–10 larvae per leaflet) are the appearance of dry or necrotic patches on older leaves. In a severe infestation, the leaves of the palm have a burnt appearance (Fig. 5.13C), and silken and excrement galleries are observed, leading to a reduced fruit yield (Perera *et al.*, 1988; Venkatesan *et al.*, 2009).

Morphological description

Adult females can reach a length of 20 mm; they are light grey in colour with antennae and their front wings measure 20–29 mm in length with three faint spots. The males are smaller and have a thin abdomen that ends in a prominent tuft of scales, with fringes of hair on the posterior and anal margin of the wings, providing an easy means of identifying the sexes (Cock and Perera, 1987; Mariau, 2001).

Biological aspects and epidemiology

Females lay their eggs near the apical zone of older leaves (Mariau, 2001), forming a silk web where the larvae and insects develop (Fernando and Chandrasiri, 2010). They feed on the phloem of the leaves where they establish (Cock and Perera, 1987). Although *O. arenosella* adults are slow fliers, they have been shown to travel long distances in strong winds (Howard *et al.*, 2001). In addition, it has been found that the dynamics of this insect population correlates with certain climatic events (Perera *et al.*, 1988).

Phytosanitary management

For management of *O. arenosella*, the recommendations are (i) use of light traps to physically eliminate infestations; (ii) cutting and burning the first affected leaves of the palm; (iii) use of insecticides, although this can be problematic in the long term (Chakravarthy and Doddabasappa, 2009); (iv) use of biopesticides (Kendra, 2010); and (v) use of biological controls of natural enemies such as parasitoids (Venkatesan *et al.*, 2009).

5.5.2 *Brontispa longissima*

First report and distribution

The coconut leaf beetle (*Brontispa longissima* L.; Fig. 5.14A) is a very harmful pest affecting coconut production. It is native to the Aru Islands, Indonesia, Papua New Guinea and the Bismarck Archipelago where it was first reported in the period 1929–1937 (CABI, 2020). It has spread since then to more than 20 countries in the Asia-Pacific and Middle East regions (Aguirre Uribe, 1998). The main host of *B. longissima* is coconut, but it is capable of infesting several palm species (Cabelin and Ceniza, 2021).

Symptoms

Damage by *B. longissima* appears as red-brown narrow decayed strips, running parallel to the midrib, on leaflets of the unopened spear. The damage is caused by the grazing of both adults and larvae. The strips become larger when the leaves unfold and look blotchy and grey-brown in colour, and may even look burnt when symptoms are severe (Fig. 5.14B, C). Palm development and growth can be delayed, affecting the production of fruit, and in some cases causing the death of the palm (Appanah *et al.*, 2007). All ages of palm are affected, and severe damage can be seen on recently planted palms or seedlings in nurseries (Zhang *et al.*, 2015).

Morphological description

Adult males are smaller than females. Their shape is elongated and ovoid, dorsoventrally flattened, 8–12 mm long and 2.00–2.25 mm wide,

Fig. 5.14. (A) *Brontispa longissima* adult insect. (B, C) Damage caused to an individual coconut plant (B) and to the crown (C). Photos courtesy of C. Lee (A), H. Tsatsia and G. Jackson (B) and G. Jackson (C).

with antennae of approximately 2.75 mm. Their colour can be different shades of brown to reddish-brown and even black (Fig. 5.14A; Lever, 1979).

Biological aspects

The egg-to-adult cycle takes 5–9 weeks. The adult beetles are nocturnal and fly well but can only make short flights, often of only a few hundred metres, so their natural spread is slow. They propagate mainly through the movement of infested palms and palm tissues. The main damage is caused when adult *B. longissima* feed on young, unopened leaflets, causing accumulated damage, exceeding that caused by the larvae, which also feed on the leaflets (Cabelin and Ceniza, 2021).

Phytosanitary management

Cultural control practices are based on the removal of infested palm leaves or full palms if necessary (Liu, 1994; Appanah *et al.*, 2007). Chemical control by applying insecticides can be undertaken, and satisfactory control can be achieved at low cost using a low volume of fine spray, applied from above, to the central spike of each palm. With regard to biological control, certain parasitoids and predatory insects have been used successfully in several countries to help control the beetle (Appanah *et al.*, 2007).

5.6 Other Pests and Diseases

There are several other diseases and pests that affect coconut and other palms that have not been included here. For further information, we recommend the reader to consult the list of coconut diseases reported by Ploetz *et al.* (1999), and the websites www.cocopest.org/index.asp (accessed 16 August 2023) for additional pests of coconut and www.cabidigitallibrary.org/ (accessed 16 August 2023) for both pests and diseases.

5.7 Conclusion

This chapter has described a number of pests and diseases that affect coconut with recommendations for their management, but in most cases the recommendations are for dealing with a problem once it has already arrived in a country. What we need, as much as possible, is to act in advance to avoid the arrival of pests and diseases like those described in this chapter. If we take as an example LY, which currently occurs in Honduras (Central America) and Guadalupe (Caribbean), in both cases the disease is spreading from north to south, and if this continues, it will eventually affect the rest of the coconut-producing countries in Central and South America. This will include Brazil, the largest coconut-producing country in the Americas, a country that has developed a very important industry based on packed coconut water that is rapidly expanding into the North American and European markets, setting a trend that will be followed for other coconut products. Unfortunately, 60% or more of the coconut varieties planted in Brazil are LY susceptible and arrival of the disease would be devastating. The same catastrophe could play out in many other coconut-producing countries that are presently without LY.

An even worse catastrophe could happen if the phytoplasmas present in other parts of the world arrive in the Americas. For instance, the phytoplasma that is responsible for Saint Paul Wilt disease, presently in Africa, has been shown to kill Malayan Dwarf coconut varieties (Quaicoe *et al.*, 2009), which are currently the LY-resistant varieties planted in the Americas. Thus, there is a significant risk of movement of phytoplasma diseases in both directions, from west to east and east to west, and this could easily happen with the movement of plant material (Fig. 5.15).

Similar intercontinental exchanges could also happen for other high-risk pathogens and pests, and therefore we need to prepare for this eventuality, as well as strengthening the quarantine measures between neighbouring countries and within countries. In addition, more research is needed to improve coconut disease and pest management. Although some organizations such as the Organismo Internacional Regional de Sanidad Agropecuaria (OIRSA, El Salvador) and CGIAR (formerly known as the Consultative Group on International Agricultural Research) promote regional and global alliances to prevent and manage devastating plant pest and disease calamities, unfortunately there is currently no regional or global initiative specifically focused on coconut.

Fig. 5.15. This is a conversation that actually happened somewhere in a coconut-producing country in South America. Figure courtesy of N. Marmolejo.

It is very important that the coconut community starts developing organized regional and global alliances. For this to be successful, it will be necessary for the leadership of national coconut organizations to participate. Fortunately, the International Coconut Community (ICC, Indonesia) and the International Coconut Genetic Resources Network (COGENT, Indonesia) are working on promoting good practices for safe coconut germplasm exchange, field screening markers for resistant varieties and developing tools for molecular detection of agents. Deeper cross-border partnerships and research investment through regional and global alliances are likely be the only way the industry can successfully deal with the current suite of pests and diseases and address emerging coconut industry threats.

5.8 References

Aguirre Uribe, L. (1998) Official Mexican Norm NOM-067-FITO-1995, in which you have the procedures for production and phytosanitary certification for hybrid seed of coconut against 'amarillamiento letal'. Memoria, Mexico.

Al-Dobai, S., Elkahky, M. and Faleiro, R. (eds) (2017) *Proposed Multi-disciplinary and Multi-regional Strategy for the Management of Red Palm Weevil. Proceedings of the Scientific Consultation and High-Level Meetings on Red Palm Weevil Management,* 29–31 March, Rome. Available at: www.fao.org/publications/card/fr/c/CA1541EN/ (accessed 15 August 2023).

Appanah, S., Sim, H.C. and Sankaran, K.V. (eds) (2007) Developing an Asia-Pacific strategy for forest invasive species: the coconut beetle problem – bridging agriculture and forestry. RAP Publication 2007/02. Food and Agriculture Organization of the United Nations, Regional Office for Asia and the Pacific, Bankok. Available at: www.fao.org/documents/card/en?details=bcff7ed8-1127-5f29-9968-31334a3364ce%2f (accessed 15 August 2023).

Ashburner, G.R., Cordova, I.I., Oropeza, C.M., Illingworth, R. and Harrison, N.A. (1996) First report of coconut lethal yellowing disease in Honduras. *Plant Disease* 80: 960.

Ashry, I., Mao, Y., Al-Fehaid, Y., Al-Shawaf, A., Al-Bagshi, M. *et al.* (2020) Early detection of red palm weevil using distributed optical sensor. *Scientific Reports* 10: 3155.

Asipuela-Haro, R., Torres, Á. and Martínez, R. (2017) Transmission of the sudden wilt disease in the O × G oil palm (*Elaeis oleifera × Elaeis guineensis)* by the bugs *Lincus curvatus* Campos and Roell and *Macropygium reticulare* F. (Hemiptera: Pentatomidae). *Revista de Protección Vegetal* 32, 1–9.

Bedford, G. (2014) Advances in the control of rhinoceros beetle, *Oryctes rhinoceros* in oil palm. *Journal of Oil Palm Research* 26, 183–194.

Bedford, G. (2022) *Oryctes rhinoceros* (coconut rhinoceros beetle). In: *CABI Compendium*. CAB International, Wallingford, UK. Available at: https://doi.org/10.1079/cabicompendium.37974 (accessed 15 August 2023).

Boccardo, G., Beaver, R.G., Randles, J.W. and Imperial, J.S. (1981) Tinangaja and bristle top, coconut diseases of uncertain etiology in guam, and their relationship to cadang-cadang disease of coconut in the Philippines. *Phytopathology* 71, 1104–1107.

Brasier, C.M. and Griffin, M.J. (1979) Taxonomy of *Phytophthora palmivora* on cocoa. *Transactions of the British Mycological Society* 72, 111–143.

Cabelin, G. and Ceniza, M.J.C. (2021) Comparative biology of coconut leaf beetle (Clb) *Brontispa longissima* (Gestro) (Coleoptera: Chrysomelidae) on selected coconut varieties. *CLSU International Journal of Science and Technology* 5, 8–20.

CABI (2020) *Brontispa longissima* (coconut hispine beetle). In: *CABI Compendium*. CAB International, Wallingford, UK, p. 10059. Available at: https://doi.org/10.1079/cabicompendium.10059 (accessed 15 August 2023).

CABI (2022) *Rhynchophorus palmarum* (South American palm weevil). In: *CABI Compendium*. CAB International, Wallingford, UK. Available at: https://doi.org/10.1079/cabicompendium.47473 (accessed 15 August 2023).

Camino, M., Hernández, R., Gutiérrez, O., Castrejón, G., Arzuffi, B. *et al.* (2000) Pruebas con la feromona de agregación (rhynchophorol: Rhynko-Lure) producida por el macho de *Rhynchophorus palmarum* en la Costa Grande de Guerrero, México. *ASD Oil Palm Papers* 20, 9–12.

Castañeda, J.G.D.L.M., Mora, I.Y.L., Cupul, W.C. and García, J.M.P. (2022) Efficacy and cost efficiency of traps for mass capture of *Rhynchophorus palmarum* L. in a Mexican *Cocos nucifera* L. orchard. *Tropical and Subtropical Agroecosystems* 25: 108.

Chakravarthy, A.K. and Doddabasappa, B. (2009) *Testing of Garlic-based Biopesticide on Insect Pests of Coconut (*Cocus nucifera *L.): A Report*. University of Agricultural Sciences, Gandhi Krishi Vignan Kendra, Bangalore.

Chase, A.R. and Broschat, T.K. (1991) *Diseases and Disorders of Ornamental Palms*. APS Press, St Paul, Minnesota.

Chinchilla, C. M., and Escobar, R. (2007) El anillo rojo y otras enfermedades de la palma aceitera en centro y suramérica. *ASD Oil Palm Papers* 30, 1–27.

Chung, G.F. (1997) The bioefficacy of the aggregation pheromone in mass trapping of rhinoceros beetles (*Oryctes rhinoceros* L.) in Malaysia. *Planter* 73, 119–127.

Cock, M.J.W. and Perera, P. (1987) Biological control of *Opisina arenosella* Walker (Lepidoptera, Oecophoridae). *Biocontrol News and Information* 8, 283–310.

Coconut Development Board (2022) Cultivation practices for coconut. *Indian Coconut Journal* LXV, 34–36.

Cordova, I., Jones, P., Harrison, N.A. and Oropeza, C. (2003) *In situ* PCR detection of phytoplasma DNA in embryos from coconut palms with lethal yellowing disease. *Molecular Plant Pathology* 4, 99–108.

de la Torre, A.R.C., de la Torre, A.J.A. and Moya, O.M. (2010) Biología, hábitos y manejo de *Rhynchophorus palmarum* L. (Coleoptera: Curculionidae). *Revista Mexicana de Ciencias Agrícolas* 1, 613–624.

Dollet, M. and Lopez, G. (1978) Study of the association between flagellate protozoa and *Marchitez sorpresiva* in South American oil palms. *Oleagineux* 33, 209–217.

Dollet, M., de Franqueville, H. and Ducamp, M. (2012) Bud rot and other major diseases of coconut, a potential threat to oil palm. *Oilseeds and Fats, Crops and Lipids* 19, 102–107.

Dzido, J.L., Sánchez, R., Dollet, M., Julia, J.F., Narváez, M. *et al.* (2020) *Haplaxius crudus* (Hemiptera: Cixiidae) transmits the lethal yellowing phytoplasmas, 16SrIV, to *Pritchardia pacifica* Seem. and *H. Wendl* (Arecaceae) in Yucatan, Mexico. *Neotropical Entomology* 49, 795–805.

Earthstoriez (2018) India on the coconut palm. Available at: www.earthstoriez.com/india-coconut-myth-folklore/ (accessed 31 August 2023).

Eden-Green, S.J. (1995) History, world distribution and present status of lethal yellowing-like diseases of palms. In: *Proceedings of an International Workshop on Lethal Yellowing-Like Diseases of Coconut,* Elmina, Ghana, November 1995. Natural Resources Institute, London, pp. 9–25.

EPPO (2008) Data sheets on quarantine pests. *Rhynchophorus ferrugineus. EPPO Bulletin* 38, 55–59.

EPPO Global Database (2005) *Rhynchophorus palmarum*. European and Mediterranean Plant Protection Organization (EPPO), Luxembourg. Available at: https://gd.eppo.int/taxon/RHYCPA/datasheet (accessed 15 August 2023).

EPPO Global Database (2017) *Opisina arenosella (NEPASE): Distribution*. European and Mediterranean Plant Protection Organization (EPPO), Luxembourg. Available at: https://gd.eppo.int/taxon/NEPASE/distribution (accessed 15 August 2023).

Faleiro, J.R. (2006) A review of the issues and management of the red palm weevil *Rhynchophorus ferrugineus* (Coleoptera: Rhynchophoridae) in coconut and date palm during the last one hundred years. *International Journal of Tropical Insect Science* 26, 135–154.

Fenwick, D.W. (1968) Red-ring disease of the coconut palm. In: Smart, G.C. Jr and Perry, V.G.V. (eds) *Tropical Nematology*. Center for Tropical Agriculture. University Agriculture Press, Gainesville, Florida, p. 38–48.

Fernando, L.C.P. and Chandrasiri, K.A.S. (2010) More evidence on the attraction of males to the female sex pheromone of *Opisina arenosella* Walker. *COCOS* 12, 79–81.

Garcia, A., Ribeiro, G. and Rodrigues, A. (1999) Ocorrência das principais doenças do coqueiro (*Cocos nucifera* L.) em Rondônia e medidas de controle. *Fitopatologia Brasileira* 24, 182–185.

Giblin-Davis, R.M. (2001) Borers of palms. In: Howard, F.W., Moore, D., Giblin-Davis, R.M. and Abad R.G. (eds) *Insects on Palms*. CAB International, Wallingford, UK, pp. 267–305.

Griffith, R. (1983) Red ring research 1956–1983. *Oleagineux* 38, 423–429.

Griffith, R. (1987) Red ring disease of coconut palm. *Plant Disease* 71, 193–196.

Griffith, R., Giblin-Davis, R.M., Koshy, P.K. and Sosamma, V.K. (1990) Nematode parasites of coconut and other palms. In: Luc, M., Sikora, R.A. and Bridge, J. (eds) *Plant Parasitic Nematodes in Subtropical and Tropical Agriculture*. CAB International, Wallingford, UK, pp. 493–527.

Gronenborn, B., Randles, J.W., Knierim, D., Barrière, Q., Vetten, H.J. *et al.* (2018) Analysis of DNAs associated with coconut foliar decay disease implicates a unique single-stranded DNA virus representing a new taxon. *Scientific Reports* 8: 5698.

Hadidi, A., Flores, R., Randles, J.W. and Semancik, J. (2003) *Viroids*. CSIRO Publishing, Collingwood, Australia.

Hadidi, A., Sun, L. and Randles, J.W. (2022) Modes of viroid transmission. *Cells* 11: 719.

Hallett, R.H., Perez, A.L., Gries, G., Gries, R., Pierce, H.D. *et al.* (1995) Aggregation pheromone of coconut rhinoceros beetle, *Oryctes rhinoceros* (L.) (Coleoptera: Scarabaeidae). *Journal of Chemical Ecology* 21, 1549–1570.

Hanold, D. and Randles, J.W. (1991) Coconut cadang-cadang disease and its viroid agent. *Plant Disease* 75, 330–335.

Hodgson, R.A.J., Wall, G.C. and Randles, J.W. (1998) Specific identification of coconut tinangaja viroid for differential field diagnosis of viroids in coconut palm. *Phytopathology* 88, 774–781.

Holmes, F.O. (1925) Non-pathogenicity of the milkweed flagellate in Maryland. *Phytopathology* 15, 294–296.

Howard, F.W. (1987) *Myndus crudus* (Homoptera: Cixiidae), a vector of lethal yellowing of palms. In: *Proceedings of the 2nd International Workshop on Leafhoppers and Planthoppers of Economic Importance*. Brigham Young University, Provo, UT, pp. 117–129.

Howard, F.W., Moore, D., Giblin-Davis, R.M. and Abad, R.G. (2001) Defoliators of palms. In: Howard, F.W., Moore, D., Giblin-Davis, R.M. and Abad, R.G. (eds) *Insects on Palms*. CAB International, Wallingford, UK, pp. 33–108.

Huger, A.M. (2005) The *Oryctes* virus: its detection, identification, and implementation in biological control of the coconut palm rhinoceros beetle, *Oryctes rhinoceros* (Coleoptera: Scarabaeidae). *Journal of Invertebrate Pathology* 89, 78–84.

Hurtado, R.E. and Rincón, V.O. (2009) La geomática como instrumento para modelar y hacer seguimiento a la pudrición del cogollo (PC) y al insecto plaga: *Rhynchophorus palmarum*. Zona Occidental-Tumaco (Colombia). *Palms* 30, 21–35.

Iyer, R., Gunasekharan, M. and Hegde, V. (2018) Coconut: maladies and remedies. In: Nampoothiri, K.U.K., Krishnakumar, V., Thampan, P.K. and Achuthan Nair, M. (eds) *The Coconut Palm (*Cocos nucifera *L.) – Research and Development Perspectives*. Springer, Singapore, pp. 489–518.

Jaffé, K., Sánchez, P., Cerda, H., Hernández, J.V., Jaffé, R. *et al.* (1993). Chemical ecology of *Rhynchophorus palmarum*: attraction to host plants and to a male-produced aggregation pheromone. *Journal of Chemical Ecology* 19, 1703–1720.

Kaakeh, W. (2006) Toxicity of imidacloprid to developmental stages of *Rhynchophorus ferrugineus* (Curculionidae: Coleoptera): laboratory and field tests. *Crop Protection* 25, 432–439.

Kendra, G.K.V. (2010). *Testing of Garlic Based Bio-pesticide on Insect Pests of Coconut (Cocos nicufera L). A Report*. University of Agricultural Sciences, Bangalore, India.

Koltunow, A.M. and Rezaian, A. (1989) A scheme for viroid classification. *Intervirology* 30, 194–201.

Labouisse, J.P., Sileye, T., Bonnot, F. and Baudouin, L. (2011) Achievements in breeding coconut hybrids for tolerance to coconut foliar decay disease in Vanuatu, South Pacific. *Euphytica* 177, 1–13.

Lédo, A. da S., Passos, E.E.M., Fontes, H.R., Ferreira, J.M.S., Talamini, V. *et al*. (2019) Avanços na propagação do coqueiro. *Revista Brasileira de Fruticultura* 41, e159–e173.

Lever, R.J.A.W. (1979) *Pests of the Coconut Palm*. Food and Agriculture Organization of the United Nations, Rome.

Liu, S. (1994) *The Application of Fungicide Resistant Entomopathogenic Green Muscardine Fungus in Taiwan: Biological control of Coconut Leaf Beetle (*Brontispa longissima*) and Diamondback Moth (*Plutella xylostella*)*. Technical Bulletin Vol. 138, Food and Fertilizer Technology Center for the Asian and Pacific Region.

Löhr, B., Negrisoli, A. and Molina, J.P. (2019) *Billaea rhynchophorae*, a palm weevil parasitoid with global potential. *Journal of Plant Protection* 37, 101–108.

Louise, C., Dollet, M. and Mariau, D. (1986) Recherches sur le Hartrot du cocotier, maladie à Phytomonas (Trypanosomatidae) et sur son vecteur *Lincus* sp. (Pentatomidae) en Guyane. *Oléagineux* 14, 437–448.

Manjeri, G., Muhamad, R. and Tan, S.G. (2014). *Oryctes rhinoceros* beetles, an oil palm pest in Malaysia. *Annual Research and Review in Biology* 4, 3429–3439.

Mariau, D. (2001) *La Fauna de la Palma de Aceite y del Cocotero: los Insectos y Àcaros Plagas y sus Enemigos Naturales*, Vol. 15. Editions Quae, Versailles, France.

Martyn, E.E. (2021) Red ring disease of coconuts in Trinidad and Tobago. *Tropical Agriculture* 98, 218–225.

McCoy, R.E. and Martinez-Lopez, G. (1982) *Phytomonas staheli* associated with coconut and oil palm diseases in Colombia. *Plant Disease* 66, 675–677.

Mirisola Filho, L.Â. (2002) *Cultivo de Coco Anão*. Aprenda Fácil Editoria, Viçosa, Brazil.

Mohamed, N.A. and Imperial, J.S. (1984) Detection and concentration of coconut cadang-cadang viroid in coconut leaf extracts. *Phytopathology* 74, 165–169.

Moscoso-Ramírez, P.A., Ortíz-García, C.F., Palma-López, D., Ruíz-Beltrán, P. and Sánchez-Soto, S. (2002) Incidencia de enfermedades letales en progenitores e híbridos de cocotero en Tabasco, México. *Revista Fitotecnia Mexicana* 25, 327–332.

Mou, D.F., Humphries, A.R., Soto, N., Helmick, E.E., Ascunce, M.S. *et al*. (2020) A survey of auchenorrhynchan insects for identification of potential vectors of the 16SrIV-D phytoplasma in Florida. *Florida Entomologist* 103, 344–352.

Narváez, M., Vázquez-Euán, R., Harrison, N.A., Nic-Matos, G., Julia, J.F. *et al*. (2018) Presence of 16SrIV phytoplasmas of subgroups A, D and E in planthopper *Haplaxius crudus* Van Duzee insects in Yucatán, Mexico. *3 Biotech* 8: 61.

Narváez, M., Nic-Matos, G. and Oropeza, C. (2022) *In vitro* transmission of 16SrIV phytoplasmas to coconut plants by *Haplaxius crudus* in Yucatan, Mexico. *3 Biotech* 12: 5.

Oropeza, C., Cordova, I., Puch-Hau, C., Castillo, R., Chan, J.L. and Sáenz, L. (2017) Detection of lethal yellowing phytoplasma in coconut plantlets obtained through *in vitro* germination of zygotic embryos from the seeds of infected palms. *Annals of Applied Biology* 171, 28–36.

Oropeza, C., Sáenz, L., Narváez, M., Nic-Matos, G., Córdova, I. *et al*. (2020) Dealing with lethal yellowing and related diseases in coconut. In: Adkins, S., Nguyen, Q. and Biddle, J. (eds) *Coconut Biotechnology: Towards the Sustainability of the 'Tree of Life'*. Springer International Publishing, Cham, Switzerland, pp. 169–197.

Ovando-Cruz, M.E., Serrano-Altamirano, V., Gálvez-Marroquín, L.A., Ariza-Flores, R., Martínez-Bolaños, M. *et al*. (2019) Evaluación de trampas para *Rhynchophorus palmarum* L. (Coleoptera: Curculionidae) en la costa de Oaxaca, México. *Agro Productividad* 12, 3–8.

Palma-Cancino, P.J., Jose, A.S.G., Maria, N., German, N.M., Yasmin, C.M. *et al*. (2020) First report of mortality in *Phoenix canariensis* associated with subgroup 16SrIV-D phytoplasmas in Coahuila, Mexico. *African Journal of Biotechnology* 19, 846–857.

Pedaute, R.F. and Duran-Vila, N. (2016) Viroides: aspectos generals. In: Flores, R. and Serra-Soriano, R. (eds) *Enfermedades de Plantas Causadas por Virus y Viroides*. Bubok, Sociedad Española de Fitopatología, Spain, pp. 53–91.

Perera, P., Hassell, M.P. and Godfray, H.D.J. (1988) Population dynamics of the coconut caterpillar, *Opisina arenosella* Walker (Lepidoptera: Xyloryctidae), in Sri Lanka. *Bulletin of Entomological Research* 78, 479–492.

Plavsic-Banjac, B., Hunt, P. and Maramorosch, K. (1972) Mycoplasmalike bodies associated with lethal yellowing disease of coconut palms. *Phytopathology* 62, 298–299.

Ploetz, R., Harrison, N. and Jones, P. (1999) Diseases of coconut palm (*Cocus nucifera* L.). APS Press, St Paul, Minnesota. Available at: www.apsnet.org/edcenter/resources/commonnames/Pages/CoconutPalm.aspx (accessed 15 August 2023).

Ponnamma, K.N., Lalitha, N. and Khan, A.S. (2002) Bioefficacy of the pheromone-rhinolure in the integrated pest management for rhinoceros beetle, *Oryctes rhinoceros* L. – a major pest of oil palm. In: *Proceedings of the 15th Plantation Crops Symposium Placrosym XV,* 10–13 December, 2002, Mysore, India. Central Coffee Research Institute, Karnataka, India, pp. 525–530.

Prades, A., Salum, U.N. and Pioch, D. (2016) New era for the coconut sector. What prospects for research? *Oilseeds and Fats, Crops and Lipids* 23: D607.

Quaicoe, R.N., Dery, S.K., Philippe, R., Baudouin, L., Nipah, J.O. *et al.* (2009) Resistance screening trials on coconut varieties to Cape Saint Paul wilt disease in Ghana. *Oléagineux* 16, 132–136.

Quillec, G. and Renard, J.L. (1984) *Phytophthora* rot of coconut palm. *Oleagineux* 39, 143–147.

Ramachandran, C.P., Ponnamma, K.M., Koya, K.M. and Kurian, C. (1979) The coconut leaf-eating caterpillar, *Nephantis serrinopa* Meyrick – a review. *Philippine Journal of Coconut Studies* 4, 9–17.

Randles, J.W. and Hatta, T. (1979) Circularity of the ribonucleic acids associated with cadang-cadang disease. *Virology* 96, 47–53.

Randles, J.W., Miller, D. C., Morin, J.-P., Rohde, W. and Hanold, D. (1992) Localisation of coconut foliar decay virus in coconut palm. *Annals of Applied Biology* 121, 601–617.

Restrepo, L.G., Rivera, F. and Raigosa, J.D. (1982) Ciclo de vida, hábitos y morfometria de *Metamasius hemipterus* Olivier y *Rhynchophorus palmarum* L. (Coleoptera: Curculionidae) en caña de azucar (*Saccharum officinarum* L.). *Acta Agronómica* 32, 33–44.

Rifai, M.A. (1969) A revision of the genus *Trichoderma*. *Mycological Papers* 116, 1–56.

Rodriguez, M.J.B., Vadamalai, G. and Randles, J.W. (2017) Economic significance of palm tree viroids. In: Hadidi, A., Flores, R., Randles, J.W. and Palukaitis, P. (eds) *Viroids and Satellites*. Elsevier, London, pp. 39–49.

Sáenz, A. (2005) Aspectos generales e importancia del agente causal de anillo rojo. *Revista Palmas* 26, 59–70.

Santi, I.S., Prayoga, A., Setyawan, H. and Atmaka Aji, W. (2022) Implementation of integrated pest control to reduce rhino beetle (*Oryctes rhinoceros*) attacks in oil palm plantations. *Tropical Plantation Journal* 1, 68–74.

Suma, P., La Pergola, A., Longo, S. and Soroker, V. (2014) The use of sniffing dogs for the detection of *Rhynchophorus ferrugineus*. *Phytoparasitica* 42, 269–274.

Tanaka, S., Harrison, R.L., Arai, H., Katayama, Y., Mizutani, T. *et al.* (2021) Confirmation of *Oryctes rhinoceros* nudivirus infections in G-haplotype coconut rhinoceros beetles (*Oryctes rhinoceros*) from Palauan PCR-positive populations. *Scientific Reports* 11: 18820.

Thévenin, J.-M., Motulo, H.F.J., Kharie, S. and Renard, J.L. (1995) Lutte chimique contre la pourriture du coeur à *Phytophthora* du cocotier en Indonésie. *Plantations, Recherche, Développement* 2, 41–50.

Tropicsafe (2022) Project – insect-borne prokaryote-associated diseases in tropical and subtropical perennial crops. EU-Horizon-2020. No. 727459. Available at: www.tropicsafe.eu/ (accessed 16 August 2023).

Tuckek, C. (1926) *Phytophthora* bud rot of coconut palms in Porto Rico. *Journal of Agricultural Research* 32, 471–498.

Vadamalai, G., Thanarajoo, S.S., Joseph, H., Kong, L.L. and Randles, J.W. (2017) Coconut cadang-cadang viroid and coconut tinangaja viroid. In: Hadidi, A., Flores, R., Randles, J.W. and Palukaitis, P. (eds) *Viroids and Satellites*. Elsevier, London, pp. 263–273.

Varsani, A., Martin, D.P., Randles, J.W., Vetten, H.J., Thomas, J.E. *et al.* (2021) Taxonomy update for the family Alphasatellitidae: new subfamily, genera, and species. *Archives of Virology* 166, 3503–3511.

Vásquez-Ordóñez, A.A., Löhr, B.L. and Marvaldi, A.E. (2020) Comparative morphology of the larvae of the palm weevils *Dynamis borassi* (Fabricius) and *Rhynchophorus palmarum* (Linnaeus) (Curculionidae: Dryophthorinae): two major pests of peach palms in the Neotropics. *Papéis Avulsos de Zoologia* 60: e202060(s.i.).27.

Venkatesan, T., Jalali, S.K. and Srinivasamurthy, K. (2009) Competitive interactions between *Goniozus nephantidis* and *Bracon brevicornis*, parasitoids of the coconut pest *Opisina arenosella*. *International Journal of Pest Management* 55, 257–263.

Watanapongsiri, A. (1966) A revision of the genera *Rhynchophorus* and *Dynamis*. PhD thesis, Department of Agriculture, Bangkok.
Wefels, E., Morin, J.P. and Randles, J.W. (2015) Molecular evidence for a persistent-circulative association between coconut foliar decay virus and its vector *Myndus taffini*. *Australasian Plant Pathology* 44, 283–288.
Weintraub, P.G., Trivellone, V. and Krüger, K. (2019) The biology and ecology of leafhopper transmission of phytoplasmas. In: Bertaccini, A., Weintraub, P.G., Rao, G.P. and Mori, N. (eds) *Phytoplasmas: Plant Pathogenic Bacteria – II: Transmission and Management of Phytoplasma-associated Diseases*. Springer, Singapore, pp. 27–51.
Wierzbicki, R. and Eiras, M. (2021) Pospiviroides de importância para hortaliças e ornamentais: panorama dos aspectos fitopatológicos e dos riscos para os sistemas agrícolas brasileiros. *Revisão Anual de Patologia de Plantas* 27, 51–76.
Zhang, X., Tang, B. and Hou, Y. (2015) A rapid diagnostic technique to discriminate between two pests of palms, *Brontispa longissima* and *Octodonta nipae* (Coleoptera: Chrysomelidae), for quarantine applications. *Journal of Economic Entomology* 108, 95–99.

6 Physiological and Nutritional Disorders of Coconut

Shunmugiah Veluchamy Ramesh[1]*, Kukkehalli Balachandra Hebbar[1], Ravi Bhat[2], Periasamy Subramanian[2], P. Abhin Sukumar[1], Sundaravelpandian Kalaipandian[3] and Steve W. Adkins[3]

[1]Division of Physiology, Biochemistry and Post-Harvest Technology, Indian Council of Agricultural Research (ICAR), Central Plantation Crops Research Institute (CPCRI), Kasaragod, Kerala, India; [2]Division of Crop Production, ICAR-CPCRI, Kasaragod, Kerala, India; [3]School of Agriculture and Food Sustainability, The University of Queensland, Gatton, Australia

Abstract

Coconut (*Cocos nucifera* L.) is an important plantation crop growing in tropical regions of the world and commanding an immense commercial potential. Considering the perennial nature of the crop, it is important to manage physiological and nutritional disorders when they appear. Most disorders are due to unfavourable interactions at the soil, water and palm interface, and at times, these interactions are aggravated by other biotic and abiotic factors. In general, a good supply of nutrients can address most soil deficiencies and thereby overcome any symptoms associated with nutritional disorders. Instances of an association with biotic agents or other factors are also reported, such as those caused by lethal yellowing disease. In this context, this chapter discusses and presents an overview of the nutritional requirements of the coconut palm and its various growth disorders. It concludes with the physiological optimization of growth and development of coconut and presents future perspectives.

6.1 Introduction

Coconut (*Cocos nucifera* L.) is a perennial and commercially important oil crop grown in humid tropical environments that feeds millions of people. Coconut is one of the most important palm crops in the world and is known as the 'Tree of Life', because of its myriad of food, nutritional, therapeutic and cosmetic uses (Foale, 2003). Coconut products include coconut oil, tender coconut water, raw kernel, copra, coconut shell, coconut cake, coir pith, wood and leaf-based products (DebMandal and Mandal, 2011). The various parts of the coconut palm have many practical uses, ranging from food to coir for stuffing pillows and making ropes, mats, beds and cots, and wood to make utensils, furniture, rosary boxes, brushes and fuel (Watt, 1889). Copra, the dry kernel of the coconut fruit, is commonly used for oil extraction, and contains 65–75% oil (NMCE, 2007). Besides being used for oil extraction, the kernel is also used as a raw material for chutney, curries, sweets, toffee and other foods. The unopened floral spathe is tapped to extract inflorescence sap (neera) or used to make fermented toddy, which can be processed into jaggery, sugar and vinegar (Hebbar *et al.*, 2022). Analysis of tender coconut water suggests that it

* Email: ramesh.sv@icar.gov.in

DOI: 10.1079/9781789249736.0006

contains several inorganic ions, including potassium (K; 290 mg 100 mL^{-1}), calcium (Ca; 44 mg 100 mL^{-1}), sodium (Na; 42 mg 100 mL^{-1}), magnesium (Mg; 10 mg 100 mL^{-1}) and phosphorus (P; 9.2 mg 100 mL^{-1}), all of which are vital for human health (Effiong *et al.*, 2010).

The ideal monthly average rainfall for coconut fruit production is 150 mm. Coconut palms can be found in regions with an annual rainfall ranging from 1300 to 2500 mm (Rajagopal *et al.*, 1990). However, coconut is frequently exposed to water-deficit conditions, either from soil or atmospheric conditions, and owing to its perennial nature, it has a relatively long productive lifespan (Rajagopal and Kasturi Bai, 2002). A mature coconut palm produces a bunch of fruit every month and hence on average a palm produces more than 12 bunches of fruit each year (each bunch contains 8–10 fruit) under ideal climatic conditions (DebMandal and Mandal, 2011). A major factor driving coconut productivity is its biological efficiency, which is principally determined by its efficiency in dry-matter production and the appropriate partitioning of the dry matter into economically important palm parts. Genetic makeup, soil characteristics, agro-climatic conditions and management strategy greatly influence coconut productivity (Naresh Kumar *et al.*, 2018).

6.1.1 Overview of the nutritional requirements of coconut

To develop comprehensive methods for estimating crop fertilizer requirements, two basic tests, a soil test and a plant test, have been developed (Nathanael, 1961). Since Justus von Liebig (1840) proposed his mineral theory of crop nutrition, soil and plant scientists have created and used a variety of tests to determine a crop's nutritional requirements. The oldest of these, a field chemical test, has been used increasingly to determine an appropriate fertilizer application to improve crop productivity. However, soil nutrient profiles can be unhelpful because the symptomatology (due to a nutrient deficiency) does not always show a similarity to the nutrient profile; hence, such tests are considered to have little relevance when addressing deficiencies in palm nutrition. Furthermore, the optimal soil nutrient levels for any palm species have yet to be well determined. A more reliable method for identifying nutrient deficiencies in perennial crops is through a leaf sap test. Such a diagnostic test offers more accurate knowledge of the nutritional condition of the plant in relation to the fertility of the soil, especially for tree crops such as coconut (Selvamani and Duraisami, 2014). From field experiments on hybrid coconut palms, Mohandas (2012) designed experiments to maintain soil health and promote consistent fruit production. It was found that 50% of the necessary nitrogen (N) should be provided by organic manure, such as vermicompost or composted coir pith, while the remaining 50% should be provided by inorganic fertilizers. In coastal coconut soils with low organic matter, organic manures that contain 50% of the required N can also help by adding further organic matter to the soil.

6.1.2 Physiological and nutritional disorders in palms

Physiological disorders can be described as 'variations in a plant's normal physiological processes caused by inadequate environmental conditions, such as nutritional imbalance, or genetic inheritance' (Meera *et al.*, 2017). Physiological disorders of plants considerably affect the development and production of high-quality fruit (Mishra *et al.*, 2016). Extremes in environmental factors such as ambient temperature, light intensity, soil moisture, aeration and nutrient content can disrupt the metabolic processes of plants, which in turn causes various disorders (Kumar and Kumar, 2016). At the farm level, it is difficult to control the abiotic circumstances that cause physiological disorders. Yield and quality improvement can be achieved by choosing the right varieties to plant and by the crop protection and agronomic practices used (Meera *et al.*, 2017). Some of the commonly occurring physiological and nutritional disorders in coconut and their features are shown in Table 6.1.

For a variety of physiological disorders, the symptoms may appear to be disease related, but they are usually preventable by modifying the environment. Micronutrient deficiencies in fruit crops cause substantially more developmental abnormalities than macronutrient shortages

Table 6.1. Common physiological and nutritional disorders of coconut (*Cocos nucifera* L.) caused by various agents, biotic and abiotic factors, and nutritional deficiencies.

Disorder	Characteristic symptoms	Causative agent(s)	Remedial measures	Reference(s)
Barren fruit	Oblong fruit, defective or deformed kernel and shell Low husk content	Coreid bug (*Paradasynus rostratus*) Eriophid mites (*Aceria guerreronis*) Rodents Defective pollination	2,4-Dichlorophenoxyacetic acid spray Bee keeping Thinning of fruit and application of potassium chloride (1 kg) and sodium borate (200 g)	Sandhu and Gill (2013); Sharma (2006)
Button shedding	Premature fruit fall	Multiple factors (e.g. abiotic, nutritional deficiencies)	Root feeding of 1-naphthaleneacetic acid (~220 μM) Mulching and summer irrigation	Sudhakara (1990)
Tapering stem wilt or pencil tip	Diminished vigour and lanky palms Decreasing collar girth causing pencil-point symptom	Nutrient deficiency and genetic makeup of palms	Root feeding of micronutrient (iron, manganese, copper) mixture	Thirumalaiswamy *et al.* (1994)
Coconut rapid decline	Reduced trunk girth and leaf chlorophyll Low nut yield	Copper and zinc micronutrient deficiencies	Micronutrient mixture application	Ranasinghe *et al.* (2002)
Leaf scorch decline	Decreased leaf area, photosynthesis and transpiration High levels of abscisic acid in xylem	Adverse abiotic factors	Soil structural improvement and appropriate nutrition supply	Ranasinghe (2005)
Finschhafen disorder	Yellowing or bronzing of leaflets	Leafhopper (*Zophiuma butawengi*)	Encouraging natural enemies	Gitau *et al.* (2011)
Lethal yellowing disorders	Yellowing, wilting and death of palms	Phytoplasmas	Strict quarantine and removal of infected palms Growing resistant varieties	Eziashi and Omamor (2010); Gurr *et al.* (2016)
Bronze leaf wilt disorder	Yellowish-bronze leaves Palm heart decay	Unfavourable soil and water relations	Mineral nutrition Drainage	Briton-Jones (1940); Bain (1937)
Rubbery copra	Chlorotic leaves Necrosis of leaflets Leathery dried kernel	Sulfur deficiency	Application of sulfur (900 g per palm)	Southern (1969)
Crown choke/crown rot disorder	Abnormal fronds Arrested growth of bud Choked crown	Boron deficiency	Boron nutrition	Chakrabarthy *et al.* (1970); Brunin and Coomans (1973)

Continued

Table 6.1. Continued.

Disorder	Characteristic symptoms	Causative agent(s)	Remedial measures	Reference(s)
Nutritional disorders	Uniform chlorotic leaves Arrested growth and leaf shedding	Nitrogen deficiency	Foliar spray of 2% urea or root feeding (1% urea)	
	Leaf crinkling Leaf hooks Necrotic inflorescence and fruit	Boron deficiency	Borax/sodium tetraborate 0.2%	
Abnormalities due to environmental or external factors	Necrosis of plant parts	Low temperature or cold injury	Nursery flooding to avoid freezing damage Nutrient and irrigation scheduling to improve cold hardiness	Rashid *et al.* (2015)
	Low biomass Photorespiration increases Reproductive phases severely affected	High temperature		Hebbar *et al.* (2020)
	Wilting, improper inflorescence development and low economic yield of palms	Water-deficit stress	Proper irrigation scheduling	Rajagopal and Kasturi Bai (1999); Gomes and Prado, (2007); Hebbar *et al.* (2017)
	Longitudinal trunk splits Phenotypic abnormalities	Excessive water uptake Salinity or salt spray from oceans	Drainage	Hebbar *et al.* (2021)

(Zheng *et al.*, 2020). A mineral element is considered essential when a shortage of it inhibits a plant from completing its reproductive phase. Deficiency symptoms can be rectified by providing the appropriate mineral to the plants when that mineral is directly involved in a key metabolic pathway (Arnon, 1954). Palms are commonly affected by poor mineral nutrition in the field. K, N, Mg and manganese (Mn) are the most common nutritional deficiencies seen in palms. Other essential nutrients, such as Ca, boron (B), zinc (Zn) and copper (Cu) deficiencies are occasionally seen in the field if such elements are not present in the fertilizers used, although these deficiencies are uncommon (Bataglia and Furlani, 2010). Nutrient deficiencies can be caused by several factors such as a lack of nutrients in the soil, poor soil aeration, nutrient imbalance, high soil pH or planting palms at high densities. Although certain nutrient deficiencies in palms can be identified based on their visual symptoms, leaf nutrient analysis can be used to confirm the diagnosis. Iron (Fe) and B deficiency are the exceptions because their symptoms are often not related to leaf Fe or B concentrations. To perform leaf nutrient analysis, four to six leaflets from the youngest leaf are required.

6.2 Physiological Disorders

6.2.1 Barren fruit

Barren fruit, also known as seedless or imperfect fruit, are coconut fruit that lack or have a defective kernel (Sharma, 2006). The husk growth in a barren fruit is normally complete, but the quantity produced is significantly lower than in normal fruit. In general, barren fruit are oblong in shape, whereas the typical healthy fruit of the same trees are round. When the embryo is present in the barren fruit, it is often in various stages of degeneration. The shell and kernel of the barren fruit are usually deformed. This condition can be caused by a fungus that infects the embryo, causing the kernel to degenerate and completely spoil the water inside. It can also be caused by attack by insects such as the coreid bug (*Paradasynus rostratus* L.) or eriophid mites (*Aceria guerreronis* Keifer, 1965), or by rodents. Defective pollination and fertilization can also be a causative factor for barren fruit. High temperatures can cause drying of the stigmatic surface, leading to ineffective pollination. This disorder can be managed by keeping bees and spraying 2,4-dichlorophenoxyacetic acid (~270 μM), a synthetic auxin to improve fruit set and yield. Thinning of fruit and the application of an appropriate amount of nutrients (1 kg potassium chloride and 200 g sodium borate, in addition to the recommended dose of fertilizer), or application of common salt has also been found to be beneficial (Sandhu and Gill, 2013).

6.2.2 Button shedding

Premature button shedding is a frequent concern, with crop loss occurring due to the premature loss of fruit. It usually occurs when the fruit approaches maturity but before the kernel has formed. Button shedding occurs for a variety of reasons. Excessive soil acidity or alkalinity, inadequate soil drainage, extended drought, genetic factors, a lack of nutrients, inappropriate pollination, attack from insect pests and the adoption of poor agronomic management techniques are all causal factors for button shedding. Abnormal soil pH, either acidity or alkalinity, can be rectified by adding lime or gypsum to the soil, respectively. Another reason thought to be responsible for button shedding is nutrient deficiency in the soil. The application of nitrogenous manures improves fruit production in both temperate and tropical palms. The application of fertilizers containing K could also help to reduce this disorder. Root feeding of the palms with 1-naphthalene-acetic acid (~220 μM) not only reduces button shedding but also enhances yield (Sandhu and Gill, 2013). Another measure, covering the soil with green fronds, ensures that evapotranspiration is reduced and thereby soil moisture is retained. Sustainable production can also be restored with summer irrigation with adequate drainage in the winter.

6.2.3 Tapering stem wilt

Many plantations with coastal sandy and laterite soils have been found to have tapering stem wilt, often known as pencil-tip disease. This disease has an adverse effect on the crop by lowering vitality and productivity and causing premature mortality. Nutrient deficiencies in the soil or inadequate transfer of available nutrients by the roots are some of the reasons identified for tapering stem wilt (Cooke, 1950). In research conducted more recently, it has also been discovered that it can be produced by genetic factors of that palm or can be caused by a physiological impediment to translocation (Thirumalaiswamy *et al.*, 1994). Diseased palms have a lanky and ungainly appearance, along with diminished vigour and restricted growth. First, the basal leaves turn chlorotic, and later, the leaves turn pale and shrink in size. Following this, the number of fronds and spathes decreases substantially. Affected trees rarely yield fruit, and even if fruit are produced, they typically remain immature and malformed in appearance. Along with these symptoms, the collar girth gradually decreases, resulting in the manifestation of the characteristic 'pencil-point' appearance. Growth of the palm is arrested when the disease reaches its terminal stage (Thirumalaiswamy *et al.*, 1994).

6.2.4 Coconut rapid decline

Coconut rapid decline is a physiological disorder of unresolved aetiology that has been observed in Sri Lanka (Ranasinghe *et al.*, 2002). Morphologically, this disorder is characterized by a reduction in trunk girth, number of fronds and the chlorophyll content of those fronds. However, fronds show an enhanced stomatal diffusive resistance. Coconut rapid decline causes a remarkable decrease in the reproductive parts of the palm, such as the size of the inflorescence and the number of female flowers. It ultimately reduces fruit yield and palm productivity (Ranasinghe and Wijerathene, 2005). Physiologically, it has been found that palms affected by this disorder have obstructions in the transport of water from the roots to the upper parts of the palm, so water remains in the roots (Ranasinghe and Wijeratne, 2005). Preliminary studies on affected palms have revealed that Cu and Zn micronutrient deficiencies are also present, and the application of a micronutrient mixture with these nutrients resulted in improvements in terms of frond characteristics and a reduction in stomatal diffusive resistance (Wijebandara and Ranasinghe, 2004).

6.2.5 Leaf scorch decline

Another physiological disorder known as leaf scorch decline is due to an unknown cause and has been reported in Sri Lanka (Ranasinghe, 2005). It is characterized by a significant decrease in leaf area, decreased rates of photosynthesis and transpiration, reduced leaf water potential, and low leaf content of Zn and chlorophyll. In addition, palms may have a high concentration of abscisic acid in their xylem sap and a high stomatal diffusion resistance in comparison with healthy palms (Ranasinghe, 2005). There are no known ways to alleviate this disorder.

6.2.6 Finschhafen disorder

Coconut Finschhafen disorder is known to occur in young and adult palms of coconut in Papua New Guinea's West New Britain Province. This disorder has been associated with infestations of the coconut leafhopper (*Zophiuma butawengi* Heller). It is characterized by yellow or bronze colouring of frond leaflets, especially at their tips and slowly progressing back to the leaf stalk (Gitau *et al.*, 2011). To alleviate this disorder, pest management is required.

6.2.7 Lethal yellowing disease of coconut

Lethal yellowing disease of coconut is a serious complex of diseases caused by a range of phytoplasmas and has led to losses of millions of coconut and related palms worldwide (Eziashi and Omamor, 2010; Gurr *et al.*, 2016). Lethal yellowing disease is known by a variety of names. In Florida, the Caribbean Basin, the Dominican Republic, Kenya and Mozambique, the disease is known as coconut lethal yellowing. In Nigeria, it is called Awka disease, while in Ghana, the Republic of Côte d'Ivoire, Nigeria and Togo, it is called Cape St Paul wilt, Kaincopé disease or Kribi disease. The disease is called root wilt disease in India, coconut yellow decline in

Malaysia, Kalimantan or Natuna wilt in Indonesia, Weligama coconut leaf wilt disease in Sri Lanka and Bogia coconut disease in Papua New Guinea. These diseases are all characterized by yellowing, wilting and death of palms. Despite advancements in nucleic acid detection of phytoplasmas, no suitable crop management option has been identified to control these diseases.

It is thought that the disease is spread by insects, but management strategies based on vector control have been hampered due to a poor understanding of the vector species involved. In any new outbreak, one or two palms in the region show characteristic symptoms associated with the disease, and this is followed by a secondary spreading stage where palms up to 100 m away begin to display symptoms. This stage is followed by spread of the disease by as much as 100 km. Vector transmission studies, although cumbersome, have identified some planthopper species (*Haplaxius crudus* Van Duzee, previously called *Myndus crudus*) in Florida (Howard *et al.*, 1983) and *Proutista moesta* (Westwood) in India (Rajan, 2013) as potential spreaders of the disease. Hence, strict vigilance to avoid the introduction of diseased palms, rapid removal of infected palms, and replanting with relatively resistant or tolerant varieties are suggested as remedial measures.

6.2.8 Bronze leaf wilt disorder

Adverse soil conditions in coconut-growing regions of Trinidad and St Lucia due to drought, water logging or impermeable soil strata are known to cause a physiological disorder called bronze leaf wilt disorder (Briton-Jones, 1940). Further exploration of the disorder has revealed that unfavourable soil parameters and nutrient deficiency in soils, combined with improper soil–water–palm relations are the causative factors of this disorder. Hence, correctional measures that supply appropriate mineral nutrition and maintain a favourable soil physical health can be used to recover palms to regular growth (Briton-Jones, 1940).

6.2.9 Rubbery copra

Rubbery copra is a disorder in coconut attributed to sulphur (S) deficiency in soils of Papua New Guinea and Madagascar (Southern, 1969; Ollagnier and Ochs, 1972). Coconut palms grown in such S-deficient soils display characteristic chlorotic (orange-yellow) leaf symptoms, which in later stages causes severe necrosis of leaflets and shedding of fronds (Southern, 1969; Ollagnier and Ochs, 1972). Interestingly, the kernels from these palms upon drying show leathery, flexible copra, which is characterized by very low oil content, and oils with a relatively high iodine content but low saponification value. To alleviate this disorder, application of S-containing fertilizers (900 g S per palm^{-1}), once every 2 years, is recommended (Southern, 1969).

6.2.10 Crown choke/crown rot disorder

The phenomenon of coconut crown choke or crown rot disorder has been reported in Assam in India (Chakrabarthy *et al.*, 1970) and in Côte d'Ivoire (Brunin and Coomans, 1973) and is attributed to a deficiency in B nutrition. Characteristic symptoms of the disorder include shorter abnormal fronds that pack around the palm apex, deformed pinnae, hooked fronds arresting bud growth and eventually the death of the palm. The abnormal fronds surrounding the apical bud give the choked appearance of the palm. A supply of B, as sodium pentaborate, at the rate of 15 g per palm at the time of planting and every 6 months thereafter (Brunin and Coomans, 1973), or sodium tetraborate (20 g per palm) (Chakrabarthy *et al.*, 1970) or sodium borate (250–500 g per palm) through irrigation (Cecil and Pillai, 1978), is an effective treatment.

6.3 Physiological Abnormalities due to Environmental or Climatic Factors

6.3.1 Low-temperature conditions

Plant injuries brought on by low temperatures (chilling temperatures) are known as chilling injuries. One of the key environmental issues limiting coconut productivity is low temperatures (Sun *et al.*, 2021). Depending on its intensity, frost damage can result in fruit tissue browning, distortion, puckering and damage to the point where the fruit is completely spoiled (Kumar and

Kumar, 2016). The primary sign is widespread and uniform necrosis of the leaflets. The rachis, petiole and even the trunk tissue turn necrotic in more severe cases (Broschat, 2006). The tissues get weaker in freezing temperatures because they cannot perform their regular metabolic and cellular functions. Most tropical and subtropical fruits are cold sensitive (Gross *et al.*, 2002). Palms vary greatly in their cold tolerance, both within and among species. Palms affected by low-temperature stress are also prone to pathological infections such as bud rot and after being affected by cold often display evidence of micronutrient limitations such as Mn, because of reduced root activity at the cooler temperatures (Broschat, 2006).

Cold damage can be diagnosed based on symptoms and prevailing external temperatures to which the palms have been exposed. Management includes selecting palms that are suited to the colder environment, and during very cold periods covering young palms, or if in pots bringing them inside throughout the cold period. Palm nurseries can also be flooded to prevent freeze damage. Coconut trunk injury can be caused by prolonged cooling temperatures. Low temperatures, especially those below 15°C, have been demonstrated to lead to severe leaf blotches and irregular fruit growth (Rashid *et al.*, 2015). Appropriate fertilization, nutrient scheduling and cold-mitigating actions can be used to improve the cold hardiness of coconut and other palms (Broschat, 2010).

6.3.2 High temperatures

High temperature is a major environmental stress factor that inhibits plant development, metabolism and productivity. Numerous biochemical processes involved in plant growth and development are very sensitive to high temperatures. Temperature influences the process of photosynthesis by modulating the activities of the enzymatic machinery involved in the process, and by altering the phenomenon of electron transport (Sage and Kubien, 2007). High temperatures can cause both photorespiration and dark respiration to increase, which lowers the overall biomass production (Hebbar *et al.*, 2017). A modest decrease in stomatal density has also been observed at high temperatures. Temperature has a significant impact on coconut reproductive activities such as pollen germination (Ranasinghe *et al.*, 2010; Hebbar *et al.*, 2018), pollen-tube development and fruit set (Hebbar *et al.*, 2020).

6.3.3 Water-deficit stress

Drought stress has an impact on coconut cultivation and production in practically all coconut-producing countries, because it is primarily a rainfed crop (Coomans, 1975). Prolonged drought in the summer has been observed to cause wilting in large areas of coconut. Some palms revive when the rain returns, but the yield can be reduced for at least 3 years after a significant drought (Rajagopal *et al.*, 1996; Hebbar *et al.*, 2017). The physiological impacts of water deficit on coconut and its influence on the overall growth of the palm and fruit productivity have been well documented (Gomes and Prado, 2007). Drought can severely reduce photosynthesis, so it is essential to maintain a good soil and leaf water balance to safeguard against this and to ensure the survival of coconut seedlings during summer (Kasthuri Bai *et al.*, 2006).

6.3.4 Excessive water uptake resulting in trunk splitting

Palms with large volumes of water in their trunks may develop deep longitudinal splits. The trunks of these trees are often wet and support the growth of other life forms such as algae, epiphytic plants, lichens and mosses. It has also been observed that excessive moisture absorption by soft-walled parenchyma cells in the middle region of the trunk and expanding beyond the hard vascular and fibre cells that predominate in the peripheral tissue could be an anatomical reason for the trunk splitting. This condition is most common in locations with excessive rainfall, although it can also occur in drier areas during prolonged rainfall events. Minor stem cracking does not normally damage the structure of a palm, but cracks that extend halfway through or more could undermine the trunk strength.

6.3.5 Root suffocation in palms

Palms that are planted too deeply, on poorly drained soils, in landscape settings or in field sites can suffer from root suffocation. Fe deficiency is a common symptom of the early signs of root suffocation, which include the production of chlorotic new leaves. In severe situations, the leaves may wilt and dry out, and the trunk may shrink. When palms are planted deeper than they were originally growing, the oxygen content of the soil can be reduced, which may result in root suffocation. Decreased soil aeration in waterlogged soils can cause root rotting, which can eventually kill the root system and the palm.

6.3.6 Salinity stress resulting in foliar salt injury

High-speed winds, which blow salt from the sea, leading to deposits on palms can cause scorching (Hebbar *et al.*, 2017). Plants exposed to wind-blown salt may accumulate Na^+, which slows their growth and results in morphophysiological abnormalities (Parida and Jha, 2013). Nevertheless, coconut palms have evolved multiple defence mechanisms against salt stress (Parida and Jha, 2013). Osmotic adjustment is one such mechanism, wherein cytoplasmic accumulation of compatible low-molecular-weight solutes (proline and sugars) and vacuole compartmentalization of potentially harmful ions maintains cell turgor. Enhanced concentrations of organic molecules and the maintenance of high concentrations of Na^+ in the roots ensures a relatively low Na^+:K^+ ratio in the leaves. This suggests that the root system of coconut seedlings plays a significant role in the phenomenon of salinity tolerance (Lima *et al.*, 2017). Furthermore, seawater substitution studies for hydroponically grown coconut seedlings have shown a high accumulation of Na^+ in the roots and shoots, combined with little transport to the leaves, are the underlying factors that maintain a balanced K^+:Na^+ ratio in the leaves. This physiological process was also suggested to be involved in the salinity tolerance mechanism of coconut seedlings (Hebbar *et al.*, 2021).

6.4 Nutritional Disorders

Coconut cultivation can experience deficiency symptoms from macro- (e.g. N, P and K) and micronutrients in plantations where the palms have not been supplied with appropriate fertilizer, and more specifically in palms that have not been supplied with organic manures.

6.4.1 Nitrogen deficiency

Insufficient N in the soil is the main reason for N deficiency in plants. Generally, the main N-deficiency symptom is the uniform yellowing of leaves and leaf veins (Nautiyal *et al.*, 2000). Cereal crops have distinctive yellowing at the tips of lower leaves that resembles a 'V' shape. In coconut, uniform chlorosis, a light green discoloration or yellowing of the oldest leaves, is the first sign of a N deficit condition. The lower leaves begin to yellow at the tip and progress upwards from there (Fig. 6.1). Younger leaves will also discolour as the condition intensifies. When the N deficiency is severe, leaf shedding occurs, and growth stops. The golden-yellow colour characteristics of older leaves turn to reddish grey before dying. A N shortage is easily identified and diagnosed based on the visual symptoms; however, leaf nutrient estimation provides a clearer picture. At times, a N shortage can be mistaken for Fe- or S-deficiency symptoms; however, the chlorotic symptoms observed during Fe and S deficiencies are generally in vigorous young leaves. In the case of N deficiency, the older leaves show a chlorotic appearance. As a corrective measure, foliar application of 2% urea three times every 2 weeks, soil applications of 1–2 kg of urea per palm, or root feedings of 200 mL of 1% urea twice a year can be applied.

6.4.2 Potassium deficiency

As one of the essential nutrients for palm health, K is needed in rather high concentrations. K helps plants carry out a variety of functions such as photosynthesis, enzyme activation and osmoregulation (Nelson and Patnude, 2012). In coconut, the oldest leaves show symptoms first, which eventually spread to younger leaves. On

Fig. 6.1. Coconut palms displaying the characteristic nitrogen-deficiency symptoms of yellowing of the lower whorl of leaves.

leaflets, translucent yellow or orange scattered rust-coloured dots appear (yellowing from the leaf margin). Older leaves exhibit curled leaflet tips as well as necrotic spots (Fig. 6.2). The leaflet turns slightly yellowish, and this becomes more visible near the leaf tip. The reddish colour spots gradually enlarge and become bigger so that brown patches form. Leaflets show necrotic edges that eventually wither. Tapering of the crown is commonly found in K-deficient palms.

6.4.3 Magnesium deficiency

In palms, Mg deficiency results from the excessive leaching of Mg^{2+} ions from sandy and related soils characterized by poor cation-exchange capacity. In addition, excessive concentrations of Ca, N and K in the soils can also predispose palms to Mg deficiencies (Broschat, 2005). Mg deficiency in coconut is characterized by the yellowing of leaflets on mature fronds. The leaflet becomes pale yellow with a green band on either side of the frond. The tips of the leaves gradually dry out and appear scorched. The palm's crown turns a pale-yellow colour at the mild stage of Mg deficiency. Yellowing of the fronds becomes obvious when the deficiency intensifies over time. Mg shortages are typically diagnosed by symptomatic observations. The severity of the discoloration is a sign of the extent of the Mg shortage, which is visually different from a K deficiency (Fig. 6.3). While Mg-deficient leaves have noticeably green leaf centres and intense

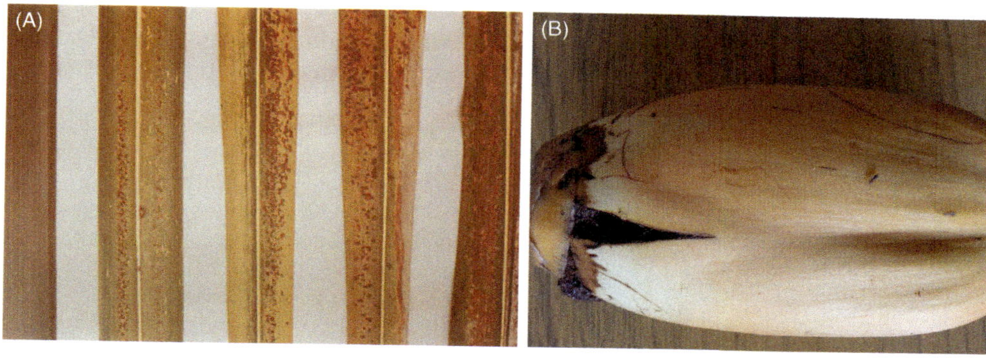

Fig. 6.2. Potassium deficiency symptoms in coconut: (A) leaves showing a slight yellowish colour, and yellow or orange scattered rust-coloured dots, and (B) splitting of fruit.

Fig. 6.3. Magnesium deficiency in coconut palms, the leaves show yellowing symptoms and green leaf centres.

lemon-yellow to orange borders, K-deficient leaves typically have orange to bronze tones that progressively fade to green towards the base of the leaf. To manage Mg deficiency, a hydrated calcium sulfate application of 2–5 kg per palm $year^{-1}$ and root feeding (0.2% solution) is recommended.

6.4.4 Phosphorus deficiency

Initially, P deficiency causes purple coloration in the leaves, which gradually turns yellowish and causes premature leaf death. A reduction in growth rate, leaves that stay upright and early leaf shedding are the predominant symptoms of the deficiency. Leaf size, leaf number and total growth all decrease. Often during P deficiency, root growth is also found to be restricted. Other than stunted growth and a decrease in yield, there are no obvious visual signs for P deficiency.

P deficiency is prevalent in a wide range of soil types, and hence as a management strategy, foliar application of 2% diammonium phosphate twice every 2 weeks or soil application of 5 kg of farmyard manure per palm is recommended. The application of 1% diammonium phosphate twice a year to the roots is also recommended.

6.4.5 Boron deficiency resulting in crown choking disorder

The element B, which is poorly understood, is the most common micronutrient shortage that has an impact on the production of many crops (Dordas and Brown, 2005). B deficiency, resulting from a lack of B in the soil, can be caused due to high soil pH and soil drying. Heavy leaching can also cause a temporary B deficiency. Newly developing leaves always show symptoms of B deficiency. In coconut palms, the characteristic B deficiency symptoms include 'hook leaves' wherein the leaflets are joined and the developing leaf wrinkles to produce sharply curved leaflet tips. The appearance of leaves remains serrated and zigzag shape. Leaf gas exchange is severely impeded in coconut due to B deficiency. B deficiency reduces stomatal conductance and leaf water loss, and hence increases the leaf-to-air vapour difference (Pinho *et al.*, 2010). Normally with this deficiency, the lower basal region of the frond does not produce any leaflets. B deficiency affects the photosynthetic capability of coconut palms because it decreases photosystem II electron transport by 12.5%, photosynthesis by 35.7% and stomatal conductance by 45.6% (Pinho *et al.*, 2010).

Typical symptoms of B deficiency appear when new leaves emerge several months after the occurrence of the deficiency, although a temporary B shortage might influence leaf development even during its mature leaf phase (Broschat, 2007). Additional signs of B deficiency in coconut palms include: (i) folioles with white to translucent spots; (ii) necrosis of tissues in the surrounding spots; (iii) corrugation, with a shape resembling the white spots and lines; (iv) young leaves that remain concealed within the mature leaves; (v) non-existence of a spear; (vi) the apex of the folioles showing necrotic symptoms; and (vii) corrugation on the leaf lamina and midrib. Brownish stains appear on the palm heart, and the two sides of each rachis often have necrosis (Pinho *et al.*, 2015). The symptoms of B deficit are extremely distinct and are usually easy to diagnose; however, Mn-deficiency symptoms can sometimes be confused with those of B deficiency. Leaf analysis is not particularly helpful, because B deficiencies are often momentary in nature. In addition, as B remains stationary inside the palm, the deficiencies only influence the developing leaf primordium in the bud region (Fig. 6.4).

6.4.6 Manganese deficiency

Inadequate amounts of Mn in the soil can cause Mn deficiency, and a high soil pH can considerably lower the availability of Mn. Cold soil temperatures can also result in a temporary Mn deficiency in soils where Mn levels are only marginally adequate by lowering root activity levels. Mn

Fig. 6.4. (A) Boron deficiency showing improper unfurling of leaves and button shedding. (B) Hen and chicken symptoms associated with boron deficiency in fruit.

shortage is extremely frequent in alkaline soils, and can also happen in containers with cooler soil temperatures and inadequate drainage (Broschat, 2014). In coconut, the youngest leaves appear chlorotic with longitudinal necrotic streaks. As the deficiency intensifies, newly developing leaflets, except for their basal sections, seem necrotic and withered. This withering causes the leaflets to curl around the rachis, furnishing a browned appearance to the leaf (frizzle top). In Mn-deficient palms, growth is arrested severely, and the freshly emerging leaves show necrotic petiole stubs. The diagnosis of this condition may only be determined by visual symptoms, but as symptoms of B insufficiency can be identical, leaf nutrient analysis is also recommended. The symptoms of late-stage K deficiency are almost impossible to differentiate from those of Mn deficiency at a distance, and thus close inspection is necessary to check for the basal symptom distribution and distinctive longitudinal streaking characteristic of Mn deficiency. Management of this deficiency can be achieved by applying manganese sulfate to the soil at a rate of 25 kg ha^{-1}.

6.4.7 Iron deficiency

Palms that have been transplanted to a great depth or are growing in soils with inadequate aeration frequently experience Fe deficiency. Deep planting and waterlogged soils effectively suffocate the roots and decrease their capacity to absorb nutrients such as Fe. Yellowing or chlorosis in between the veins of young leaves is the predominant sign of an Fe shortage. Chlorotic young leaves that have uniform necrotic tips and smaller leaves are signs of increasing deficiency as the condition worsens. Root damage due to root-rot diseases will similarly cause Fe-deficiency symptoms as the root surface area available for Fe uptake is reduced. Fe deficiency occurs far more frequently in container production than in settings involving landscape or field nurseries. As they usually develop slowly, palms can remain in a container for more than a year and can be transplanted into the field whenever required (Broschat, 2011). Management can be achieved through the application of Fe sulfate at the rate of 0.25–0.50 kg per palm $year^{-1}$.

6.4.8 Sulphur deficiency

Yellowish-green or yellowish-orange leaflets are typical symptoms of S deficiency. The older leaves still remain green. As the stem ages, the leaves droop. The number and size of leaves are diminished in aged palms. Due to rachis fragility, a covering of deceased fronds can occasionally form around the stem. The fruit can also fall off prematurely. The copra may be rubbery and of low market value.

6.4.9 Zinc deficiency

The production of tiny leaves with a reduced leaf size of 50% is a sign of Zn deficiency. The length and width of the leaflets can decrease as they become chlorotic. Flowering is postponed in an acute deficit. Button shedding can also be caused by Zn deficiency. Zn deficiency is mainly a problem when coconut is grown in salty soils.

6.4.10 Calcium and copper deficiencies

Younger leaves show marginally narrow white bands and interveinal chlorosis as a result of Ca deficiency. In addition, the appearance of rust on the leaf border followed by the leaf rolling up can be symptomatic. Ca deficiency appears in palms grown in acidic soils. Palms experiencing Cu deficiency show bluish-copper leaves due to turgor depletion, and the terminal leaves roll. The leaves look bleached grey and the palms do not flower.

6.5 Physiological Optimization of Growth and Development of Coconut

To be productive in unfavourable environmental conditions, coconut growth and development need to be physiologically optimized. This is crucial to achieve a sustainable output when palms are planted in challenging environments (Naresh Kumar *et al.*, 2012). High temperatures have a harmful effect on the reproductive phase of coconut development, as they affect pollen germination and pollen-tube growth (Hebbar *et al.*, 2018, 2020). As coconut is mostly grown in coastal areas, seawater intrusion and sea-level rises brought on by climate change are potential threats to the growth and development of the coconut. They have some resilience, in that coconut seedlings can withstand up to 8.32 dS m^{-1} electrical conductivity without suffering any harmful impacts on their physiology or biomass and growth. Coconut palms may have a salt-tolerance mechanism that involves Na^+ accumulation in the roots and shoots and K^+ accumulation in the leaves (Hebbar *et al.*, 2020). Although coconut may indeed be more resilient than other plants to environmental stresses, particularly water and salt inundation, the resilience of the current commercial coconut varieties has limits. In the face of more extreme weather events and temperatures due to climate change, the identification of more water-efficient types and varieties resistant to drought, heat and salinity are needed. Additionally, the forecasting of future locations suitable for coconut farming will aid in developing a comprehensive grasp of future breeding needs and adaption tactics.

6.6 Concluding Remarks and Perspectives

Further understanding of nutritional inadequacies and physiological disorders in coconut will undoubtedly help to increase productivity. Different cultural and nutritional practices are needed to mitigate the detrimental effects of known and unknown nutrient disorders. By having well-documented measures and approaches, the physiological abnormalities brought on by diverse environmental stresses can be minimized, and productivity and consistency of coconut productivity throughout the year can be enhanced.

6.7 References

Arnon, D.J. (1954) Criteria of essentiality of inorganic micronutrients for plants. In: Wallace, T. (ed.) *Trace Elements in Plant Physiology*. Chronica Botanica, Waltham, Massachusetts.

Bain, F.M. (1937) *Bronze Leaf Wilt Disease of the Coconut Palm*. US Government Printing Office, Trinidad.

Bataglia, O.C. and Furlani, P.R. (2010) Nutrition and fertilization of palms. *Acta Horticulturae* 1003, 57–62.

Briton-Jones, H.R. (1940) *The Diseases of the Coconut Palm*. Bailliere, Tindall and Cox, London.

Broschat, T.K. (2005) Magnesium deficiency in palms. University of Florida Institute of Food and Agricultural Sciences, Gainsville, Florida. Available at: http://edis.ifas.ufl.edu/EP266 (accessed 16 August 2023).

Broschat, T.K. (2006) Physiological disorders of landscape palms. University of Florida Institute of Food and Agricultural Sciences, Gainsville, Florida. Available at: https://edis.ifas.ufl.edu/publication/EP263 (accessed 16 August 2023).

Broschat, T.K. (2007) Boron deficiency symptoms in palms. *Palms* 51, 115–126.

Broschat, T.K. (2010) Fertilization improves cold tolerance in coconut palm. *Horticulture Technology* 20, 852–855.

Broschat, T.K. (2011) Iron deficiency in palms. University of Florida Institute of Food and Agricultural Sciences, Gainsville, Florida. Available at: https://edis.ifas.ufl.edu/publication/EP265 (accessed 16 August 2023).

Broschat, T.K. (2014) Manganese deficiency in palms. University of Florida Institute of Food and Agricultural Sciences, Gainsville, Florida. Available at: https://edis.ifas.ufl.edu/publication/EP267 (accessed 16 August 2023).

Brunin, C. and Coomans, P. (1973) Boron deficiency in young cocofruit in the Ivory Coast. *Oleagineux* 28, 229–234.

Cecil, S.R. and Pillai, N.G. (1978) Role of boron in coconut nutrition. *Indian Coconut Journal* 9, 1–3.

Chakrabarthy, B.K., Nath, B.K., Dhar, B.K. and Goswami, R.N. (1970) Note on crown rot disease of coconut. *Indian Journal of Agricultural Science* 40, 502–594.

Cooke, F.C. (1950) Tapering disease of coconut. *Ceylon Coconut Quarterly* 1, 17–21.

Coomans, P. (1975) Influence des facteurs climatiques sur les fluctuations saisonnieres et annuelles de la production du cocotier. *Oleagineux* 30, 153–159.

DebMandal, M. and Mandal, S. (2011) Coconut (*Cocos nucifera* L.: Aceraceae): in health promotion and disease prevention. *Asian Pacific Journal of Tropical Medicine* 4, 241–247.

Dordas, C. and Brown, P.H. (2005) Boron deficiency affects cell viability, phenolic leakage, and oxidative burst in rose cell cultures. *Plant and Soil* 268, 293–301.

Effiong, G.S., Ebong, P.E., Eyong, E.U., Uwah, A.J. and Ekong, U.E. (2010) Amelioration of chloramphenicol induced toxicity in rats by coconut water. *Journal of Applied Science Research* 6, 331–335.

Eziashi, E. and Omamor, I. (2010) Lethal yellowing disease of the coconut palms (*Cocos nucifera* L.): an overview of the crises. *African Journal of Biotechnology* 9, 9122–9127.

Foale, M. (2003) *Coconut Odyssey: The Bounteous Possibilities of the Tree of Life*. Australian Centre for International Agricultural Research, Canberra.

Gitau, C.W., Gurr, G.M., Dewhurst, C.F., Mitchell, A., Fletcher, M.J. *et al.* (2011) *Zophiuma lobulata* (Hemiptera: Lophopidae) causes Finschhafen disorder of coconut and oil palms. *Annals of Applied Biology* 158, 139–148.

Gomes, F.P. and Prado, C.H. (2007) Ecophysiology of coconut palm under water stress. *Brazilian Journal of Plant Physiology* 19, 377–391.

Gross K.C., Wang C.Y. and Saltveit M. (2002) The commercial storage of fruits, vegetables, and florist and nursery crops. Agricultural Research Service, US Department of Agriculture, Available at: www.ba.ars.usda.gov/hb66/index.html (accessed 16 August 2023).

Gurr, G.M., Johnson, A.C., Ash, G.J., Wilson, B.A., Ero, M.M. *et al.* (2016) Coconut lethal yellowing diseases: a phytoplasma threat to palms of global economic and social significance. *Frontiers in Plant Science* 7: 1521.

Hebbar, K.B., Berwal, M.K., Arivalagan, M. and Chaturvedi, V.K. (2017) Physiological and biochemical response of coconut to climate change variables. In: Hebbar, K.B., Naresh Kumar, S. and Chowdappa, P. (eds) *Impact of Climate Change on Plantation Crops*. Astral International, New Delhi.

Hebbar, K.B., Rose, H.M., Nair, A.R., Kannan, S., Niral, V. *et al.* (2018) Differences in *in vitro* pollen germination and pollen tube growth of coconut (*Cocos nucifera* L.) cultivars in response to high temperature stress. *Environmental and Experimental Botany* 153, 35–44.

Hebbar, K.B., Apshara, E., Chandran, K.P. and Prasad, P.V. (2020) Effect of elevated CO_2, high temperature, and water deficit on growth, photosynthesis, and whole plant water use efficiency of cocoa (*Theobroma cacao* L.). *International Journal of Biometeorology* 64, 47–57.

Hebbar, K.B., Santhosh, A., Sukumar, A.P., Neethu, P., Ramesh, S.V. *et al.* (2021) Effect of sea water substitution on growth, physiological and biochemical processes of coconut (*Cocos nucifera* L.) seedlings – a hydroponic study. *Scientia Horticulturae* 280: 109935.

Hebbar, K.B., Ramesh, S.V., Ghosh, D.K., Beegum, P.P., Pandiselvam, R. *et al.* (2022) Coconut sugar – a potential storehouse of nutritive metabolites, novel bio-products, and prospects. *Sugar Technology* 24, 841–856.

Howard, F.W., Norris, R. and Thomas, D. (1983) Evidence of transmission of palm lethal yellowing agent by a planthopper, *Myndus crudus* (Homoptera, Cixiidae). *Tropical Agriculture* 60, 168–171.

Kasturi Bai, K.V., Rajagopal, V. and Naresh Kumar, S. (2006) Chlorophyll fluorescence transients with response to leaf water status in coconut. *Indian Journal of Plant Physiology* 11, 410–414.

Kumar, R. and Kumar, V. (2016) Physiological disorders in perennial woody tropical and subtropical fruit crops – a review. *Indian Journal of Agricultural Sciences* 86, 703–717.

Lima, B.L.D.C., Lacerda, C.F.D., Ferreira, M., Ferreira, J.F.D.S., Bezerra, A.M. *et al.* (2017) Physiological and ionic changes in dwarf coconut seedlings irrigated with saline water. *Revista Brasileira de Engenharia Agrícola e Ambiental* 21, 122–127.

Meera, S., Farheen, H., Jignasa, R., Preeti, G. and Adodaria, B.K. (2017) Physiological disorders and quality improvement by appropriate remedies in fruit crops. *Trends in Biosciences* 10, 6979–6981.

Mishra, D.S., Tripathi, A. and Nimbolkar, P.K. (2016) Review on physiological disorders of tropical and subtropical fruits: causes and management approach. *International Journal of Agriculture, Environment and Biotechnology* 9: 925.

Mohandas, S. (2012) Effect of NPK fertilizer levels on mineral nutrition and yield of hybrid (tall × dwarf) coconut. *Madras Agricultural Journal* 99, 87–91.

Naresh Kumar, S., Anil Kumar Singh, P.K., Aggarwal, V.U.M. and Venkateswarlu, B. (2012) *Indian Agriculture: Impact, Adaptation and Vulnerability-Salient Achievements from ICAR Network Project, 2012*. IARI Publication, New Delhi.

Naresh Kumar, S., Hebbar, K.B., Kasturi Bai, K.V. and Rajagopal, V. (2018) Physiology and biochemistry. In: *The Coconut Palm (*Cocos nucifera *L.) – Research and Development Perspectives*. Springer, Singapore, pp. 443–488.

Nathanael, W.R.N. (1961) Coconut nutrition and fertilizer requirements – the plant approach. *Ceylon Coconut Quarterly* 12, 101–120.

National Multi-commodity Exchange of India Ltd (NMCE) (2007) *Report on Copra*. NMCE, Mumbai, India.
Nautiyal, C.S., Bhadauria, S., Kumar, P., Lal, H., Mondal, R. *et al.* (2000) Stress induced phosphate solubilization in bacteria isolated from alkaline soils. *FEMS Microbiology Letters* 182, 291–296.
Nelson, S. and Patnude, E. (2012) *Potassium Deficiency of Palms in Hawai'i*. University of Hawaii at Manoa, Honolulu, Hawaii.
Ollagnier, M. and Ochs, R. (1972) Sulphur deficiencies in the oil palm and coconut. *Oleagineux* 27, 193–198.
Parida, A.K. and Jha, B. (2013) Physiological and biochemical responses reveal the drought tolerance efficacy of the halophyte *Salicornia brachiata*. *Journal of Plant Growth Regulation* 32, 342–352.
Pinho, L.G., Campostrini, E., Monnerat, P.H., Netto, A.T., Pires, A.A. *et al.* (2010) Boron deficiency affects gas exchange and photochemical efficiency (JPI test parameters) in Green Dwarf coconut. *Journal of Plant Nutrition* 33, 439–451.
Pinho, L.G., Monnerat, P.H., Pires, A.A., Freitas, M.S.M. and Marciano, C.R. (2015) Diagnosis of boron deficiency in Green Dwarf coconut palm. *Agricultural Sciences* 6: 164.
Rajagopal, V. and Kasturi Bai, K.V. (2002) Drought tolerance mechanism in coconut. *Bulletin BuroTrop* 17, 21–22.
Rajagopal, V., Kasturi Bai, K.V. and Voleti, S.R. (1990) Screening of coconut genotypes for drought tolerance. *Oleagineux* 45, 215–223.
Rajagopal, V., Jacob, M., Mathew, J. and Shivashankar, S. (1996) Impact of dry spells on the ontogeny of coconut fruits and its relation to yield. *Plantation Research and Development* 3, 251–255.
Rajan, P. (2013) Transmission of coconut root (wilt) disease through plant hopper, *Proutista moesta* Westwood (Homoptera: Derbidae). *Pest Management in Horticultural Ecosystems* 17, 1–5.
Ranasinghe, C.S. (2005) Changes in the physiological performance of leaf scorch decline (LSD) affected coconut (*Cocos nucifera* L.) palms. *Experimental Agriculture* 41, 255–265.
Ranasinghe, C.S. and Wijerathene, A.J. (2005) Vegetative, reproductive, and physiological changes in coconut palms affected by coconut rapid decline. *COCOS* 17, 1–10.
Ranasinghe, C.S., Wijesekara, R. and Wimalasekara, R. (2002) Rapid decline syndrome of coconut (*Cocos nucifera* L.) – preliminary report of a new condition. *Palms* 46, 167–171.
Ranasinghe, C.S., Waidyarathna, K.P., Pradeep, A.P.C. and Meneripitiya, M.S.K. (2010) Approach to screen coconut varieties for high temperature tolerance by *in-vitro* pollen germination. *COCOS* 19, 13–26.
Rashid, M.M., Hasan, M.S., Rahman, M.H., Ahmmed, A.N.F. and Khalequzzaman, K.M. (2015) Physiological and structural abnormalities of coconut plants (*Cocos nucifera* L.) at different stages of growth and plantation sites under low temperature condition in Bangladesh. *The Agriculturists* 13, 113–119.
Sage, R.F. and Kubien, D.S. (2007) The temperature response of C3 and C4 photosynthesis. *Plant, Cell & Environment* 30, 1086–1106.
Sandhu, S. and Gill, B.S. (2013) *Physiological Disorders of Fruit Crops*. New India Publishing Agency, New Delhi.
Selvamani, V. and Duraisami, V.P. (2014) Identifying and mapping soil fertility constraints for coconut in Coimbatore and Tiruppur Districts of Tamil Nadu State, India. *Journal of Plantation Crops* 42, 348–357.
Sharma, R.R. (2006) Physiological disorders in tropical and subtropical fruits: causes and control. In: *Fruit Production Problems and Solutions*. International Book Distributing Co., Lucknow, India, pp. 301–25.
Southern, P.J. (1969) Sulphur deficiency in cocofruit. *Oleagineux* 24, 211–220.
Sudhakara, K. (1990) Button shedding and premature nutfall in coconut. *Journal of Plantation Crops* 18, 66–77.
Sun, C., Zhang, R., Yuan, Z., Cao, H. and Martin, J.J.J. (2021) Physiology response and resistance evaluation of twenty coconut germplasm resources under low temperature stress. *Horticulturae* 7: 234.
Thirumalaiswamy, K., Vijayaraghavan, H. and Savery, M.A.J.R. (1994) Management of tapering disorder in coconut. *CORD* 10: 34.
von Liebig, J. (1840) *Die organische Chemie in ihrer Anwendung auf Agrikulktur und Physiologie*. Friedrich Vieweg Sohn, Brunswick, Germany.
Watt, G. (1889) *A Dictionary of Economic Products of India*. Superintendent of Government Printing, Kolkata, India.
Wijebandara, D.M.D.I. and Ranasinghe, C.S. (2004) Response of rapid decline affected coconut (*Cocos nucifera* L.) palms to micro-nutrients and common salt. *COCOS* 16, 11–21.
Zheng, Y., Ma, Y., Liu, W. and Qiu, F. (2020) Plant nutrition and physiological disorders in fruit crops. In: *Fruit Crops*. Elsevier, Amsterdam, pp. 47–58.

7 Harvest and Postharvest Management of Coconut

Ravi Pandiselvam[1*], Anandu Chandra Khanashyam[2], R. Dakshayani[3], Fernanda Caro Beveridge[4], Steivie Karouw[5] and M.R. Manikantan[1]

[1]Indian Council of Agricultural Research – Central Plantation Crops Research Institute, Kasaragod, Kerala, India; [2]Department of Food Science and Technology, Kasetsart University, Chatuchak, Bangkok, Thailand; [3]Indian Institute of Food Processing Technology, Thanjavur, Tamil Nadu, India; [4]School of Agriculture and Food Sustainability, The University of Queensland, Gatton, QLD, Australia; [5]Indonesian Board of Coconut Research, Manado, North Sulawesi, Indonesia

Abstract

Coconut is a high-quality crop providing a livelihood for millions of people. This chapter discusses the major harvest and postharvest technologies that are used in coconut production. The fruit is versatile and widely consumed, and is generally marketed at two stages of development, mature or immature. Therefore, fruit maturity is one of the most important parameters to be considered when harvesting. For immature (tender) fruit production, fruit of 6–8 months of age are harvested, whereas 'snowball' fruit are 8–9 months old, and for kernel-based products like copra, the fruit are 11–12 months old at harvest. Machines can help in reducing the time and drudgery involved in fruit processing. After primary processing, a variety of products such as coconut water, virgin oil, crude oil, milk and desiccated coconut can be prepared from the fruit. The process for gaining some of these important value-added coconut products is discussed in this chapter.

7.1 Introduction

The coconut (*Cocos nucifera* L.), known as 'Kalpavriksha' in India or the 'Tree of Life' in most other parts of the world, is a perennial palm that is widely cultivated in tropical countries for its fruit, sap, fibre, shell, water, meat, oil and timber. In 2015, it was estimated that coconut palms were cultivated on 12 million ha of land, spread over 94 countries, providing security and a livelihood for millions of people, both directly and indirectly (Rethinam, 2019). Most countries have focused their coconut farming on copra production (to obtain crude endosperm oil and dried kernel). India, for example, harvests over 13 billion coconuts per year, but its postharvest processing is limited largely to copra, coconut oil, coir and coir-based goods. From the country's harvest, over 40% is made into copra, with about 80% of this being processed into oil. A good proportion of the remaining coconut fruit (48%) is consumed in its raw form domestically, while a further 10% is consumed in its immature form for drinking (Ghosh, 2015). As coconut is such a versatile crop, it is feasible to make value-added goods from almost every part

* Email: anbupandi1989@yahoo.co.in

DOI: 10.1079/9781789249736.0007

of the palm. Recently, more efforts have been made in the production of more lucrative coconut products, such as for medicinal uses and high-nutritional-value food products (Foale, 2003), together with coconut wood for building materials and furniture (Nguyen *et al.*, 2015).

Postharvest handling is one of the major processing steps to be performed for effective utilization of the harvested commodity, to maximize value, minimize waste and protect the quality of the product until it reaches the consumer. If there is a lack of knowledge or suitable infrastructure for the postharvest processing of coconut, the industry can face considerable losses after harvest. Being an important horticultural crop grown in many tropical regions, coconut has made its mark with several products and by-products, namely coconut fibre (used for mats, ropes and coir), shell (activated carbon and shell products) and seed-based food products (nata de coco, coconut milk, coconut oil, young tender fruit and desiccated coconut powder). Additionally, other uses have gained attention for their ability to value add or for waste minimization. The widespread use of coconut products has led to ongoing research into improving postharvest processing, handling, value addition and storage of coconut products worldwide, all of which will be discussed further in this chapter.

In 2019, the market for coconut and coconut-based products was valued at US$11.05 billion. However, both the suppliers and the processors of coconut raw materials have faced major difficulties because of the coronavirus disease 2019 (COVID-19) pandemic. Although the market was severely impacted by the pandemic, it has raised awareness about the value of having a healthy diet for maintaining a strong immune system. Additionally, in recent years, the industry has seen a rise in consumer interest in dairy alternatives made from plants, such as those based on coconut milk and cream. The demand for coconut products is anticipated to continue to increase because of rising consumer health concerns as well as the adoption of vegan diets (Mordor Intelligence, 2022). In addition to the increase in food applications, growth in the coconut oil market is also projected to be driven by the large demand in the cosmetics industry and its rising acceptance in food applications. Moreover, due to its nutritional qualities, coconut water is becoming increasingly popular as a natural energy drink among active and health-conscious consumers. According to projections from market research (Fortune Business Insights, 2021), the global market for coconut products will increase at a compound annual growth rate of 10.5% from 2020 to 2027, rising from US$11.60 billion in 2020 to US$23.39 billion by 2027.

Regardless of the opportunities highlighted above, a lack of knowledge, practical skills or facilities with coconut postharvest technologies continues to result in severe losses, beginning with the harvesting of fruit and then during seasoning, drying and storage. Even if there is an increase in coconut production, yield and area, the tedious nature of manual coconut harvesting, the scarcity of climbers, and the operational and safety issues associated with manual climbing are likely to drive mechanization of coconut harvesting and processing.

Labour availability is one of the key resources needed for sustainable agriculture. According to Pathiraja *et al.* (2010) in Sri Lanka, labour costs account for 29% of the total production costs. Similarly, Kalathiya *et al.* (2007) found that the cost of manures, fertilizers and labour together accounted for 50% of the cost of coconut cultivation. A study by Anithakumari (2013) on Indian coconut farmers reported that 78% of respondents said that there was a labour shortage and that compared with human labour, farmers that used mechanization in land preparation claimed to save 60–75% on costs. The study also analysed the need for technology adaptation in the coconut industry and reported that most respondents (76.3%) use equipment such as the keramithra, a two-bladed tool used in the de-husking of coconuts. Power tillers were used by only 2.6% of farmers, whereas 12.8% used sprayers and 65.7% still use ladders to harvest the fruit. It is likely that increased mechanization and technology adoption by farmers in the coconut industry will be needed to meet consumer demand.

A variety of climbing tools for coconut collection have recently been created by research institutions, individuals and universities, and they all claim to be secure and simple to use. While Lam and Xu (2011) designed an autonomous robot to climb the tree like an inchworm by employing minimal sensors, Mani and

Jothilingam (2014) developed and manufactured a semi-autonomous tree climber and harvester named the 'Cocobot', which incorporated a triangular structure with an arm and a fruit cutter. Suparat and Maneewarn (2015) proposed an image-processing technique for developing a robotic unit to spray insecticide on to the coconut canopy. While few of these technologies have transitioned beyond the prototype, it is anticipated that some or all will be important for the industry in the coming years.

The demand for high-quality, nutrient-dense foods by consumers has sparked interest in creating innovative methods for processing coconut into food. Industry and academic research has developed a non-thermal technology to help the manufacturing of safe, high-quality, processed coconut foods. Non-thermal methods such as ozonation, high hydrostatic processing, ultrasonic processing, pulsed ultraviolet-light processing, cold plasma processing, pulsed electric field processing and ultrafiltration can all be used to achieve sterility and promote a longer shelf life for liquid products such as coconut sap, tender coconut water and mature coconut water (Prithviraj *et al.*, 2021; Sharma *et al.*, 2021).

7.2 Harvesting of Coconut

7.2.1 Maturity and time of harvest

Coconut fruit take roughly 11–13 months from flowering to full maturity. The fruit consist of a waxy exocarp, which is green, yellow or reddish-brown in colour before maturing to a brown or grey colour. The exocarp covers the mesocarp or husk, which is the fibrous middle portion of the fruit. The endocarp, or shell, is rigid, hard and a deep brown colour, and is followed by a thin brown testa and then the edible endosperm (Siriphanich *et al.*, 2011). At roughly 5–6 months after flowering, the fruit reaches its maximum size. Only the husk and shell develop swiftly during the first 6 months, while the inner seed cavity is filled with liquid endosperm. The hardening and thickening of the husk along with the development of a solid endosperm happens in the following 6–8 months. Between month 6 and month 12, the formation of the solid endosperm starts from the stylar end as a jelly-like substance and progresses towards the pedicel end as it thickens and turns white with a thickness of 10–15 mm as it grows (Pandiselvam *et al.*, 2022). The harvesting of coconut fruit is undertaken at various stages of maturity depending on the intended use. When coconuts are 6–8 months old, they can be harvested as immature, tender young fruit that are collected mainly for their coconut water and tender soft endosperm, whereas 8–9-month-old coconut fruit are picked to make 'snowball' tender coconuts, a coconut where the husk and shell are removed. These are round and white in colour with water still held within. Fruit that are 9–10 months old are used to make coconut chips, while fully grown coconuts (11–12 months old) are harvested for the manufacturing of copra and other kernel-based products (Manikantan *et al.*, 2018).

7.2.2 Harvesting methods

Coconut harvesting is a skilful and tedious job, usually undertaken by professional climbers. Climbing is normally carried out with the aid of ladders 3–4 m long. The climbers usually employ a rope ring around their ankles and/or arms to aid in climbing. Husks tied to the palm trunk are used as climbing stairs in some regions. Another method employed by climbers is the use of a rope loop around the climber and the palm trunk to assist in climbing. Pole harvesting is another common practice that is seen in some parts of India and Sri Lanka, where the fruit is harvested with the aid of a sharp knife attached to the end of a long pole. Although it is becoming less common, coconut harvesting using specially trained animals such as monkeys is sometimes practised in parts of South-east Asia such as Malaysia, Indonesia and Thailand. In places where traditional palm climbers are difficult to find, palm-climbing devices or mechanical harvesters are beginning to be used.

Many types of climbing aids, including paddle, electric, tractor and manually operated, have been developed and tested for their suitability for fruit harvesting. One of the commercially available paddle-type climbing aids is the Chemberi Joseph model. This comprises two assemblies, a top and a bottom assembly, both consisting of a steel rope wire looped around the palm. The climber operates the assembly by using their hands and feet to raise the assembly along the trunk. A similar device with better safety features has been developed by the Indian Council of Agricultural Research – Central Plantation Crops Research Institute (ICAR-CPCRI) (Fig. 7.1; Singh *et al.*, 2012). Another push-up type of coconut harvester, developed by the Tamil Nadu Agricultural University (TNAU), and described as the TNAU model, has the climber sitting on a steel frame attached to the trunk with the help of belts. The climber pushes up the frame by hand, while the feet are inserted into the lower frame attached to rollers, which in turn are lifted by the legs. The rubber roller provides grip and reduces the damage caused by the machine to the palm trunk.

7.3 Postharvest Handling of Coconut

7.3.1 Storage conditions for fresh coconuts

The harvested coconut fruit are generally collected and transported to a shaded gathering spot soon after harvest. Fruit that are unmarketable owing to significant insect damage, such as decay, or those with an unsatisfactory fruit size are rejected. The grading of coconut is usually based on size, uniformity, shape and skin blemishes. Seasoning is the technique of storing the collected fruit in piles in the shade for a few days before further processing and is practised in countries such as India and Sri Lanka, where the fruit may need to be harvested early. Seasoning can aid in easier de-husking and cleaner shelling, and reduces the water content and increases kernel thickness. Film-wrap and vacuum-packaging technologies can be used to extend the shelf life of fresh tender coconuts during a short period of storage (Jangchud *et al.*, 2007). The same study showed that film-wrapping and vacuum packaging of burnt aromatic coconuts could increase their shelf life to 18 and 28 days, respectively, compared with 14 days for non-treated fruit. Temperature also plays a significant role in the shelf life of fresh fruit. Although coconut is a tropical fruit, the optimal storage temperature of the fruit is 0–1°C, and at this temperature, the fruit can be successfully stored for 2 months (Ghosh, 2015). Studies have shown that reducing the temperature from 30°C to 5°C significantly increases the storage life of Makapuno coconut fruit from 3 to 42 days (Kietsuda *et al.*, 2014). Waxing of the endocarp can also be undertaken to increase the fruit storage life. Waxing reduces weight loss and cracking of the de-husked fruit during transport (Ghosh, 2015). The recommended relative humidity for processed coconut fruit (trimmed or de-husked) storage is 80–85%. Storage at lower relative humidity can result in weight and transpiration losses, whereas storage at higher relative humidity can lead to mould growth.

Fig. 7.1. A safety device developed by the Indian Council of Agricultural Research – Central Plantation Crops Research Institute to aid in he climbing of coconut palms.

7.3.2 Storage conditions for fresh kernels

The edible part of the coconut fruit consists of the solid endosperm or kernel and the liquid endosperm or water. The coconut kernel has long been used for a variety of purposes, including both raw and cooked consumption. Fresh kernel gratings are mostly used in the preparation of curries, chutneys, bread, cookies, toffee, desserts, salad toppers and other culinary applications. The fresh kernel is difficult to preserve due to its high fat, sugar and moisture content. Using hurdle technology (an approach to inactivate microorganisms during food preservation) and refrigeration, researchers demonstrated that they could successfully preserve kernel gratings for up to 1 month, when used in combination with humectants, preservatives, acidulants and antioxidants (Gamlath *et al.*, 2004). Additionally, hurdle technology was used to increase the shelf life of kernel gratings from 1 month at ambient temperatures to 3 months at refrigerated temperatures (Gunathilake, 2005). Successful pasteurization of fresh-cut coconut kernel was demonstrated by Ferrentino *et al.* (2015) using a combination treatment of high-pressure CO_2 and high-power ultrasound for 1 month.

7.3.3 Storage conditions for copra

In many countries, the postharvest processing and commercialization of coconut fruit is still in its infancy. Copra is the dried endosperm of the coconut. Almost 40% of the coconuts produced in India, and more than 50% of the coconuts produced around the world, are converted into copra, of which around 80% is processed into oil (Ghosh, 2015). Copra is usually cured and dried to a moisture content of below 7% and is cut into uniform pieces that can be stored in a well-ventilated room for long periods of time. Improper drying of copra can cause fungal and types of other microbial attack, insect infestation, production of unpleasant odours, poor colour and a higher free fatty acid content. Studies have shown that when underdried copra samples (at 12% moisture content) were stored under optimal conditions for 5 months, they lost up to 10% of their weight and had a higher percentage loss of free fatty acids (Ghosh, 2015). In small warehouses, after drying, copra can be weighed and stored in coarse sacking or jute bags until milled, whereas in large warehouses, copra is stored in piles of 1000–2000 t where aeration and rotation of the product must be undertaken (Malabrigo, 1977).

7.4 Value Addition of Coconut

Coconut is a unique crop that can claim hundreds of applications that no other horticultural oil crop can match (Fig. 7.2). Coconut goods and by-products have an extremely wide variety of commercial applications. Coconut is both a food and an oilseed crop. It also serves as a source of fibre, wood and fuel. Desiccated coconut is one of the most common value-added products from the coconut kernel. This is where the kernel is shredded and dried to less than 3% moisture content. It is a widely used commercial product in confectionery and culinary sectors around the world, as one of the primary components in fillings for chocolate, sweets and other confections. It is also used raw as a garnish for cakes, cookies and ice cream, and is toasted in snacks. Coconut chips are another important value-added product from the kernel. These are a ready-to-eat snack generally produced by osmotic dehydration of the kernel from 9–10-month-old coconut fruit. The kernel can also be used to produce coconut milk, which is an oil–protein–water emulsion extracted from the kernel through pressing. Coconut milk is used for the preparation of fish and meat meals, curries, sweets, desserts, puddings, drinks, cakes, cookies, coconut jam, ice creams and other foods. Coconut milk can be spray dried to obtain coconut milk powder, which retains its natural flavour and texture, and has a better shelf life than fresh milk. After milk extraction, the kernel residue can be dried and powdered to obtain a protein-rich coconut flour. Another important value-added product of coconut is virgin coconut oil, which is produced from the processing of fresh coconut kernel.

The liquid endosperm or tender coconut water is a nourishing isotonic beverage rich in minerals, amino acids and carbohydrates. Packaged tender coconut water, minimally processed tender coconut and snowball tender coconuts are the major value-added products created from young, tender coconut fruit. After extraction and additive treatments, the tender coconut water can be packed in pouches or aluminium cans or using Tetra Pak technology to produce packed coconut water. The minimal processing of tender coconuts, which includes the partial or complete removal of the husk, is undertaken to improve handling and reduce bulkiness. Snowball

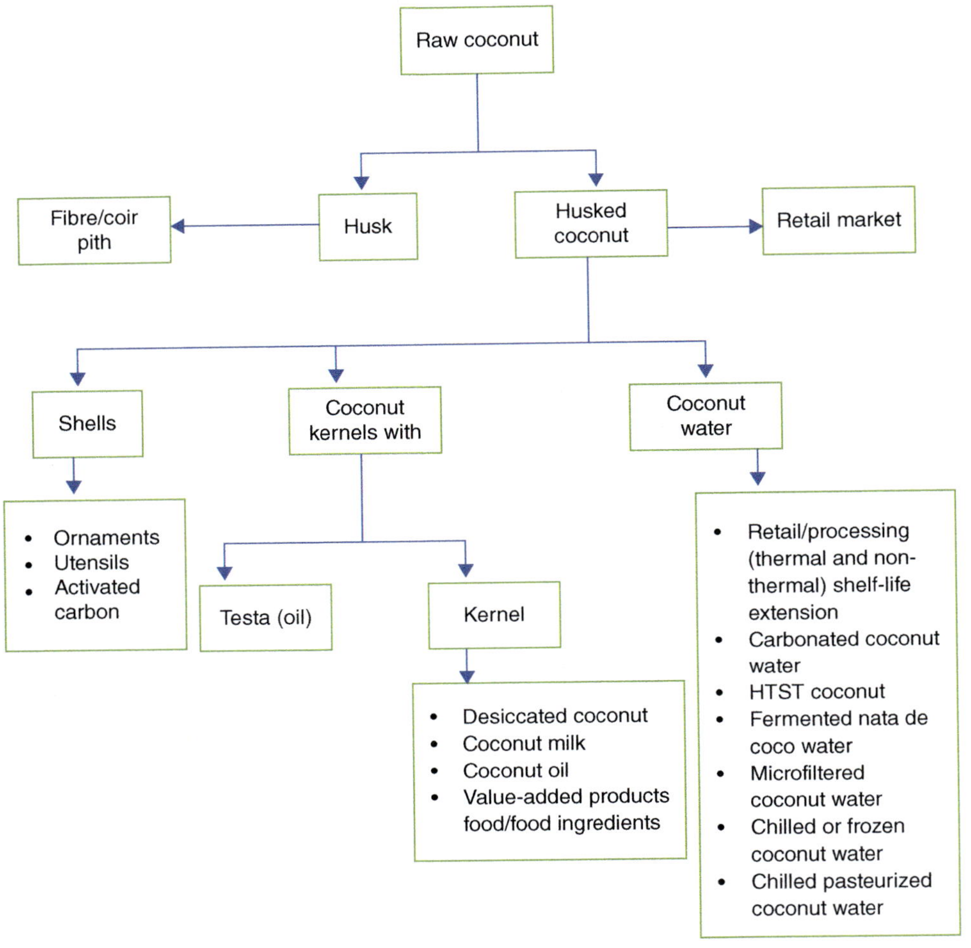

Fig. 7.2. Postharvest handling of coconut. HTST, high temperature, short time.

tender coconuts are an example of minimally processed coconut. Another important value-added product from coconut water is nata de coco. The Gram-negative, cellulose-forming bacterium *Acetobacter aceti* subsp. *xylinium* is used in its preparation to form a delicious gelatinous substance. The culture solution is made by combining coconut water, sugar and acetic acid in a predetermined ratio, followed by inoculation with *A. aceti* as a liquid culture. After 2–3 weeks of incubation, it is harvested, cut and immersed in flavoured syrup and used in ice creams, beverages, pickle, etc. The processing of coconut sap, tender coconut water and mature coconut water using various thermal and non-thermal techniques to preserve quality has been shown to increase the shelf life of these products. The majority of research is now focused on developing energy-efficient approaches and new products.

7.5 Processing Machinery

7.5.1 Mature coconut

De-husking

The first and foremost operation in coconut processing is de-husking, which is time and energy consuming. Operations involving mechanical and manual de-husking are described below. Recognizing the need to improve de-husking, a manual, twin-bladed de-husking tool was developed by Waters (1949). This tool consisted of two thin blades in a juxtaposed arrangement that when impaled on to the coconut fruit could remove a portion of the husk. For complete nut removal, the approach needed to be repeated three to four times. This process has not been widely adopted owing to the chances of injury to the person handling the tool. An alternative, the coconut spanner, developed in India, consisted of a long handle and a sharpened tongue, which worked by piercing, loosening and separating the coconut husk. It needed to be used three to four times to remove the nut from the husk, and every time the worker had to bend and reposition the coconut on the stand and then stand to use the tool again. This approach required prolonged manpower and therefore was not popular. Another tool made by Hill (1983) comprised two wedge blades (one movable and one fixed) attached to a central crossbar, the end of which slid on the spaced rod forming a frame. The fruit was mounted in a bowl in a vertical position and the blade in contact with the crossbar could be moved vertically down, which resulted in the blade piercing the coconut husk, with the separate immovable and movable blades making de-husking easier. The drawback of this equipment was that it was tiresome to place the fruit into the bowl and there were difficulties in piercing the fruit. Kerala Agricultural University (KAU) in India developed a coconut de-husking machine that had a foot pedal that operated movable and stationary cutting blades. When the foot pedal was depressed, the movable and stationary blades separated and would rip apart the coconut husk, which remained pierced on to the blades. This process had to be repeated three to four times for complete removal of the husk. This method possessed very few advantages and thus was not commonly accepted. Another team from KAU developed a coconut de-husking tool that was easier to operate. It had a hinge pin, a movable wedge-like blade, a stationary wedge, a wedge seat, a lever and a pedestal with a base. The coconut fruit was stabbed by hand upright on to two juxtaposed wedge-shaped blades. By moving the lever upwards, the movable wedge-shaped blade on the load arm of the lever moved away from the immovable wedge-shaped blade and loosened a sector of husk from the fruit. To be separated completely, the above steps had to be repeated two or three times, but only took around 8–20 s to do. The approach was not commercialized because scaling up was difficult. However, because the tool did not require maintenance and was simple to operate, it was adopted at a domestic level. ICAR-CPCRI developed a manually operated de-husker with three sharp, juxtaposed blades, which initially faced upwards. The blades were raised up and outwards by swivelling around a pivot point located at the bottom of the device. The fruit was then impaled on the blades by hand and a foot-lever returned the blades to their original position, thereby removing sections of the husk. The approach was tiresome and hence this method was not widely adopted. To overcome the disadvantages in these earlier designs, mechanical, power-operated de-huskers have now been developed (Nampoothiri *et al.*, 2019)

A power-operated semi-automatic coconut de-husking machine was designed and developed at ICAR-CPCRI in 1996. It comprised a rotating platform with a dozen fruit holders, a piercing system equipped with a shock absorber to avoid fruit breakage owing to the load applied, and a set of scooping blades for dividing and easing the husk from the fruit. Another powered, hydraulically operated machine was designed by Fletchers Engineering Ltd, Australia. This machine had a hydraulic system, an electric motor, an arm-like structure and a fruit-holding mechanism. The electric motor provided the hydraulic power for operating the arm-like structure, which was involved in the opening of the husk. The holding mechanism helped to keep the fruit in position with the eye end facing up. During operation, the coconut holder was lifted from its position to facilitate the de-husking process. The limitation of this device was its need for power and the requirement for a skilled labourer to place the fruit into the correct position for de-husking. Azmi *et al.* (2015) developed a coconut de-husking machine that consisted of a series of rollers with spikes, a chain drive, a presser, a clearer, and shaft and belting system. The two rollers, operating in opposite directions, passed the coconut through the centre of the device, which forced it into the spacing between the rollers where peeling spikes were inserted. These spikes were sharp and equally spaced, and facilitated fruit penetration and gripping, and in the longer term resulted in de-husking of the fruit.

The apparatus was used on a small scale in rural areas, with the major drawback being that it could only be used for small-scale applications. A mechanical coconut de-husker developed by Abi Varghese had three parts, namely an inlet throat, a husking mechanism and an outlet. The entire coconut fruit was placed vertically into a feed chute where it became entrapped between the rollers. The right-hand roller had sharp, slightly curved, hook-like knives that helped in the removal of the husk, while the left-hand roller was spring-loaded to press the fruit towards the right spiked roller. As the rollers rotated, the husk was removed, and the de-husked coconut fruit fell out below the rollers. Many other mechanical de-huskers have been developed, including a pneumatically powered unit, while the Caribbean Industrial Research Institute in Trinidad and Tobago has developed an electric motor-powered coconut de-husker.

Splitting and de-shelling

Splitting of coconut fruit by manual means can result in injury for the operator. Thus, Roshni *et al.* (2009) developed a power-operated coconut punch-cum-splitter for splitting fruit and extraction of coconut water.

For value addition and kernel storage, de-shelling is the next operation that needs to take place. Rey (1955) developed a knife-shaped shallow spoon that moves backwards and forwards during the spinning of a cam rod, which results in the scooping out of the coconut kernel from the shell of a manually cut fruit. A later design used pressurized water jets to detach the kernel from the shell (Brandis and Glaser, 1973). This process was wasteful of water and required intensive labour to operate. More recently, Tonpe *et al.* (2014) developed a de-shelling machine that overcame many of the demerits of these earlier manually operated devices. The machine was tested for its performance based on its de-shelling capacity and de-shelling efficiency. Its average de-shelling capacity was reported to be 195 fruit h^{-1} and the machine could de-shell fruit without breakage.

ICAR-CPCRI has developed a semi-automatic de-shelling machine for coconut that has become popular across India (Fig. 7.3). It consists of two concentrical blades with 24 teeth rotating at 24 rpm, a stationary shaft on which coconuts are placed and a motor. The coconuts are pressed towards the rotating blades by keeping them on a stationary shaft, which leads to de-shelling by the impact force created using a rotating blade. It has wide application in virgin coconut oil, coconut milk and coconut chip processing industries.

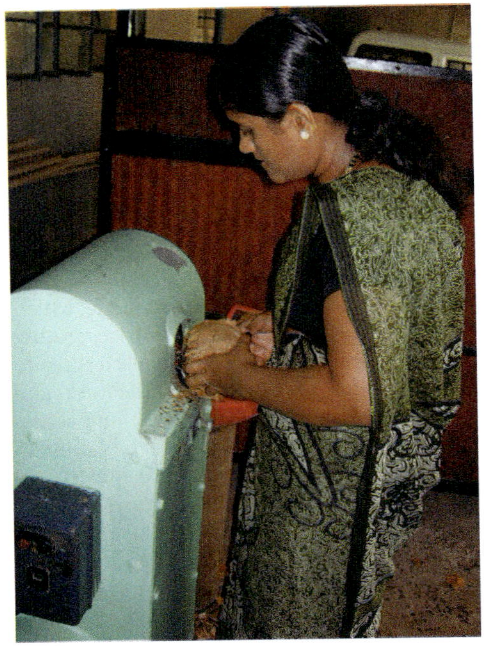

Fig. 7.3. A coconut de-shelling machine developed by the Indian Council of Agricultural Research – Central Plantation Crops Research Institute.

7.5.2 Copra processing

As described earlier, copra drying is performed on fresh coconut kernel for about 7–9 days. Such prolonged treatments can lead to contamination of the kernel, microbial spoilage, dirt and rancidity. To overcome these disadvantages, various mechanical drying practices have been developed. Mohanraj and Chandrasekar (2008) designed a forced convection solar drier for copra consisting of a solar air heater, a drying chamber, a transparent cover (glass), a thick copper heat-absorbing plate and a centrifugal fan. The air to be heated goes between the transparent cover and the heat-absorbing plate, which then produces a warming effect that is passed into the chamber to dry the copra. The moisture produced within the unit is removed by the centrifugal fan. A study using a solar-ambient hybrid source heat pump dryer for copra drying in hot and humid weather conditions was undertaken by Mohanraj (2014), who showed that the moisture content could be reduced from 52.0% to 9.2% when using a specific energy load of 0.79 kg kW h^{-1}. Lesidan (2020) developed a semi-direct dryer for flat terrains. This dryer consists of a drying bed, a plenum chamber, a tunnel and firing chamber. Coconut husks were used as the fuel for the drying process. The time required for drying to establish a moisture content of 12% was 24 h. Smoke-dried copra processing is commonly practised in the Pacific Region, India and the Philippines, as smoke can have an inhibitory effect on growth of mould. The major shortcomings of smoke drying are the formation of polyacrylamides and the brown colour given to the treated copra.

7.5.3 Coconut kernel processing

For slicing of the coconut kernel, ICAR-CPCRI developed a technology based on that of a pedal-powered sewing machine. The process comprises a stainless-steel slicing blade fixed firmly on to a circular disc, a specially designed curved feeder to feed the coconut kernel on to the slicing blade and an outlet to collect the respective slices. Likewise, a pedal-operated mechanism has also been developed in which the pedal power is transferred to the cutting blade by a belt and pulley system. A further coconut slicing machine has been developed by ICAR-CPCRI that has improved efficiency and speed of slicing.

Grating of coconut kernel for the preparation of various foods is a time-consuming process, so this has led to the development of several manual and mechanical graters. Raj *et al.* (2016) developed a grating machine that consisted of a clamp to fix the unit to a table, a holder to grip the kernel in position, a motor to run the machine, and a blade and scrapper to remove the kernel. The coconut holder can move sideways, with the help of a handle, to allow the complete grating of the kernel, irrespective of the shape and size of the fruit.

7.5.4 Young, tender coconut

Cutting/punching machine

Tender coconut water, a well-known refreshing drink, is an isotonic beverage drunk in many parts of the world. It is not only used as a thirst-quenching liquid but also as a mineral-rich drink. Due to its importance, a tender coconut cutting/punching device to extract coconut water has been developed. Patil and Chendake (2017) developed both a sitting and a standing model that can punch holes and cut open young coconut fruit. The punched hole allows the water to be collected or drunk using a straw. The fruit is then cut open by the cutting machine, usually held on a wooden platform.

Trimming machine

Tender coconuts have a huge consumer demand due to their sweet water and smooth, soft kernel. The tender coconut husk weight increases transportation costs, and therefore husk trimmers have been developed. Husk trimming can be undertaken manually using a knife or mechanically with a specialized trimming machine. The mechanical trimming machines consist mainly of a fruit holder, a fruit shoulder and body trimming knife, and a base cutting knife (Pandiselvam *et al.*, 2019, 2020).

Coconut sap collection and processing

In Sri Lanka and India, sap collection is usually undertaken by inflorescence tapping. The spadix is said to be ready for tapping when the unopened female flowers start to create a bulge at the end of the spathe. The spathe is then manipulated for 3 weeks by daily beating with a tapping rod to increase sap flow. During this time, a 4–5 cm cut is made at the tip of the spathe and an earthen pot is positioned for collection of the sap. It takes around 15–21 days for tapping to take place and for sap collection to be undertaken; however, sometimes it continues for a period of 6–8 months, depending on the location. The sap is then further processed into coconut jaggery, coconut vinegar or coconut treacle.

Storage

For short-term fruit storage, heaping of the harvested fruit is normally undertaken, but this can result in fruit losses. Moisture control plays a major role in the effective storage of coconut fruit. Haseena *et al.* (2010) investigated the influence of ambient temperature and relative humidity on the storage of tender coconut fruit, either with or partly without their husks in place (60% of the husk removed). Storing fruit at 27 ± 2°C and 70 ± 5% relative humidity for 13 days resulted in a fruit weight loss of 19.6%, while the volume of coconut water decreased from 232 to 205 mL, and the total sugar content rose from 3.4 to 4.8 g 100 mL^{-1}. This increase in sugar content might be due to the conversion of starch and polysaccharides to soluble sugars, as well as dehydration of the fruit.

The tender coconut water shelf life, determined from fruit that were stored at ambient or refrigerated conditions for 20 days, showed that refrigeration could increase the shelf life beyond that achieved under ambient storage. Mature de-husked coconuts can be waxed or film wrapped to help prevent water loss and to increase fruit shelf life. When stored at 3–6°C with 90–95% relative humidity, young tender coconuts can be stored for 3–4 weeks without any quality loss. The general practice for young, tender coconuts was to trim the husk to the maximum possible level and then to place in cold storage. However, this process can lead to fruit browning and development of mould on the fruit. Trimming and chemical treatment with 2% sodium metabisulfite can prevent browning, and these coconuts can be stored up to 7 days longer than those that are not treated (Paull and Ketsa, 2015).

During copra storage, pest and mould infestations are also common. The major pests affecting stored coconut fruit are the red-legged beetle (*Necrobia rufipes* (Fabricius, 1781)) and the saw-toothed grain beetle (*Oryzaephilus surinamensis* L.), which can induce up to 15% losses during storage. Fumigation can help prevent such losses but is rarely practised.

7.6 Conclusion

Coconut is regarded as one of the most beneficial plants to mankind, as almost all parts of the palm are useful and have domestic or industrial value. It is called the 'wonder fruit' and 'nature's supermarket' and provides the consumer with numerous products such as young tender coconut water, kernels, oil, virgin oil, milk, coir, pith, shell charcoal and carbon, timber and fuel. It provides sustenance and a livelihood for rural populations living in coconut-producing countries and for millions of small-holder farmers, processors and dealers. Despite the cultural and economic value of coconut, the industry faces many issues. These issues and opportunities are discussed in the other chapters of this book. For the harvest and postharvest management of coconut, prominent issues are a severe lack of skilled labour for harvesting, plant protection tasks and processing of products. Capacity building and technical developments in harvesting and post-harvesting operations have the potential to act as a foundation for rural development in regions where coconut is grown. Efforts in product diversification, the implementation of strict quality standards for coconut goods and higher productivity are some of the strategies that can help ensure the coconut industry's long-term sustainable future. Moreover, increasing mechanization with the help of robotics and automation is likely to play a major role in improving the lives and incomes of those working in the coconut industry, but further work is needed to make these technologies safe, accessible and economical.

7.7 References

Anithakumari, P. (2013) Labour availability in coconut cultivation and impact on technology adoption as perceived by coconut farmers. *Journal of Plantation Crops* 41, 186–190.

Azmi, H., Sanuddin, A.B., Zakimi, M.Z., Jamali, M.S., Radhwan, H. *et al.* (2015) Akademia baru design and development of a coconut de-husking machine (machine component design). *Journal of Advanced Research Design* 4, 9–19.

Brandis, C. and Glaser, F. (1973) Specification for separations of kernels from coconuts. Philippine Patent No. 9865.

Ferrentino, G., Komes, D. and Spilimbergo, S. (2015) High-power ultrasound assisted high-pressure carbon dioxide pasteurization of fresh-cut coconut: a microbial and physicochemical study. *Food and Bioprocess Technology* 8, 2368–2382.

Foale, M. (2003) *The Coconut Odyssey: The Bounteous Possibilities of the Tree of Life*. Australian Centre for International Agricultural Research, Canberra.

Fortune Business Insights (2021) Coconut derivatives market. Fortune Business Insights. Available at: www.fortunebusinessinsights.com/industry-reports/coconut-derivatives-market-100540 (accessed 17 August 2023).

Gamlath, S., Dassanayaka, L. and Gunatilake, P. (2004) Preservation of fresh coconut gratings by hurdle technique. *Food Australia* 56, 140–142.

Ghosh, D.K. (2015) Postharvest, product diversification and value addition in coconut. In: Sharangi, A.B. and Datta, S. (eds) *Value Addition of Horticultural Crops: Recent Trends and Future Directions*. Springer India, New Delhi, pp. 125–165.

Gunathilake, K. (2005) Application of hurdle technique to preserve fresh scraped coconut at ambient and refrigerated storage. *Journal of the National Science Foundation of Sri Lanka* 33, 265–268.

Haseena, M., Bai, K.V.K. and Padmanabhan, S. (2010) Post-harvest quality and shelf-life of tender coconut. *Journal of Food Science and Technology* 47, 686–689.

Hill, E.D. (1983) Coconut Husk Removing Tool. United States Patent Office, Serial No. 4383479.

Jangchud, K., Puchakawimol, P. and Jangchud, A. (2007) Quality changes of burnt aromatic coconut during 28-day storage in different packages. *Food Science and Technology* 40, 1232–1239.

Kalathiya, K.V., Dave, S.K. and Mehta, D.R. (2007) Cost and return in different sizes of coconut holding in Valsad District (South Gujarat). *Indian Journal of Agricultural Research* 41, 228–231.

Kietsuda, L., Diane, M.B., Orrawan, P. and Siriphanich, J. (2014) Postharvest quality and storage life of 'Makapuno' coconut (*Cocos nucifera* L.). *Scientia Horticulturae* 175, 105–110.

Lam, T.L. and Xu, Y. (2011) Climbing strategy for a flexible tree climbing robot – treebot. *IEEE Transactions on Robotics* 27, 1107–1117.

Lesidan, M. (2020) Design and development of semi-direct copra dryer for flat terrain. *CORD* 36, 16–22.

Malabrigo, P. (1977) Drying, storage, and preparation of copra for extraction of oil. *Journal of the American Oil Chemists Society* 54, a485–a488.

Mani, A. and Jothilingam, A. (2014) Design and fabrication of coconut harvesting robot: COCOBOT. *International Journal of Innovative Research in Science, Engineering and Technology* 3, 1351–1354.

Manikantan, M.R., Pandiselvam, R., Beegum, S. and Mathew, A.C. (2018) Harvest and postharvest technology. *The Coconut Palm (*Cocos nucifera *L.) – Research and Development Perspectives.* Springer, Singapore, pp. 635–722.

Mohanraj, M. (2014) Performance of a solar-ambient hybrid source heat pump drier for copra drying under hot-humid weather conditions. *Energy for Sustainable Development* 23, 165–169.

Mohanraj, M. and Chandrasekar, P. (2008) Drying of copra in a forced convection solar drier. *Biosystems Engineering* 99, 604–607.

Mordor Intelligence (2022) Coconut market size & share analysis – Growth trends & forecasts (2023–2028) Available at: www.mordorintelligence.com/industry-reports/coconut-products-market (accessed 17 August 2023).

Nampoothiri, K.U.K., Krishnakumar, V., Thampan, P.K. and Achuthan Nair, M. (eds) (2019) *The Coconut Palm (*Cocos nucifera *L.) – Research and Development Perspectives*. Springer, Singapore.

Nguyen, Q.T., Bandupriya, H.D., Lopez-Villalobos, A., Sisunandar, S., Foale, M., and Adkins, S.W. (2015) Tissue culture and associated biotechnological interventions for the improvement of coconut (*Cocos nucifera* L.): a review. *Planta* 242, 1059–1076.

Pandiselvam, R., Manikantan, M.R., Subhashree, N., Mathew, A.C., Balasubramanian, D. *et al.* (2019) Correlation and principal component analysis of physical properties of tender coconut (*Cocos nucifera* L.) in relation to the development of trimming machine. *Journal of Food Process Engineering* 42: e13217.

Pandiselvam, R., Manikantan, M.R., Balasubramanian, D., Beegum, P.S., Mathew, A.C. *et al.* (2020) Mechanical properties of tender coconut (*Cocos nucifera* L.): implications for the design of processing machineries. *Journal of Food Process Engineering* 43: e13349.

Pandiselvam, R., Khanashyam, A.C., Manikantan, M.R., Balasubramanian, D., Beegum, P. *et al.* (2022) Textural properties of coconut meat: implication on the design of fiber extraction and coconut processing equipment. *Journal of Natural Fibers* 19, 11092–11104.

Pathiraja, P.M.E.K., Fernando, M.T.N., Abeysekara, A.W.A.D.R. and Subasinghe, S.D.J N. (2010) An assessment of labour availability in major coconut growing areas in coconut triangle. *COCOS* 19, 13–26.

Patil, S.B. and Chendake, A.D. (2017) Development of multipurpose punching and cutting machines. *International Journal of Agricultural Science and Research* 7, 177–188.

Paull, R E. and Ketsa, S. (2015) *Coconut: Postharvest Quality-maintenance Guidelines*. University of Hawai'i at Manoa, Honolulu, Hawaii.

Prithviraj, V., Pandiselvam, R., Babu, A.C., Kothakota, A., Manikantan, M.R. *et al.* (2021) Emerging non-thermal processing techniques for preservation of tender coconut water. *Food Science and Technology* 149: 111850.

Raj, R.S.P., Raj, S.B. and Ashana, A.N. (2016) Design of an innovative coconut grating machine using tinkercad. *India International Journal of Research in Mechanical Engineering* 4, 178–182.

Rethinam, P. (2019) International scenario of coconut sector. In: *The Coconut Palm (*Cocos nucifera *L.) – Research and Development Perspectives*. Springer, Singapore, pp. 21–56.

Rey, H.O. (1955) Device for removing meat from coconut. Philippine Patent No. 793.

Roshni, T., Jippu, J., Ratheesh, C.S., Sachin, J. and Sreevisakh, K.L. (2009) Development of a household coconut punch-cum-splitter. *CIGR Journal* 10: 1188.

Sharma, R.K., Chauhan, O.P. and Xavier, J.R. (2021) Technological innovations in food processing and value addition to coconut. *Journal of Food and Agriculture Research* 1, 69–85.

Singh, D., Kushwaha, H., Mathew, A., Manojkumar, T., Hebbar, K. *et al.* (2012) Performance evaluation of paddle type and push up type coconut climbing devices. *Journal of Plantation Crops* 40, 224–226.

Siriphanich, J., Saradhuldhat, P., Romphophak, T., Krisanapook, K., Pathaveerat, S. *et al.* (2011) Coconut (*Cocos nucifera* L.). In: *Postharvest Biology and Technology of Tropical and Subtropical Fruits*. Woodhead Publishing, Sawston, UK, pp. 8–35.

Suparat, N. and Maneewarn, T. (2015) Spraying analysis for a coconut climbing robot. In: *2015 7th International Conference on Information Technology and Electrical Engineering (ICITEE)*. IEEE, Piscataway, NJ, pp. 388–393.

Tonpe, K.K., Sakhare, V.P. and Sakhale, C.N. (2014) Design and performances of coconut de-shelling machine. *International Journal of Engineering Research and Applications* 4, 39–44.

Waters, C.P. (1949) Coconut husk removing tool. United States Patent Office. Serial No. 674,305.

8 Breeding and Genetics for Coconut Improvement

Lalith Perera[1]*, Manimekalai Ramaswamy[2], Pavani Dulanja Dissanayake[1], Chongxi Yang[3] and Sundaravelpandian Kalaipandian[3]
[1]Coconut Research Institute, Lunuwila, Sri Lanka; [2]Indian Institute of Agricultural Research, Sugarcane Breeding Institute, Coimbatore, India; [3]School of Agriculture and Food Sustainability, The University of Queensland, Gatton, QLD, Australia

Abstract

Coconut is known as the 'Tree of Life' and is cultivated in more than 90 countries, with consumption increasing around the world, even in non-coconut-producing countries. Although coconut is in high demand, production is not increasing for a number of reasons, including increasing palm senility, natural calamities, and abiotic and biotic threats. It is challenging to undertake coconut breeding as the palm takes many years to mature. Only a few breeding methods are generally used, including mass selection and inter- and intrahybridization. Two major coconut types are recognized, based on trunk height, tall and dwarf varieties. Various hybrids have been developed by crossing between and within these two types. This chapter focuses on the utilization of coconut germplasm to develop high-yielding, abiotic and biotic stress-tolerant varieties by combining traditional and biotechnological tools for coconut genetic improvement.

8.1 Introduction

Coconut (*Cocos nucifera* L.) is an extremely important cash crop. It is traditionally connected to the socio-economic life of people coming from coconut-growing countries. It has countless uses, including income generation, food production and other social necessities, mainly in rural communities (Harries, 1995). Burkill (1966) described the coconut palm as being among the greatest gifts of nature to humankind because of its multitude of uses. At least 96% of coconut cultivation in most coconut-producing countries is undertaken by smallholder farmers (Persley, 1992). Coconut cultivation is distributed mainly in tropical and subtropical regions, in a zone from the equator to latitudes 20°N or 20°S and to altitudes of about 1200 m above sea level. It is cultivated in more than 90 countries on about 12 million ha (Dissanayake and Perera, 2020).

Coconut grows well in moderate to well-aerated soils under high temperatures of 27–30°C, with moderate to high relative humidity. Asia-Pacific countries are the largest coconut producers in the world. Palm production and land productivity show significant differences among countries. Although Indonesia and India have the greatest production of coconut fruit (Table 8.1), the productivity of these two countries is lower compared with other countries such as Brazil, Vietnam and Mexico (Table 8.2).

* Email: lalithperera1234@yahoo.com

DOI: 10.1079/9781789249736.0008

Table 8.1. The top five coconut-producing countries in 2020 in terms of copra production year^{-1}. Adapted from FAOSTAT, 2020.

Country	Coconut production (million t)
Indonesia	16,824,848
India	14,695,000
The Philippines	14,490,923
Brazil	2,458,839
Sri Lanka	2,233,600

Table 8.2. Productivity of coconut production across various countries in 2013. Adapted from Nampoorthiri *et al.*, 2019.

Country	Productivity (t ha^{-1})
Brazil	11.0
Vietnam	9.6
Mexico	6.5
Indonesia	6.1
India	5.5
Malaysia	5.4
Sri Lanka	5.2
Thailand	4.8
The Philippines	4.3

8.2 Types of Coconut

The two major types of tall coconut palm are *niu kafa* and *niu vai*, from the Samoan names for these two forms (Harries, 1978). The *niu kafa* form evolved naturally and produces fruit that are large, angular in shape and thick husked, with a low water content, and are generally slow to germinate. Such characteristics are important in aiding the dissemination of fruit via ocean currents. In contrast, the *niu vai* form produces smaller, spherical-shaped, thin-husked fruit with an increased endosperm volume and are early germinating. Often this form carries a degree of disease resistance and is believed to have developed through human selection over numerous generations of cultivation. Harries (1978) suggested that further selection and dissemination by humankind produced several varieties from the *niu vai* form and distributed them throughout the tropics. Child (1964) pointed out that all modern-day areas under coconut cultivation have been influenced by humankind who brought about the dissemination and then the modification of the genetic structure of the coconut populations through selection.

Bahala (1987) recognized three different forms of coconuts: the wild (*niu kafa*) and domestic (*niu vai*) forms, as referred to by Harries (1978) and a cultivated form. The most extreme *niu kafa* forms, according to Bahala (1987), are only found on uninhabited coral atolls or islands, or on remote mainland beaches. In general, no or few pure wild-type coconut palms (see Chapter 1, this volume) have continued to exist in recent times, although several of their traits have been introgressed into various present-day varieties. Thus, modern varieties are found to possess mixed and complex characteristics of both wild and domesticated forms, but the degree of mixing of characteristics varies from region to region as coconut cultivation began in each region with the predominant coconut forms that were present in those regions. As well as classifying coconut based on their proposed evolutionary history, coconut palms are also categorized into two main types based on palm height and breeding characteristics: tall (also called *typica*) and dwarfs (also called *nana*) varieties (Menon and Pandalai, 1958). Historically, the most planted and commercially important types were the tall varieties (Perera *et al.*, 2009). The major characteristics of both tall and dwarf types are highlighted in Table 8.3.

Tall types reach up to 30 m in height, are largely cross-pollinated, produce medium- to large-sized fruit and are late bearing. Dwarf types grow to 10–15 m tall and are largely self-pollinated (autogamy) and early bearing, producing smaller fruit with distinctive exocarp colours such as green, red, yellow and brown. Although autogamous, cross-pollination does occur between dwarf and tall varieties. Apart from tall and dwarf types, intermediate types also exist, called semi-tall or semi-dwarf (Liyanage, 1958; Menon and Pandalai, 1958).

Typically, coconut is a large palm, growing up to 30 m and for 60 or more years. It has composite fronds that reach up to 3.5 m in length with 60–90 cm pinnate leaflets. Like other palm species, coconut palms have a fibrous root system that is produced continuously from the base of the trunk throughout its lifespan. The rootlets contain the actively growing root tips, with a

Table 8.3. Characteristics of tall (*typica*) and dwarf (*nana*) types of coconut. Adapted from Perera *et al.*, 2009.

Trait	Tall (variety *typica*)	Dwarf (variety *nana*)
Plant height	Tall (20–30 m)	Short (10–15 m)
Bole at the base of the stem	Yes	No
First flowering	6–8 years	4–5 years
Lifespan	Long (80–100 years)	Short (40 years)
Fruit bearing	Continuous	Seasonal
Nut production per palm year^{-1}	Average 40	Average 80–100
Copra production and its quality	200 g per nut, good quality	80–100 g per nut, poor quality
Growing conditions	Variable	Sensitive to climatic changes
Breeding habit	Outbreeding	Inbreeding

single layer of epidermal cells. However, this layer gradually hardens and becomes impermeable to water and nutrient uptake when older. The coconut tree possesses a single terminal bud (the 'cabbage'), which contains the vegetative growing tip, and the tree can die if this bud is damaged. A bole is formed by tall varieties at the base of the trunk, while dwarf varieties do not produce a bole. The trunk growth of tall varieties is fast compared with dwarf varieties. The coconut fruit is botanically known as a fibrous drupe, and has both a liquid (coconut water) and a solid endosperm. The endosperm is enclosed by a hard shell (the endocarp) and then the husk (a fibrous mesocarp and thin exocarp). The shell is hard and covers the solid endosperm (kernel) and constrains the liquid endosperm (water). A smooth and thin outer layer over the husk is called the exocarp, which varies in colour, especially for dwarf varieties.

After pollination and fertilization, the fruit forms and matures in 12 months or slightly earlier in some dwarf types. Yield can be estimated before fruit maturity by counting the appearance of bunches and the fruit set in each bunch. A testa or thick brown seedcoat is present between the shell and the kernel. The testa adheres tightly to the white kernel, a tissue of approximately 12 mm thick around the central cavity holding the coconut water. Coconut diversity characterization is steered by a range of phenotypic characteristics that include plant height and fruit composition (Arunachalam and Rajesh, 2017). This chapter summarizes existing coconut genetic resources, their diversity, methods for breeding and genetic analysis of the germplasm for coconut breeding.

8.3 Flower Morphology, Anthesis and Pollination

Coconut flowering occurs continuously, and palms produce about 12 inflorescences per year. The single inflorescence develops in the axil of each new leaf at a rate of around one every month. Coconut has a monoecious inflorescence, which means that the inflorescence contains both male and female flowers, and is enclosed within the spadix. Within each inflorescence are 25–30 hard, erect, 30 cm long spikelets attached to a main floral peduncle. Each spikelet has about 200–250 male flowers, densely arranged from the tip to 7–8 cm above the base, in a clustered manner. The female flowers are globular in shape, and there are approximately 10–50, located at the base of the spikelet. The female flowers are usually much larger than the more numerous male flowers. The male flowers open and shed pollen on the same day. Usually, the male flower opening starts from the topmost spikelet in the inflorescence and proceeds downwards, but irregular opening is also observed. The male flower opening begins from the day of the appearance of the inflorescence and usually continues for about 18–22 days; this is referred to as the male phase of flowering, which occurs in both tall and dwarf types. In tall varieties, the female flowering phase starts 3 days after completion of the male phase and lasts for only 2–3 days. The female flowers of the previously opened inflorescence are still receptive when a new inflorescence begins to open. Hence, tall types naturally promote cross-pollination. However, sometimes vigorously growing young palms of certain varieties produce several inflorescences close together in time, which then allows inter-spadix pollination to occur in the same coconut palm, and this promotes self-pollination, even in the typically cross-pollinating tall types such as Malayan Tall and Rennell Island Tall varieties. However, some tall varieties are 100% allogamous, such as the West African Tall. In dwarf types, the female flowering phase starts at the beginning of the male flowering phase and continues for about 12–15 days, facilitating an overlapping of the two phases. Hence, dwarf coconuts are commonly self-pollinated in nature.

Meléndez-Ramírez *et al.* (2004) performed a study on visiting pollinators of coconut flowers and reported honeybees (*Apis* spp.) to be the most common visitors. *Apis mellifera* L., *Apis cerana indica* F. and *Apis dorsata* F. play a major role in visiting both male and female flowers (Thomas and Josephrajkumar, 2013). As these insects carry pollen on their ventral surface, they are the most efficient pollinators of coconut. In the absence of *Apis* spp., other insect species carry out pollination. For example, Kevan (1993), as well as personal observation made by the authors in 2019 at Maafahi island in the Maldives, found that pollination occurred in the absence of members of the genus *Apis*. Although there are many insects found on coconut flowers, including ants, many play no role in bringing about pollination. The importance of wind pollination, although generally negligible in many countries, appears to be important in the Maldivian coconut groves (Kevan, 1993). The tiny pollination mites (*Neocypholaelaps* spp.) also aid pollination in parts of Sri Lanka. After natural pollination occurs, around 50–70% of the fertilized female flowers abort within 3 months. The percentage loss depends on the weather and the availability of photosynthate within the growing bunches. Only the unaborted flowers form fruit, and they usually mature in around 10–12 months.

8.4 Genetic Resources Utilization

8.4.1 Diversity, wild relatives and progenitors

C. nucifera L. is the only species within the genus *Cocos*, and no close relatives of the palm have been found around the globe. The morphology and breeding habits have been used to describe various coconut genetic resources, including varieties and ecotypes (Narayana and John, 1949; Menon and Pandalai, 1958; Bourdeix *et al.*, 2005). A significant diversity is present within this species for various traits including fruit size, shape and colour, as well as thickness of the husk, shell and endosperm, and water volume. Many fruit shapes can be present, from spherical to pear-shaped, from short to long, and angular to various degrees (Foale, 1991). A study found that the diversity of morphological traits was greater in South-east Asian varieties compared with South Asian, African and South American varieties (Benbadis, 1992).

8.4.2 Germplasm conservation

Erosion of coconut genetic resources is now a serious problem around the globe due to many factors including industrialization, urbanization, infrastructure development, planting of other cash crops, the increased occurrence of natural calamities, and abiotic and biotic threats. Additionally, coconut replanting has focused on planting high-yielding varieties and hybrids with a narrow genetic base into regions with naturally greater genetic diversity. Due to the recognition of these challenges, collection and conservation programmes for coconut genetic resources have been initiated at various coconut centres around the world. These germplasm collections have largely been stored as living field collections, which are expensive to keep and laborious to maintain. Alternative conservation methods, such as cryopreservation (extreme cold long-term storage of viable plant tissues), would ideally be employed as back-ups to these collections. A cryopreservation technique has been developed for both embryos (Sisunandar *et al.*, 2010) and meristems (Bazrafshan, 2022). However, cryopreservation is not yet in use for routine coconut germplasm conservation.

To improve the capability of national conservation programmes for coconut germplasm and to utilize them for crop-improvement programmes, the International Plant Genetic Resources Institute (IPGRI) decided to establish a separate organization for coconut. Through the efforts of IPGRI, the International Coconut

Genetic Resources Network (COGENT) was established, and COGENT then formed international coconut gene banks in India, Indonesia, Papua New Guinea, Brazil and the Ivory Coast (IPGRI Annual Reports, 1997, 1998). These five gene banks are in addition to the previously established smaller national gene banks in India, the Philippines, Tanzania and the Ivory Coast. A COGENT-developed Coconut Genetic Resources Database lists a worldwide collection of conserved germplasm of 1621 accessions, comprising at least 1200 tall and 421 dwarf coconut varieties. These accessions have been conserved based on their phenotypic traits and for their ecological and distributional value. These coconut collections do contain duplicates, as well as genetically close accessions. If countries have a narrow genetic base for their coconut collections and are wanting to strengthen them, they are encouraged to import coconut genetic resources from other gene banks around the world.

The safe transfer of genetic resources between countries is important as it could accidentally lead to the introduction of new pests and diseases. Seednuts (whole, intact fruit) may not be appropriate to transfer from country to country as they may carry pathogens such as viruses, viroids and phytoplasmas. Embryo transfer is considered a much safer method for germplasm movement, but the steps for embryo germination and recovery require tissue-culture facilities, and these may not be available in institutions receiving international germplasm. Other factors limiting the successful use of embryo culture can depend on the genotype, duration and conditions of storage and transport, and availability of expertise and infrastructure within the recipient facilities.

8.4.3 Germplasm evaluation for genetic diversity analysis

The evaluation of germplasm is also referred to as germplasm characterization. Both terms refer to the identification of traits possessed by the conserved material. The evaluation can be for morphological characteristics, for yield-related characteristics, tolerance/susceptibility to environmental characters, or for pests and disease resistances. Germplasm conservation aims to retain the materials in a state ready for their immediate use in breeding programmes. To do this, a clear understanding of the specific characteristics of a variety is extremely important. The various accessions held are characterized to determine these specific characteristics. Germplasm characterization can be done using different methods. The most common method is to document the morphological characteristics, although with the advancement of molecular methods, these methods are now being used in combination with morphological characterization of coconut germplasm.

Coconut genetic diversity has now been assessed and reported in various studies undertaken around the world using molecular markers (Perera *et al.*, 1999, 2000, 2001, 2003; Manimekalai and Nagarajan, 2006). These studies suggest that the world's tall coconut varieties are genetically very diverse, while dwarf varieties exhibit low genetic diversity. This could be because of their recent divergence from the tall varieties, as well as their self-pollinating nature. For all coconut types, the various phylogenetic analyses have revealed two main clusters, the first consisting of the tall types from South-east Asia, the Pacific region and the west coast of Panama. Interestingly, the dwarf varieties formed a sub-cluster within this first cluster. The second cluster had tall varieties from South Asia, East Africa and the West African region. The first group of coconut showed a similar geographical distribution as the *niu vai* type proposed by Harries (1978), while the second group showed a similar geographical distribution to the *niu kafa* type. All dwarf types from various geographical regions were grouped into one cluster in the South-east Asia and the Pacific group. Together with these results and the lower allelic richness in dwarf types, this indicates that all dwarf varieties could have evolved from a common origin, which could be from the South-east Asian or Pacific tall varieties. The *niu vai* type of coconuts were domesticated in the South-east Asia and Pacific regions and only later introduced to other regions.

8.5 Coconut Breeding

8.5.1 Breeding objectives

The major aim of coconut breeding is to increase the overall productivity of coconut by enhancing economically valuable characters including early flowering time, fruit size and number, kernel size and oil content. Recently, other traits such as drought, pest and disease tolerance have become important in coconut-breeding programmes. Due to the changing climate, the frequency of drought is increasing in most coconut-producing areas. Hence, developing drought-tolerant varieties is one of the major breeding objectives in many coconut-breeding programmes. In some coconut populations, certain pests can be controlled by using physical, chemical or biological methods; however, for other pests, such as the coconut mite (*Aceria guerreronis* Keifer), no control measures are effective so breeding for tolerance to this kind of pest has also become one of the main breeding programme goals. Similarly, a phytoplasma disease called lethal yellowing poses a significant threat to productivity in certain parts of the world. Developing disease-tolerant varieties through breeding is the only possible sustainable option to manage such threatening diseases (e.g. lethal yellowing). In addition, developing cold-tolerant varieties for southern China and cyclone-tolerant varieties for the Caribbean and countries in the Pacific region through breeding is also becoming a high priority.

8.5.2 Breeding programmes

Due to the long maturity time, breeding methods for coconut are restricted to those involving the selection of phenotypically elite parental lines and the performance of inter- and intravarietal hybridization. The earliest work carried out on coconut breeding used a simple method of mass selection, and this was the only breeding method used until techniques were developed for artificial pollination. Most of the present-day coconut plantations around the world would have come from this approach, and mass selection would have been driven by the traits considered by farmers to be important for yield. Most present-day coconut plantations would have been established from seed, collected from productive regions, and depending on farmer preference for fruit size, palm stature and copra or water content, although in many cases the selection criteria have only focused on copra or oil productivity. Mass selection has generally only been done on the female parent, and the male parent was not considered in this approach to breeding, which used only half of the additive gene effects. The efficiency of the mass-selection method to select superior mother palms with desirable traits has been useful to develop the standard criteria for the selection of the best coconut palms. Estimation of the heritability of certain coconut traits is one of the most important features to help in the selection of suitable parents for crosses based on traits such as fruit weight, rate of fruit setting from fertilized female flowers, copra yield per palm year^{-1}, the number of female flowers per bunch and fruit production bunch, as shown in early studies (Liyanage and Sakai, 1960). Furthermore, some suitable agronomic traits such as short internode length (resulting in shorter stature), high endosperm weight and other desirable yield characters have also been determined as useful for the selection of high-quality palms (Liyanage *et al.*, 1988).

8.5.3 Coconut hybrids

Hybridization is the production of fruit by the crossing of two superior parent palms and is one of the current breeding strategies for coconut. Despite inherent constraints, dwarf × tall hybrids show significant genetic gains in a number of traits, including fruit yield and parental precocity. Many different combinations of dwarf × tall hybrids have been produced to date. Although tall × tall hybrids are not early flowering, they can produce higher yields if the parents come from different geographical origins (Bourdeix *et al.*, 1990).

In the undertaking of intervarietal hybridization, hand pollination is performed between two genetically distant varieties. During the artificial pollination step, male flowers from the inflorescence are emasculated, and the female flowers are covered with a bag. Once the female flowers are receptive, the pollen from the desired male parent is collected and dusted on to the female flowers. The rate of authenticity of the produced fruit depends on how closely supervised the required steps have been. The fruit produced (often referred to as seednuts) from various crosses are assessed in the field, and the superior lines from among these crosses are identified and used for further evaluation. Seednut production by hand pollination or artificial pollination methods are usually expensive and laborious to do for coconut, and the process can only produce a limited number of seednuts per cross, as the success rate of pollination is often very poor. Thus, if superior hybrids have been identified, these hybrids are often grown in isolated seed gardens, where natural pollination between the chosen parents can occur. This dramatically reduces the cost of hybrid seednut production. In these seed gardens, isolation is achieved by using a natural barrier of forest trees, which can prevent foreign pollen arriving from elsewhere. In a second approach, guard palms of the same pollen parent are planted about 12 rows deep around the seed garden to intercept foreign pollen. Such barriers are necessary if true hybrids are to be produced. The female parent in the seed garden is usually emasculated to ensure that the received pollen is only from the male parent. This seed-garden approach using barriers to pollen inflow is the most common procedure practised for the mass production of coconut hybrid seednuts.

8.5.4 Constraints to breeding

The major constraints to the traditional coconut-breeding methodologies are the long generation time, palm heterozygosity, the requirement for a large planting area, no possibility of vegetative propagation, and the low fruit number produced each year. As a typical coconut palm takes 7–10 years to mature, it is difficult to assess any reproductive traits and to undertake progeny testing until they reach this age. In addition, the lack of a vegetative propagation method means that it is impossible to multiply elite trees with the detected desirable traits. On average, a coconut palm produces only 50–80 seednuts year^{-1}, which is then reduced by having only a 20% success with hand pollination. This is not only restricting for seednut production from the desirable palms but also limits the size of the population of the full-sibling family, which is difficult to evaluate in genetic trials. In addition, due to its long maturity time, it is nearly impossible to develop inbred lines from these heterozygous palms. Hence, coconut-breeding approaches, as described earlier, have been restricted to just a few options (e.g. mass selection of elite palms, and intra- and intervarietal hybridization). The development of methods for clonal propagation through somatic embryogenesis of coconut have progressed over the past few decades; however, they are yet to be applied to commercial coconut seedling production and breeding (Benbadis, 1992; Kalaipandian *et al.*, 2021).

8.6 Biotechnology

8.6.1 Molecular marker-based genetic diversity analysis

Due to the constraints in conventional coconut-breeding technologies, biotechnological approaches are an attractive alternative for use in future breeding programmes. Molecular markers were first used in coconut breeding in the 1990s and were used to determine genetic diversity and generate genetic maps, for example for the genes for pathogen resistance. The genetic diversity of tall types was assessed in several studies by using molecular techniques, such as random amplified polymorphic DNA (RAPD; Everard, 1996), restriction fragment length polymorphism (RFLP; Lebrun *et al.*, 1999), and then more recently using simple sequence repeats (SSRs). Recently, SSR studies have revealed the most informative and precise analysis of genetic diversity in coconut (Perera *et al.*, 2003), including revealing that the genetic diversity of dwarf types is lower than that of tall types (Fig. 8.1; Perera, 2005).

8.6.2 Practical applications of molecular markers

Perera (2010) successfully utilized SSR markers to identify genetically distant coconut varieties and then to use them as parents in breeding programmes in Sri Lanka. This marker approach was also used to confirm the identity of hybrids (e.g. Sri Lankan Green Dwarf × Sri Lankan Tall, and Sri Lankan Yellow Dwarf × Sri Lankan Tall; Perera, 2010), and to identify whether accidentally mislabelled or seednut lot mixing had occurred. This study identified two SSR markers

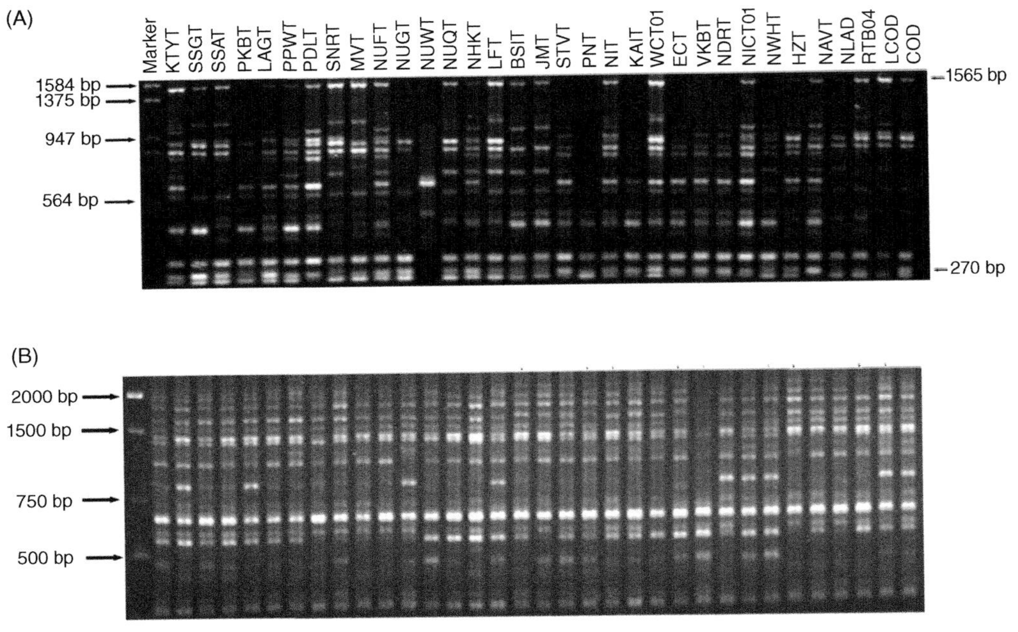

Fig. 8.1. The genetic diversity of tall and dwarf coconut types analysed using (A) random amplified polymorphic DNA and (B) inter-simple sequence repeat analysis. The marker in (A) is a λ DNA *Eco*RI/*Hin*dIII double digest. From Perera *et al.*, 2005, with permission.

useful for the identification of certain varieties such as Sri Lankan Tall, Sri Lanka Green Dwarf and Sri Lankan Yellow Dwarf, and these identified varieties were subsequently used as parents in a commercial breeding programme in Sri Lanka. Before molecular markers, hybrid seedlings were identified by assessing their appearance (e.g. the yellow petiole seen in a successful Sri Lankan Yellow Dwarf × Sri Lankan Tall Hybrid). However, this method has now been shown to be imprecise by the molecular marker work. This indicates the importance of including a molecular marker approach for the identification and selection of hybrids prior to planting. Similarly, Bandaranayake and Kearsey (2005) generated molecular markers that can specifically identify the San Ramon variety and used them to identify San Ramon fruit from within a mixed population to plant in a field gene bank. In another study, the genetically pure Sri Lankan Brown Dwarf variety was identified and selected from a mixture of dwarf varieties using molecular markers (Perera *et al.*, 2002). In Sri Lanka, a molecular marker technique was employed to purify a true-to-type dwarf yellow palm variety prior to it being used in hybridization work, which avoided genetic variation appearing in the hybrid palms (Perera *et al.*, 2015). In India, Rajesh *et al.* (2013) utilized sequence characterized amplified region (SCAR) markers to distinguish certain kinds of tall and dwarf types. SSRs were also used to test for somaclonal variation in somatic embryogenesis-derived plantlets, and to study somaclonal variation in field-planted coconuts in Sri Lanka (Fernando *et al.*, 2004; Perera *et al.*, 2008).

8.6.3 Molecular breeding

Molecular breeding refers to the utilization of molecular tools to speed up the breeding programme of a species. In coconut, several studies have been previously undertaken using molecular tools to help improve the efficiency of the breeding approach. These studies used quantitative trait locus (QTL) mapping for marker-assisted selection and gene discovery.

Linkage and QTL mapping

The first coconut genome map was generated by using inverse sequence-tagged repeat (ISTR) markers applied to an F_1 population generated from an East African Tall × Laguna Tall cross (Rohde *et al.*, 1999). In the Philippines, an F_1 mapping population was then created from a Malayan Yellow Dwarf × Laguna Tall, mapped using amplified fragment length polymorphism (AFLP), ISTR, RAPD and inter-simple sequence repeat (ISSR) markers (Herran *et al.*, 2000). In this latter study, approxiamtely 16 linkage groups were generated from 382 markers and through this, six QTLs for traits such as early germination, early flowering and high yield were found. This was a pioneering study and demonstrated the possible use of marker-assisted breeding for coconut. In another study, the same mapping population was used to identify QTLs for other traits, such as for leaf production, and trunk girth and height (Ritter *et al.*, 2000).

The selection of parents and the appropriate size of the mapping population are factors known to be important in the production of a genome map for any crop species (Bandaranayake and Kearsey, 2005). For coconut, this is critical, as a single cross can only supply a limited number of seednuts each year. Bandaranayake and Kearsey (2005) reported that about 200–400 individuals were necessary to form an appropriate mapping population. In a study in Sri Lanka, a large mapping population of 350 lines was generated from crosses involving 30 genetically similar Sri Lankan Red Dwarf individuals (from the South-east Asia and Pacific group) and a single Sri Lankan Tall individual (from the Indo-Atlantic group) to obtain a higher level of segregation of the traits of interest. Previously, SSR markers had been used to detect 30 identical palms from the Sri Lankan Red Dwarf population. The data showed that an excellent F_2 population linkage map could be generated from the F_1 population. Perera (2006) showed that a segregating mapping population, formed from a Sri Lankan Yellow Dwarf (tolerant) × Sri Lanka Tall (susceptible), could be generated to study palm susceptibility to *Aceria* spp. mite (Figs 8.2 and 8.3).

Selection of parents for resistance breeding

The use of molecular markers is a simple and reliable way to identify early pathogen disease. For example, Weligama coconut leaf wilt disease, a phytoplasma disease, was able to be detected

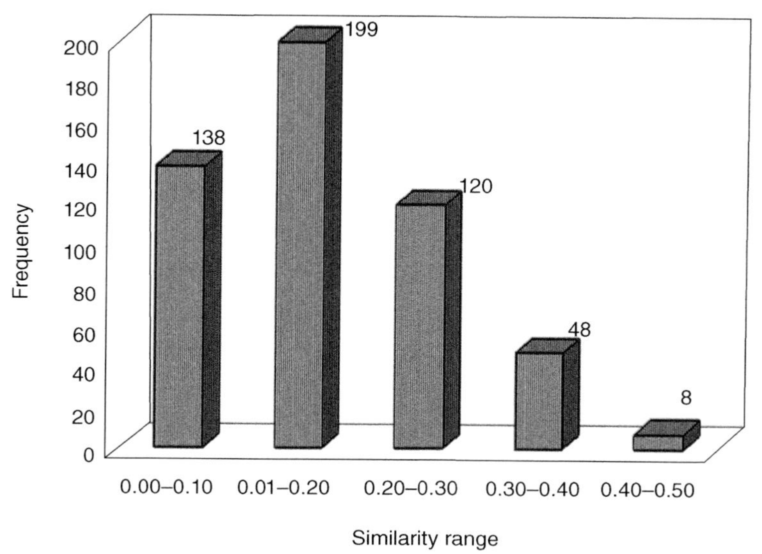

Fig. 8.2. Frequency distribution of 528 pairwise similarity coefficients of coconut accessions based on 92 simple sequence repeat markers. The number of pairwise values falling in the respective frequency range is indicated above each bar. From Perera, 2006, with permission.

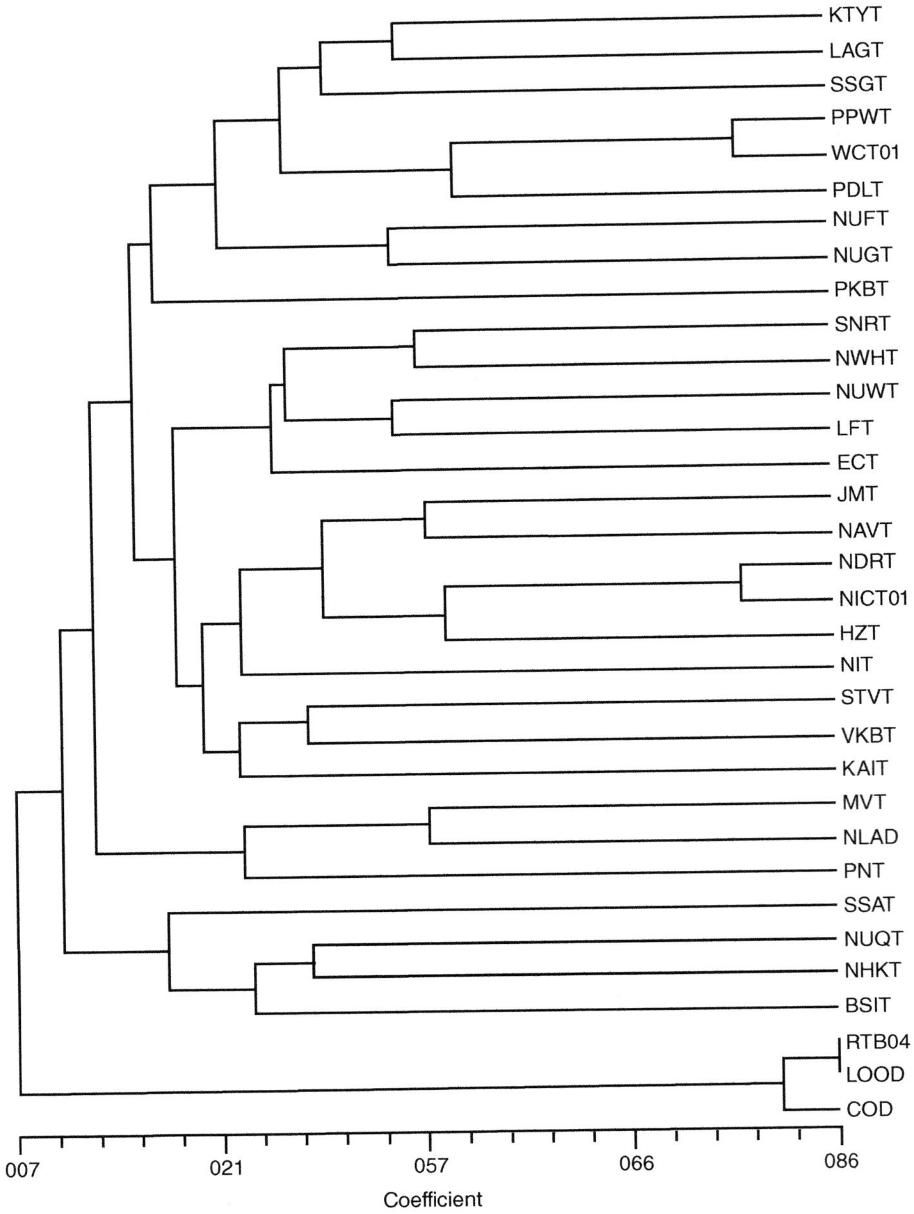

Fig. 8.3. Dendrogram of coconut germplasm accessions generated using Jaccard's similarity coefficient obtained with 92 simple sequence repeat markers. From Perera, 2006, with permission.

using a molecular marker approach (Perera *et al.*, 2012), and these markers could be used to screen progeny for the agent, in a resistance breeding and disease management programme.

Recently, two homologous genes that activate a systemically acquired pathogen resistance (a salicylic acid-dependent defence response) have been identified. Expression of these two genes, when salicylic acid was applied exogenously to *in vitro*-growing coconut plantlets, was elevated, providing resistance to lethal yellowing disease (Nic-Matos *et al.*, 2017). Such studies indicate the potential opportunities for the use of these types of techniques in disease resistant breeding programmes.

8.6.4 Association mapping studies

In simple terms, association mapping is the identification of traits linked to molecular markers. It is an alternative method to the identification of QTLs or molecular markers that are associated with desirable traits found in populations. This association mapping can be performed if larger populations and high-throughput molecular marker systems are available, and is dependent on linkage disequilibrium. Hence, the intensity of the linkage disequilibrium should be evaluated prior to performing association mapping (Batugal *et al.*, 2009). Several gene-based molecular markers, such as SSRs, were generated from transcriptome sequencing of coconut and used in association mapping studies to find the markers linked to palm height in coconut (Xia *et al.*, 2014). Recently, Zhou *et al.* (2020) conducted association studies of 80 coconut accessions from six different categories including Red Dwarf, Yellow Dwarf, Hainan Red, Hainan Tall, a Dwarf Malayan × West African Tall hybrid and an Aromatic Green Dwarf, and mapped a trait that indicates fatty acid content in the mature kernel.

8.6.5 Genomics for utilization of coconut germplasm

Genome sequencing has provided information on key genes and genetic pathways for desirable characters, including pest and disease resistance, and QTLs for yield and oil synthesis in coconut. Recent progress with next-generation sequencing technologies has enabled a better understanding of the complex coconut genome. For example, RNA sequencing has identified 57,304 unique genes in coconut (Fan *et al.*, 2013). Nejat *et al.* (2015) generated a transcriptome (the sum total of mRNA molecules expressed) and identified resistance and susceptibility genes in coconut for lethal yellowing disease. They reported 39,873 differentially expressed genes, of which 21,860 and 18,013 genes were downregulated or upregulated after infection, respectively. These genes can now be used as markers to screen for tolerant varieties of coconut germplasm.

More recently, it has been found that single-nucleotide polymorphisms (SNPs) can be used for germplasm selection, as their occurrence is frequent throughout the genome. In this way, SNPs can be used to accelerate crop-improvement programmes. There are now automated high-throughput SNP genotyping systems that enable breeding programmes to proceed faster when compared with other molecular marker approaches. Whole-genome sequencing, using recent next-generation sequencing technologies, discovered SNP pools that could be used as markers in germplasm collections. Pesik *et al.* (2017) developed single-nucleotide amplified polymorphism (SNAP) markers from SNP markers for coconut. These SNAP markers have been used for genotyping of several coconut varieties including Banten Tall, Pati Dwarf and Kopyor coconuts.

The first draft genome of the coconut palm was sequenced by Xiao *et al.* (2017) from the Hainan Tall variety. The genome consisted of 2.20 Gbp of sequence scaffolds, estimated to cover about 91% of the genome, and had 28,039 annotated protein-coding genes. The assembled genome of Hainan Tall was used to analyse the signalling-pathway genes for salt tolerance (Yang *et al.*, 2021). Lantican *et al.* (2019) reported coconut genome sequencing from a dwarf coconut variety, Catigan Green Dwarf. This study produced 7139 unique SSR markers, which were designed from genome sequences, that can now be used in the development of high-yielding, climate change-resilient, pest- and disease-tolerant varieties. In another study, the draft genome of the Chowghat Green Dwarf variety was assembled (1.93 Gb), with a total of 6828 genes being forecast based on the RNA sequence data (Muliyar *et al.*, 2020).

Recently, Wang *et al.* (2021) revealed the source of differences in characteristics between two coconut types using a multi-omics approach. They found that variation in palm height might be due to an altered gibberellin metabolism, leading to shorter (dwarf) coconut palms. Muñoz-Pérez, *et al.* (2022) used a genotyping-by-sequencing approach to generate several SNPs along the genome of a diverse collection of 112 coconut accessions. They showed that the Atlantic group of coconuts have a slightly higher genetic diversity and faster linkage disequilibrium decay. While Pacific varieties possessed a slower linkage disequilibrium decay, a low fixation index and lower nucleotide diversity.

8.7 Conclusion and Future Prospects

Coconut-breeding programmes were started in the early 1920s, and a number of traits, including yield and flowering time, were significantly improved in these programmes. As there are some inherent phenological and reproductive trait limitations in coconut, generally only mass selection and inter- and intravarietal hybridization have been used for breeding. Some countries have taken measures to improve their coconut genetic resources through safe germplasm exchange programmes. Considering the importance of the known genetic diversity within coconut, efforts have been made to conserve and utilize this diversity. Most diversity is found among the tall coconut types. Hence, the most effort has been made to maintain large numbers of tall varieties in field garden collections. Breeding for disease and pest resistance requires well-characterized genetic variability estimates. Currently, the available coconut genome sequences and genetic analysis methods offer the chance to establish high-throughput genetic marker systems to characterize the variability among the conserved genetic resources and to utilize it in coconut-breeding programmes. The use of molecular methods for coconut is only in the very early stages. Cooperative and collaborative research is needed to help generate new technologies to expand and improve coconut-breeding approaches and enable widespread application by the industry.

8.8 References

Arunachalam, V. and Rajesh, M.K. (2017) Coconut genetic diversity, conservation and utilization. In: Ahuja M. and Jain S. (eds) *Biodiversity and Conservation of Woody Plants*. Sustainable Development and Biodiversity 17. Springer, Cham.

Bahala, R.T. (1987) *Report of Training on Coconut Breeding*. FAO, Manado, Indonesia.

Bandaranayake, C.K. and Kearsey, M.J. (2005) Genome mapping, QTL analysis and MAS: importance, principle, constraints, and application in coconut. *International Plant Genetic Resources Newsletter* 142, 47–54.

Batugal, P., Bourdeix, R. and Baudouin, L. (2009) Coconut breeding. In: Jain, S.M. and Priyadarshan, P.M. (eds) *Breeding Plantation Tree Crops: Tropical Species*. Springer, Berlin, pp. 327–376.

Bazrafshan, A. (2022) *Ex situ* conservation of coconut (*Cocos nucifera* L.) germplasm using cryopreservation. PhD thesis, The University of Queensland, Brisbane, Australia.

Benbadis, B.K. (1992) Coconut and date palm. In: Hammerschlag, F.A. and Litz, R.W. (eds) *Biotechnology of Perennial Fruit Crops*. CAB International, Wallingford, UK, pp. 383–400.

Bourdeix, R., N'Cho, Y.P., Le Saint, J.P. and Sangare, A. (1990) Coconut (*C. nucifera* L.) selection strategies. I. Rundown of achievements. *Oléagineux* 45, 366–371.

Bourdeix, R., Santos, G., Labouisse, J.P. and Baudouin, L. (2005) Useful definition of terms and nomenclature. In: Batugal, P., Ramanatha Rao, V. and Oliver, J. (eds) *Coconut Genetic Resources*. International Plant Genetic Resources Institute (IPGRI), Rome, pp. 9–10.

Burkill, I.H. (1966) *A Dictionary of the Economic Products of the Malay Peninsula*. Ministry of Agriculture and Co-operatives, Kuala Lumpur.

Child, R. (1964) *Coconuts*. Longmans, London.

Dissanayake, P.D. and Perera, L. (2020) Emerging non-food applications in the coconut industry. In: Ghosh, P.K. (ed.) *The Coconut Palm (Cocos nucifera* L.*)*. Nova Science Publishers, New York, pp. 283–300.

Everard, J.M.D.T. (1996) Use of molecular markers for breeding of the coconut palm (*Cocos nucifera* L.). MSc thesis, University of New England, Armidale, Australia.

Fan, H., Xiao, Y., Yang, Y., Xia, W., Mason, A.S. *et al.* (2013) RNA-Seq analysis of *Cocos nucifera*: transcriptome sequencing and *de novo* assembly for subsequent functional genomics approaches. *PLoS One* 8: e59997.

FAOSTAT (2020) Countries by commodity. Food and Agricultural Organisation of the United Nations, Rome. Available at: www.fao.org/faostat/en/#rankings/countries_by_commodity (accessed 18 August 2023).

Fernando, S.C., Weerakoon, L.K., Perera, P.I.P., Bandupriya, H.D.D., Ambagala, I. *et al.* (2004) Genetic fidelity and *ex vitro* performance of tissue cultured coconut plants. In: Peiris, T.S.G. and Ranasinghe, C.S. (eds) *Tree of Life – New Trends in the Millennium. Proceedings of the International Conference to Mark the 75th Anniversary of Coconut Research Institute, Sri Lanka, Part II*. Ceylon Printers, Colombo, Sri Lanka.

Foale, M.A. (1991) Coconut genetic diversity – present knowledge and future research needs. In: *Proceedings of the International Workshop on Coconut Genetic Resources,* 8–11 Oct. 1991, Cipanas, Indonesia. IPGRI, Rome, pp. 46–58.

Harries, H.C. (1978) Evolution, dissemination, and classification of *Cocos nucifera* L. *Botanical Review* 44, 265–320.

Harries, H.C. (1995) Coconut (*Cocos nucifera* L.). In: Smartt, J. and Simmonds, N.W. (eds) *The Evaluation of Crop Plants*, 2nd edn. Longman, New York. pp. 389–394.

Herran, A., Estioko, L., Becker, D., Rodriguez, M.J.B., Rohde, W. *et al.* (2000) Linkage mapping and QTL analysis in coconut (*Cocos nucifera* L.). *Theoretical and Applied Genetics* 101, 292–300.

IPGRI Annual Reports (1997) *1996 Annual Report of IPGRI*. International Plant Genetic Resources Institute (IPGRI), Rome.

IPGRI Annual Reports (1998) *1997 Annual Report of IPGRI*. International Plant Genetic Resources Institute (IPGRI), Rome.

Kalaipandian, S., Mu, Z., Kong, E.Y.Y., Biddle, J., Cave, R. *et al.* (2021) Cloning coconut via somatic embryogenesis: a review of the current status and future prospects. *Plants* 10: 2050.

Kevan, P.G. (1993) Wind or insects: pollination of coconut (*Cocos nucifera* L.) in the Maldives Islands. In: Conner, L.T., Rinderer, T., Sylvester, H.A. and Wongsiri, S. (eds) *Asian Apiculture.* Wicwas Press, Cheshire, CT, pp. 398–409.

Lantican, D.V., Strickler, S.R., Canama, A.O., Gardoce, R.R., Mueller, L.A. *et al* (2019) *De novo* genome sequence assembly of dwarf coconut (*Cocos nucifera* L. Catigan Green Dwarf) provides insights into genomic variation between coconut types and related palm species. *G3: Genes, Genomes, Genetics* 9, 2377–2393.

Lebrun, P., N'cho, Y.P., Bourdeix, R. and Baudouin, L. (1999) Le cocotier. In: Hamon, P., Seguin, M., Perrier, X. and Glaszmann, J.C. (eds) *Diversité Génétique des Plantes Cultivées*. CIRAD (French Agricultural Research Centre for International Development), Montpellier, France, pp. 219–240.

Liyanage, D.V. (1958) Varieties and forms of coconut palms grown in Ceylon. *Ceylon Coconut Quarterly* 9, 1–10.

Liyanage, D.V. and Sakai, K.I. (1960) Heritability of certain yield characters of the coconut palm. *Journal of Genetics* 57, 245–252.

Liyanage, D.V., Wickramaratne, M.R.T. and Jayasekara, C. (1988) Coconut breeding in Sri Lanka: a review. *COCOS* 6, 1–26.

Manimekalai, R. and Nagarajan, P. (2006) Inter-relationship among coconut (*Cocos nucifera* L.) germplasm accessions using RAPD technique. *Genetic Resource Crop Evolution* 53, 1137–1144.

Meléndez-Ramírez, V., Parra-Tabla, V., Kevan, P.G., Ramirez-Morillo, I., Harries, H. *et al.* (2004) Mixed mating strategies and pollination by insects and wind in coconut palm (*Cocos nucifera* L. Arecaceae): importance in production and selection. *Agricultural and Forest Entomology* 6, 155–163.

Menon, K.P.V. and Pandalai, K.M. (1958) *The Coconut Palm: A Monograph*. Indian Central Coconut Committee, Erakulam, India, pp. 86–102.

Muliyar, R.K., Chowdappa, P., Behera, S.K., Kasaragod, S., Gangaraj, K.P. *et al.* (2020) Assembly and annotation of the nuclear and organellar genomes of a dwarf coconut (Chowghat Green Dwarf) possessing enhanced disease resistance. *Omics: A Journal of Integrative Biology* 24, 726–742.

Muñoz-Pérez, J.M., Cañas, G.P., López, L. and Arias, T. (2022) Genome-wide diversity analysis to infer population structure and linkage disequilibrium among Colombian coconut germplasm. *Scientific Reports* 12: 2958.

Narayana, G.V. and John, C.M. (1949) Varieties and forms of coconut. *Madras Agricultural Journal* 36, 349–366.

Nejat, N., Cahill, D.M, Vadamalai, G., Ziemann, M., Rookes, J. *et al.* (2015) Transcriptomics-based analysis using RNA-Seq of the coconut (*Cocos nucifera*) leaf in response to yellow decline phytoplasma infection. *Molecular Genetics and Genomics* 290, 1899–1910.

Nic-Matos, G., Narvaez, M., Peraza-Echeverria, S., Saenz, L. and Oropeza, C. (2017) Molecular cloning of two novel *NPR1* homologue genes in coconut palm and analysis of their expression in response to the plant defence hormone salicylic acid. *Genes and Genomics* 39, 1007–1019.

Perera, L. (2005) Genetic diversity and relationships of coconut revealed by microsatellite markers. In: Sharma, A.K. and Sharma, A. (eds) *Plant Genome, Biodiversity and Evolution.* Science Publishers, CRC Press, Boca Raton, Florida.

Perera, L. (2006) *Report of the Genetics and Plant Breeding Division*. Annual Report. Coconut Research Institute, Lunuwila, Sri Lanka.

Perera, L. (2010) Hybrid testing and variety identification of coconut (*Cocos nucifera* L.) in Sri Lanka using microsatellite markers. *CORD* 26: 5.

Perera, L., Russell, J.R., Provan, J. and Powell, W. (1999) Identification and characterization of microsatellite loci in coconut (*Cocos nucifera* L.) and the analysis of coconut populations in Sri Lanka. *Molecular Ecology* 8, 335–346.

Perera, L., Russell, J.R., Provan, J. and Powell, W. (2000) Use of microsatellite DNA markers to investigate the level of genetic diversity and population genetic structure of coconut (*Cocos nucifera* L.). *Genome* 43, 15–21.

Perera, L., Russell, J.R., Provan, J. and Powell, W. (2001) Levels and distribution of genetic diversity of coconut (*Cocos nucifera* L., var. Typica, form typica) from Sri Lanka assessed by microsatellite markers. *Euphytica* 122, 381–389.

Perera, L., Peries, R.R.A. and Padmasiri, M.H.L. (2002) Sri Lanka Brown Dwarf coconut: a potential coconut variety for future breeding. *Plant Genetic Resources Newsletter* 131, 28–34.

Perera, L., Russell, J.R., Provan, J. and Powell, W. (2003) Studying genetic relationships among coconut varieties/populations using microsatellite markers. *Euphytica* 132, 121–123.

Perera, L., Perera, S.A., Bandaranayake, C.K. and Harries, H.C. (2009) Coconut. In: Vollmann, J. and Rajcan, I. (eds) *Oil Crop Breeding*. Handbook of Plant Breeding. Springer, New York.

Perera, L., Meegahakumbura, M.K., Wijesekara, H.R.T., Fernando, W.B.S. and Dickinson, M.J. (2012) A phytoplasma is associated with the Waligama coconut leaf wilt disease in Sri Lanka. *Journal of Plant Pathology* 94, 205–209.

Perera, P.I.P., Perera, L., Hocher, V., Verdeil, J.L., Yakandawala, D.M.D. *et al.* (2008) Use of SSR markers to determine the anther-derived homozygous lines in coconut. *Plant Cell Reports* 27, 1697–1707.

Perera, S.A.C.N., Kamaral, L.C.J. and Fernando, W.B.S (2015) Molecular assessment of *Cocos nucifera* L. var. Sri Lanka Yellow Dwarf for genetic purity and *Aceria* mite tolerance. *International Journal of Molecular Evolution and Biodiversity* 5, 1–5.

Persley, G.J. (1992) *Replanting the Tree of Life: Towards an International Agenda for Coconut Palm Research*. CAB International, Wallingford, UK, in association with ACIAR and the Technical Advisory Committee of the Consultative Group on International Agricultural Research.

Pesik, A., Efendi, D., Novarianto H., Dinarti, D. and Sudarsono, S. (2017) Development of SNAP markers based on nucleotide variability of WRKY genes in coconut and their validation using multiplex PCR. *Biodiversitas* 18, 465–475.

Rajesh, M.K., Jerard, B.A., Preethi, P., Regi, J.T., Fayas, T.P. *et al.* (2013) Development of a RAPD-derived SCAR marker associated with tall-type palm trait in coconut. *Scientia Horticulturae* 150, 312–316.

Ritter, E., Rodriguez, M.J.B., Herran, A., Estioko, L., Becker, D. *et al.* (2000) Analysis of quantitative trait loci (QTL) based on linkage maps in coconut (*Cocos nucifera* L.). In: Arencibia, A. (ed.) *Plant Genetic Engineering towards the Third Millennium*. Elsevier Science, Amsterdam, pp. 42–48.

Rohde, W., Becker, D., Kullaya, A., Rodriguez, M.J.B., Herran, A. *et al.* (1999) Analysis of coconut germplasm biodiversity by DNA marker technologies and construction of a first genetic linkage map. In: Oropeza, C. *et al.* (eds) *Current Advances in Coconut Biotechnology*. Kluwer Academic Publishers, Dordrecht, The Netherlands, pp. 99–120.

Sisunandar, R.A., Turquay, P., Samosir, Y. and Adkins, S.W. (2010) Cryopreservation of coconut (*Cocos nucifera* L.) zygotic embryos does not induce morphological, cytological or molecular changes in recovered seedlings. *Planta* 232, 435–447.

Thomas, R.J. and Josephrajkumar, A. (2013) Flowering and pollination biology in coconut. *Journal of Plantation Crops* 41, 109–117.

Wang, S., Xiao, Y., Zhou, Z.-W., Yuan, J., Guo, H. *et al.* (2021) High-quality reference genome sequences of two coconut cultivars provide insights into evolution of monocot chromosomes and differentiation of fiber content and plant height. *Genome Biology* 22: 304.

Xia, W., Xiao, Y., Liu, Z., Luo, Y., Fan, H. *et al.* (2014) Development of gene-based simple sequence repeat markers for association analysis in *Cocos nucifera*. *Molecular Breeding* 34, 525–535.
Xiao, Y., Xu, P., Fan, H., Baudouin, L., Xia, W. *et al.* (2017) The genome draft of coconut (*Cocos nucifera*). *GigaScience* 6: gix095.
Yang, Y., Bocs, S., Fan, H., Armero, A., Baudouin, L. *et al.* (2021) Coconut genome assembly enables evolutionary analysis of palms and highlights signalling pathways involved in salt tolerance. *Communications Biology* 4: 105.
Zhou, L., Yarra, R. and Cao, H. (2020) SSR based association mapping analysis for fatty acid content in coconut flesh and exploration of the elite alleles in *Cocos nucifera* L. *Current Plant Biology* 21: 100141.

9 Coconut Conservation and Propagation

Hannes Wilms[1,2*], Amirhossein Bazrafshan[3], Bart Panis[1,4] and Steve W. Adkins[3]
[1]*KU Leuven, Leuven, Belgium;* [2]*Research Institute for Nature and Forest (INBO), Genetic Diversity, Geraardsbergen, Belgium;* [3]*School of Agriculture and Food Sustainability, The University of Queensland, QLD, Australia;* [4]*The Alliance of Bioversity International and CIAT, Leuven, Belgium*

Abstract
The conservation of coconut genetic resources is essential for the industry to thrive. For thousands of years, people have adapted to new situations by breeding crops; however, breeding programmes rely on the diversity present within a species – diversity that is, in the case of coconut, under threat. To conserve coconut diversity, both *in situ* and *ex situ* approaches have been developed. For *in situ* conservation, on-farm conservation has been implemented as well as protecting areas containing wild coconut palms. These methods allow some natural adaptation over time but are generally not accessible to other farmers or breeders, so gene banks play an important role. To date, field collections have been the main coconut conservation method, but these collections face challenges, so complimentary *in vitro* alternatives are being developed. This chapter tackles all coconut conservation methods and gives an overview of the benefits, drawbacks and complementary approaches.

9.1 Introduction

The coconut palm (*Cocos nucifera* L.) and the coconut industry are facing serious challenges, including climate change, extreme weather events and fatal diseases, such as lethal yellowing disease and cadang-cadang disease (Vadamalai *et al.*, 2009; Oropeza-Salín *et al.*, 2020). The industry is also experiencing increased consumer demand for different coconut products (Bandyopadhyay and Frederiksen, 1999; Baudouin *et al.*, 2008; Akhtar, 2019; Research and Markets, 2019). These challenges and opportunities present coconut farmers and breeders with many difficulties, which can be overcome by changing plantation management styles and/or by using coconut varieties adapted to the changing circumstances (Bourdeix *et al.*, 2021). Therefore, multiple breeding programmes are under way around the globe (Batugal *et al.*, 2009). Such breeding programmes, however, currently rely completely on the availability of diverse germplasm present in farmers' fields, in gene banks and in natural settings.

The coconut species as a whole is not considered endangered, and it is not indexed in the International Union for Conservation of Nature Red List (IUCN, 2022); however, like many other crops, the diversity present within the species is under threat (Bourdeix and Prades, 2017), as factors such as deforestation, urbanization and changes in agricultural practices are all resulting in genetic erosion (Kameswara Rao, 2004; Bourdeix *et al.*, 2009). Not only do farmers rely on the coconut palm, but the loss of diversity and local disappearances of the palm also negatively

* Email: Hannes.wilms@inbo.be

DOI: 10.1079/9781789249736.0009

impact ecosystem services (Garcia *et al.*, 2009). Where the palm occurs naturally, it is part of the ecosystem, and multiple wild species including birds, mammals and invertebrates make use of its presence (Chakravarthy and Thyagaraj, 2012; Basheer and Aarif, 2013; Kumara *et al.*, 2015). It is also an important part of the vegetative succession of newly formed atolls, as the coconut palm is considered a pioneering species, adapted to colonizing new islands (Bourdeix *et al.*, 2013; Harries and Clement, 2014). Therefore, it is important that a wide diversity of coconut genetic material is safely conserved. Multiple options exist to conserve coconut genetic resources. *In situ* conservation is focused on the conservation of palms in natural habitats, while other approaches, such as *ex situ* field collections, provide easier accessibility for farmers and breeders.

Within the coconut community, the most important authority regarding conservation is the International Coconut Genetic Resources Network (COGENT). This organization was established in 1992 and aims to strengthen international collaboration in the conservation and use of coconut genetic resources, to promote improvements in coconut production on a sustainable basis, and to boost the livelihoods and incomes of coconut stakeholders in developing countries (COGENT, 2022). COGENT promotes collaboration between the different conservation institutes and maintains an international database, and currently comprises 39 coconut-producing member countries; however, there are still gaps and problems to overcome, such as the lack of duplicate germplasm collections and the operational problems in the individual gene banks (Bourdeix and Prades, 2017).

Around the world, immense coconut diversity still exists. The International Coconut Community (ICC) distinguishes 360 varieties, consisting of 735 unique accessions that are conserved in gene banks of COGENT member countries (Bourdeix *et al.*, 2010), with around an extra 800 unique accessions being held in national gene banks, totalling more than 1500 conserved accessions (Bourdeix and Prades, 2017). Multiple varieties are conserved as more than one accession, as different populations of a single variety can still be different from each other. However, the real diversity is much greater, as these gene banks have not collected from all the existing coconut populations. This is exemplified by researchers and locals, who have described, for example, varieties with an edible husk, which are yet to be indexed (Bourdeix *et al.*, 2013). To fill these gaps, one can blindly collect and conserve materials, but this is inefficient, especially with limited resources. It would therefore be more efficient to uncover and subsequently focus on protection and collection from diversity hotspots. While the coconut palm is found in many locations close to the equator, most of this diversity originates from its region of domestication, the Asia-Pacific region (Gunn *et al.*, 2011). Within this region, geneticists have found multiple island hotspots that carry a high diversity of coconut palms (Raiesh *et al.*, 2008; Jerard *et al.*, 2017). Outside these regions, coconut populations are often defined by a lack of diversity, caused by the founder effect and the limited time that has passed since their arrival (Baudouin and Lebrun, 2009).

While the diversity hotspots are known, the conservation of coconut genetic resources has specific difficulties. Conservation by seed storage faces considerable logistical problems, as the seeds, when compared with other seed crops, are extremely large. An average wheat grain weighs 65 mg, while the average coconut seed weighs 680 g and is therefore 10,000 times heavier and considerably larger. Additionally, coconut seeds are recalcitrant (Royal Botanic Gardens, 2022), making long-term seed storage through drying and subsequent low-temperature storage impossible. Current coconut gene banks therefore mainly rely on field collections. However, managing such field banks is time consuming and labour intensive. Poor management practices result in unwanted crosses between close-standing palms or through contaminated pollen collections (Baudouin *et al.*, 2008), which are disadvantageous when needing to provide pure materials to third parties (Kumar *et al.*, 2021). Moreover, they are just as vulnerable to environmental extremes and pest and disease attacks as field-grown coconuts (Moorhead, 2012). A further problem is the lack of backup collections, as many varieties are present only in a single field bank (Bourdeix and Prades, 2017). At least one backup collection is considered good conservation practice. Conservation challenges, their impact and potential solutions, will be discussed in this chapter.

9.2 *In situ* Conservation

The coconut palm is currently grown in most tropical countries (FAOSTAT, 2022). However, one can hardly make the argument that the whole tropical region can be considered the palm's natural environment. The original environment was probably limited to the Asia-Pacific region (Gunn *et al.*, 2011), as the presence of coconut in many other areas, especially inland, has come about through human introduction (Nayar, 2016). In the Asia-Pacific region, the coconut palm was present before human arrival, but after human settlement, their histories become intertwined, and as the crop became domesticated, it was spread further by human intervention (Summerhayes, 2018). While the coconut palm is considered domesticated, there still is some debate as to whether domesticated coconuts have become mixed with or have replaced all natural or wild types (Harries 1978; Leach *et al.*, 2003; Foale *et al.*, 2020; Leu, 2021). Multiple scholars are in favour of the hypothesis that domesticated coconuts still coexist with wild types and point to the *niu kafa* coconut type, which resembles a 'wild' coconut ideotype. Compared with the more domesticated *niu vai* coconut type, the *niu kafa* fruit has a thicker husk and a more triangular base (Fig. 9.1) and therefore is more adapted to floating and spread in an atoll ecosystem (Harries 1978; Chan and Elevitch, 2006), although it must be noted that *niu kafa* palms are now cultivated, making the occurrence of a true wild type of coconut completely devoid of any human interference rare. None the less, there are claims of wild coconut forms present on Keeling Island, Australia (Leach *et al.*, 2003) and in the Philippines (Gruezo and Harries, 1984).

One of the most important *in situ* conservation areas may indeed be Keeling Island, containing both wild and old cultivated but abandoned populations of coconut palms (Leach *et al.*, 2003). Fortunately, in 1995, this region was turned into the Pulu Keeling National Park, with a mandate to support the conservation of coconut: 'The park supports coconut trees that are considered to be the most primitive form of the species' (Australian Government, 2015). Parts of the Philippines that Gruezo and Harries (1984) describe as containing wild coconut populations are in the Guiuan Marine Reserve, which was established in 1994 (The Phillipine Government, 1994). However, within this reserve,

Fig. 9.1. *Niu vai* and *niu kafa* type coconut fruit. (A) Whole *niu vai* (top) and *niu kafa* (bottom) fruit. (B) Cross-sections of *niu vai* (top) and *niu kafa* (bottom) fruit. Photos courtesy of Sisunandar.

no specific attention to coconut conservation is given. A more broadened scope towards the conservation of ancient coconut types would include the conservation of areas that contain abandoned populations or where coconut has been reintroduced into its natural environment. This would, of course, include many smaller areas, within some of the diversity hotspots mentioned previously, but as far as this work is concerned, only the Pulu Keeling National Park can currently be considered as a true wild-type coconut *in situ* conservation site.

The European Environment Agency (EEA) defines *in situ* conservation as 'the conservation of ecosystems and natural habitats and the maintenance and recovery of viable populations of species in their natural surroundings and, in the case of domesticated or cultivated species, in the surrounding environment where they have developed their distinctive properties' (EEA, 2023). This means that on-farm conservation of domesticated palms also falls within the domain of *in situ* conservation. On-farm conservation is the conservation of materials, especially old landraces, by the farmer. These old landraces have adapted to the ecogeographical region and were selected to sustain the palm communities' needs (Dhillon *et al.*, 2004). *In situ* conservation has allowed these populations to evolve further, adapting to new biotic and abiotic circumstances (Dulloo *et al.*, 2005). Some of these types of conservation areas are on the Andaman and Nicobar Islands, where tribes have cultivated different crops, including 41 different forms of coconut (Abraham *et al.*, 2008). Other examples can be found on the west coast of Kerala (India) in multiple farms, where researchers observed many of the forms there to have adapted to the local environment and which are still grown today (Samsudeen and Thamban, 2017). However, many of these old on-farm populations are not recorded and thus are in danger of disappearing if mismanaged or replaced by more popular varieties (Bourdeix *et al.*, 2013), although it may already be too late in some places (Bourdeix and Sourisseau, 2021). This is motivating researchers to characterize palms on multiple farm sites to find and protect newly discovered forms (Labouisse and Caillon, 2001; Samsudeen *et al.*, 2013).

While multiple efforts exist to conserve coconut genetic resources within their natural habitat, outside of their natural growing area coconut palms are not always welcome. A prominent example where the palm is considered invasive is on the Palmyra Atoll. There the introduction of coconut caused a significant decline in the presence of native plant species (Young *et al.*, 2010). To overcome this problem, island authorities plan to eradicate most of the coconut palms in an effort to rebalance the natural ecosystem (Longley-Wood *et al.*, 2022). This is in sharp contrast to other initiatives such as that in Ghana, where researchers are trying to save coconut palm populations (Campbell, 2006), or in Amazonia (Brazil) where researchers claim that the presence of coconut palms might even reduce forest degeneration practices (Salm *et al.*, 2009), although their initial results could not verify this claim (Salm and Feder, 2015). Nevertheless, the loss or gain of these palms is negligible, as these palms are often remnants of 'recent' plantations or relatively recently introduced.

The strength of *in situ* conservation is that a whole population in a specific area can be protected, not just selected examples. As such, the conservation unit, through the evolutionary processes being kept in place, can further adapt to changing circumstances, which might interest breeders (Rajora and Mosseler, 2001). *In situ* conservation is generally also less costly when compared with *ex situ* approaches, as it does not need special facilities such as dedicated fields, laboratories or cold-storage equipment (Hebbar *et al.*, 2013). *In situ* conservation approaches will also be essential if 'wild' coconut types, originating from *ex situ* collections, are to be reintroduced into their natural habitat. Without the proper protection and/or planning, reintroduction of a species will not succeed (Volis and Blecher, 2010). However, there are also some drawbacks to *in situ* conservation. The palms kept on *in situ* conservation sites are often less accessible to breeders than other conservation formats, are in danger of being replaced by other varieties, and are affected by different factors such as (invasive) insects, diseases and natural disasters; additionally, these conservation areas often have to compete with other land uses (Dhillon *et al.*, 2004; Dulloo *et al.*, 2005; Bourdeix and Prades, 2017).

9.3 *Ex situ* Conservation

Ex situ conservation is defined by the EEA as 'the conservation of components of biological diversity outside their natural habitats' (EEA, 2023). This covers a wide variety of methodologies including seed collections, field banks, botanical gardens, *in vitro* collections, pollen conservation, cryopreservation and DNA banks (Engels *et al.*, 2008). Each of these *ex situ* conservation practices come with its own advantages and drawbacks. While seed conservation is a popular and convenient method for many crops, long-term conservation of coconut seed is not possible. The coconut seed is extremely large – it is the second largest seed after the double coconut (*Lodoicea maldivica* (J.F. Gmelin) Persson). Coconut seeds are also sensitive to dehydration and cold-temperature storage, making them 'recalcitrant' (Royal Botanic Gardens Kew, 2022). Seed (contained within fruit tissues) can therefore only be stored for a short period of 2 to a maximum of 7 months, depending on the variety (Harries, 2012).

Another *ex situ* method, which plays a limited role in coconut conservation, is growing plants in botanical gardens (Engels *et al.*, 2008). Most gardens are a member of the Botanic Gardens Conservation International, which currently represents 3757 different botanical gardens from more than 100 countries. However, only 125 of these have at least one variety of coconut palm. In total, they conserve about 65 unique varieties, of which 50 are only represented in one garden (BGCI, 2022), covering less than 5% of the registered varieties (Bourdeix *et al.*, 2010). This illustrates that botanical gardens are to date only of limited importance for coconut conservation, so other *ex situ* methods will be discussed in more detail in the following sections.

9.3.1 Field gardens

In field gardens (Fig. 9.2), the palms are planted together in large groups of the same accession and often managed in the same way as a normal production field would be; even the selling of seedlings or seednuts takes place, providing funding for maintenance of the collection (Engels, 2003; Bourdeix and Prades, 2017). Aside from the general management needs such as watering, fertilizing and weeding, other activities in these field gardens include characterization of the palms, often at both the morphological and molecular levels, health indexing, agronomic evaluation and labelling of all material on site (Bourdeix and Prades, 2017).

To preserve an accession, at least 45 different individuals for dwarf varieties and 90 for tall varieties are needed (Santos *et al.*, 1996). This is because the self-pollinating dwarf palms are less diverse within a collection compared with tall varieties (Gunn *et al.*, 2011). They are planted at a density of between 110 and 250 palms ha^{-1}, depending on local cultivation practices, giving one to five accessions per hectare (Santos *et al.*, 1996), and are rejuvenated every 25–30 years (Bourdeix *et al.*, 2011). The ICC estimates that the cost of maintaining one accession is between US$762 and US$2787 per annum (ICC, 2021). Multiple accessions are generally planted in the same field garden, often close together. In this arrangement, natural pollinators can cross-pollinate and contaminate the gene pool of an accession, if the resulting seed is to be planted to rejuvenate the existing palm population (Nayar, 2018). It is estimated that gathering the coconuts needed for regeneration or duplication of an accession requires 1.5 years of preparation and costs more than US$2000 per accession (Bourdeix and Prades, 2017). The process of regeneration is made more complicated as these old and often tall palms need to be scaled, and the method requires bagging the flower and hand pollination, so no other pollen can fertilize the flowers (Bourdeix *et al.*, 2020).

While maintenance is one aspect, introducing new material into a field garden comes with other challenges. From field collection to establishing a new conserved population in a field garden might take up to 15 years. This is because many of the collected fruit will be the result of cross-pollination and thus will require genetic screening before they can be maintained in the conservation collection (Bourdeix and Prades, 2017). The main advantage of conserving a plant collection in a field garden is that all the collected accessions can be characterized and evaluated and used in demonstrations. In addition, true-to-type seednuts or pollen can be collected and made available for distribution to farmers or breeders.

Fig. 9.2. The International Coconut Genebank for South Asia and the Middle East in Kidu, India, where 455 accession are conserved. Photo courtesy of B. Panis.

Field garden conservation is currently the most popular method to conserve coconut germplasm (Bourdeix *et al.*, 2020). According to the COGENT Database, 1416 coconut accessions are now conserved this way (Bourdeix *et al.*, 2005). There are currently five international field banks, located in Brazil, Côte d'Ivoire, India, Indonesia and Papua New Guinea, and 19 national collections (ICC, 2021). However, these collections are incomplete, as 47 coconut-producing countries and territories are yet to be represented in such *ex situ* collections, with experts estimating that one-third of useful diversity is not yet protected (Bourdeix and Prades, 2017). While some popular varieties are conserved in more than 15 countries, most conserved accessions are only represented in one field garden without any backups, meaning that a destructive event caused by disease or adverse weather conditions in one field garden could result in the loss of a whole accession (Bourdeix and Prades, 2017).

In addition to the lack of backups, coconut field gardens themselves face multiple challenges, some of which may even threaten their existence. First, keeping the coconut seednuts, produced for rejuvenating the collection or used for germplasm exchange, true to type is extremely difficult as several accessions can interact in a field garden, meaning that hand pollination is the only way to be certain that specific crosses have taken place (Bourdeix *et al.*, 2020). As the palms grow, this becomes more problematic, as climbing to undertake hand pollination becomes more dangerous. This is one of the reasons why an accession is rejuvenated every 25–30 years, as the size of the palms at that age makes the process more reasonable (Bourdeix *et al.*, 2011). Second, diseases and pests are threatening many field gene banks. An example is the international South Pacific gene bank in Papua New Guinea (Moorhead, 2012), where Bogia coconut syndrome has affected multiple palms in the region threatening the 57 accessions that the gene bank holds. Another dangerous disease for germplasm exchange is lethal yellowing disease, as it not only kills the palm but can also be passed down within its fruit (Oropeza *et al.*, 2017). Finally, there is the challenge of increased land demand and competing interests in many locations. While field gardens give benefits for the industry in the longer term and at a global level, at the local level this might not always be recognized. For example, in Indonesia, multiple hectares of the international South Asian gene bank were destroyed to build a horse-racing track (Bourdeix and Prades, 2017). In Côte d'Ivoire, plans are in place to build real estate over an area of the international African gene bank. It is thought that these developments will result in the loss of more than 50 important accessions (Bourdeix *et al.*, 2020).

In addition to 'traditional' field banking, a further concept has been proposed, one that is similar to the practices of older Polynesian communities, known as 'polymotu' (Bourdeix *et al.*, 2005). Polymotu is a combination of the Greek word 'poly' and the Polynesian word 'motu' and means 'many islands'. The basic concept of polymotu is to spread genetic diversity over different small islands, each planted with one accession, and the islands can be shared between several non-out-crossing varieties (Bourdeix *et al.*, 2011). This makes polymotu somewhat of a hybrid between *in situ* and *ex situ* conservation. The main advantages of a polymotu system are its remoteness, which makes the location less vulnerable to pests and diseases, while keeping the pureness of the accession. In such a system, free pollination can occur within an accession, while in traditional field gardens, manual pollination has to be undertaken to be sure of true-to-type seednuts (Bourdeix *et al.*, 2011). However, the remoteness of the approach is also a drawback, as the materials are less accessible, for example to breeders and farmers. There have been plans to initiate new polymotu projects after a successful pilot project was undertaken in Thailand (Bourdeix *et al.*, 2011). While these projects have been implemented only recently, the basis of this approach has been tried and tested, as it is part of the traditional approach to conservation of coconut by the Polynesians (Bourdeix *et al.*, 2013).

9.3.2 Short- and medium-term *in vitro* conservation

The application of *in vitro* techniques, including short-, medium- and long-term methods, can balance some of the problems associated with *in situ* coconut conservation and *ex situ* field banks. By taking the plants out of their natural environment and growing them in a well-managed, fully controlled environment, the need for space, the impact of natural disasters, and pests and even diseases can be substantially reduced or even eliminated (Withers and Engelmann, 1997). Even in the case where the *in vitro* material becomes infected, methods are available to eradicate these infections, as has been shown by removal of banana streak virus and cucumber mosaic virus from banana (*Musa* spp.) germplasm (Helliot *et al.*, 2002), and potato viruses S, A and M from potato (*Solanum tuberosum* L.) germplasm (Bettoni *et al.*, 2022). Such eradication methods include thermotherapy, meristem tip culture, chemotherapy, cryotherapy or a combination of these methods. For coconut, such methodologies are yet to be developed, but reliable virus detection methods for several coconut-related viruses do exist (Randles *et al.*, 1999).

Until recently, micropropagation technologies for coconut material had not been fully developed and propagation could only occur via seed. In contrast to field gardens, the application of *in vitro* conservation in general comes with a greater requirement for equipment, and facilities, such as those needed for culture medium preparation, sterile transfer of *in vitro* tissues (a laminar airflow hood), and growth and multiplication facilities (a light- and temperature-controlled culture room). Together with the labour costs to maintain such an *in vitro* collection (estimated to be around 85–90% of the total costs required), this makes it, in many instances, more expensive when compared with field gardens (Pence, 2011). However, while it might be more expensive, its disease-free state and considerably smaller use of land makes the *in vitro* approach ideal for germplasm conservation and exchange (Withers and Engelmann, 1997; Kumar *et al.*, 2021).

An embryo culture protocol for coconut was first attempted in the 1950s (Cutter and Wilson, 1954), and had its first success 10 years later when a protocol was developed for the production of Makapuno seedlings, a type of coconut that lacks the ability to germinate in nature (de Guzman and del Rosario, 1964). Later, this protocol was shown to be useful for the exchange of coconut genetic resources, by exchanging either *in vitro* embryos or seedlings derived from them (Assy-Bah and Engelmann, 1992b; Nair and Karun, 1999).

Embryo culture starts with isolation and culture of the zygotic embryo from the coconut fruit and its germination *in vitro* (Engelmann *et al.*, 2011). When the embryos have fully developed into seedlings, they can be maintained in culture and used for micropropagation and cryopreservation protocols (Wilms *et al.*, 2021), or converted into field-ready plants (Samosir and Adkins, 2014). Thanks to the low amount of added plant growth regulators into the media required for these steps, there is less chance of somaclonal variation occurring than for other tissue culture methods (Krishna *et al.*, 2016). However, for maintenance of these cultures, a monthly subculture is necessary to ensure culture vigour is maintained, thus making the method expensive in terms of resources and labour. Another drawback is that although it is used as a transient conservation method, embryo culture does not result in the multiplication of germplasm. Unlike many other species, coconut does not form new vegetative meristems (which can be used for multiplication) during its natural growth. Another disadvantage compared with the *in vitro* culture of other plant species is that palms, especially coconut *in vitro* cultures, grow extremely slowly. This means that acclimatization (preparation for growing in the natural environment) and further growth of the seedlings to a size that can be transferred to field conditions is also slow and remains problematic, as some seedlings can be lost during this stage (Samosir and Adkins, 2014; Caren *et al.*, 2019). Even the most efficient protocols show that acclimatization takes multiple months to achieve.

Slow growth

Slow growth protocols, while not yet developed for coconut, can solve some of the cost-related problems associated with embryo culture. As inducing slow growth will cause the seedling to require less subculturing, this is likely to be a saving in terms of media and labour. For other species (e.g. banana), slow growth induction is used as a method for medium-term storage (Panis *et al.*, 2020). To undertake this method, the plant or plant tissues are grown *in vitro* on a deprived nutrient medium, often containing osmotically active sugars, and placed under sub-optimal temperature conditions (e.g. light or temperatures lower than those ideal for growth). These conditions vary among species but generally can be divided into two different groups: tropical and temperate plant species. While temperate crops, such as potato, can be stored at 6°C (Panis *et al.*, 2020), tropical species such as banana (de Smet *et al.*, 1995) and coffee (Dussert *et al.*, 1997) cannot tolerate temperatures below 10°C and are stored at around 16°C. There is no published research on introducing slow growth conditions for coconut *in vitro* cultures. However, parallels can be drawn from existing protocols, especially those already developed for another palm such as date palm (*Phoenix dactylifera* L.; El-Dawayati, 2017).

Somatic embryogenesis

The first attempt to develop somatic embryogenic coconut cultures was in 1993 (Verdeil *et al.*, 1994), and since then the process has been continuously improved (Kalaipandian *et al.*, 2021). While originally aimed at developing a multiplication protocol, this process was also recognized as a method to conserve coconut tissues. Like other plants, such as date palm (Hassan, 2017), embryonic cultures can be conserved in their proliferating phase, or the resulting plantlets can be kept under slow growth conditions as described above, as this latter method has a lower chance of inducing somaclonal variation than may occur with callus culture. Alternatively, these somatic embryos could also be used as explants for cryopreservation (Sharma, 2005) (see Section 9.3.3).

Unfortunately, during the production of somatic embryos, it is possible that somaclonal variation can occur, as has been shown in oil palm (*Elaeis guineensis* Jacq.), a close relative of the coconut (Weckx *et al.*, 2019), which would make this technique less appropriate for conservation than other methods, although for coconut the incidence of somaclonal variation occurring during somatic embryogenesis has yet to be observed and screening of embryogenic plantlets using simple sequence repeat markers has found no variation (Bandupriya *et al.*, 2017). Nevertheless, this must be put into perspective, as the absence of polymorphisms using simple sequence repeat markers does not prove the absence of DNA changes, because point mutations in the genome are hard to find without a whole-genome sequence. Moreover, epigenetic changes are even more difficult to detect. Therefore, trait measurements taken on the field-growing regenerates should be undertaken to confirm that genetic fidelity has been retained.

It is yet to be determined how long coconut somatic embryos could be maintained in their proliferating phase. Prolonged subculture in other species has been shown to increase the chance of somaclonal variation occurring (Krishna *et al.*, 2016). More research is needed to understand whether this is also a risk for coconut. In addition, to adapt the protocol for long-term storage, there are still numerous improvements necessary, as the material needs frequent subculturing to prevent tissue necrosis (Wilms *et al.*, 2022).

9.3.3 Long-term *in vitro* conservation: cryopreservation

Plant cryopreservation is the storage of biological material at ultra-low temperatures. An advantage of this system is that a lot of material can be stored in a limited space, but the drawback is that a whole plant needs to be able to be regenerated from a small explant, which takes more time compared with other conservation methods. This process is unique to different plant tissues, with each having its own advantages and drawbacks (Panis, 2018).

An important consideration for cryopreservation is water content. All living material contains water, but when this water is exposed to subzero temperatures, ice crystals form. When the formation of these crystals happens within the cell, the cell might rupture and undergo other forms of mechanical stress, causing cell death (Benson, 2008). However, not all water in a cell can be removed, as this too would cause cell death, meaning that for cell survival a minimum amount of cell water is essential. The rate at which a tissue can be dehydrated is species and tissue specific, as orthodox seeds are able to survive greater degrees of dehydration compared with recalcitrant seeds and biological tissues (Pritchard and Nadarajan, 2008). To avoid death by ice crystal formation or dehydration, vitrification can be applied. In a vitrified or glassy state, all water is set in an amorphous solid matrix, which, combined with the cold temperature, makes all biological and enzymatic reactions within the cryopreserved samples come to a halt. This effectively stops the tissue ageing, giving it an unlimited storage life, as long as its temperature does not rise (Panis, 2008). However, to achieve this glass state, two requirements must be met: (i) the presence of a concentrated solution; and (ii) fast freezing rates (Panis *et al.*, 2005). One method to create a concentrated solution is by removing the cell water, for which different methods (e.g. air drying, silica drying) have been developed. Meanwhile, the rapid freezing relies on the absence of temperature gradients in the tissue, and as biological tissues are poor conductors of heat, these samples need to be very small (often 1 mm^3), otherwise temperature gradients would result in ice crystal formation.

For germplasm preservation, the two most obvious explants that could be extracted from the coconut palm are zygotic embryos and plumules, with the plumule being the 1 mm^3 embryonic shoot excised from the germinating embryo. The first results for the cryopreservation of mature and immature embryos were reported in 1992 (Assy-Bah and Engelmann, 1992a,b). For mature embryos (from mature fruit), an air dehydration step, combined with a high sucrose and glycerol treatment has been used, but the recovery rate can vary greatly (33–93%). For immature embryos, pre-growth in a medium with high glucose, glycerol and/or sorbitol content was used, followed by rapid freezing. While these and subsequent reports by other groups looked promising, no field-grown coconut palms were obtained according to Sisunandar *et al.* (2010). The group thus formulated its own protocol, relying on fast (flash) drying using a silica-gel apparatus in a laminar airflow hood (Fig. 9.3), followed by fast cooling (Sisunandar *et al.*, 2010). This resulted in 16 out of 20 tested varieties showing plant regeneration rates after cryopreservation of between 10% and 40% in soil. Later, the same authors found that embryo

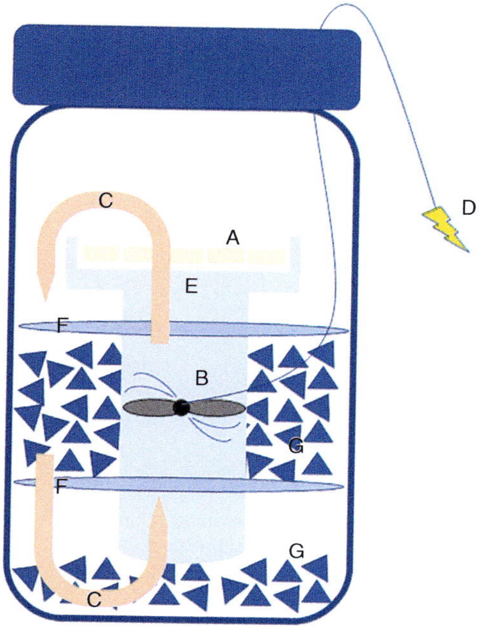

Fig. 9.3. A flash drying apparatus developed by Sisunandar *et al.* (2010) to dehydrate coconut zygotic embryos. The zygotic embryos (A) are placed on a sterile platform above an electric (D) fan (B) that forces sterile dry air (C) over the embryos (E). After passing the embryos, the air is circulated back through silica gel pearls (G), which are held in place by dividers (F). Picture by H. Wilms based on the schematics provided in Sisunandar *et al.*, 2010.

maturity played a significant role within their protocol, as mature embryos performed better than those that were immature, although plant regeneration remained at around 28% (Sisunandar *et al.*, 2014). Sajini *et al.* (2011) also used mature zygotic embryos but applied a sucrose pre-culture step followed by plant vitrification solution 3 (PVS3: 50% sucrose, 50% glycerol) treatment (instead of the air dehydration) to reduce the embryo's moisture content. This resulted in 20–25% of embryos growing into potted plants after cryopreservation. N'Nan *et al.* (2012) tried to simplify the cryopreservation protocol by combining the sucrose pre-culture and dehydration in a single step resulting in 14–75% of treated embryos germinating, indicated by emergence of the first leaf 6 months after cryopreservation, depending on the variety treated. More recently, Kumaunang *et al.* (2020) tested a sucrose pre-treatment, followed by physical dehydration for 24 h using silica gel for zygotic embryos of the Laguna Tall variety, and reported that 47% of embryos showed an emerged plumule and root formation 3 months after cryopreservation. While multiple protocols have been developed, comparing them is not always straightforward as many research groups use their own definition of the 'end point'. Results for survival or germination are not representative for growing *in vitro* or potted plants, as plants are still lost during these two essential steps (Sajini *et al.*, 2011).

The alternative candidate explant, plumules, were originally considered to be free of seed-transmissible diseases, such as lethal yellowing disease (Oropeza *et al.*, 2017). This was thought to be because many disease-causing agents (e.g. bacteria and viruses) are often absent in the meristem tissues, while potentially present in other embryo tissues (N'Nan *et al.*, 2008; Cueto *et al.*, 2013). However, this must be confirmed by a case-by-case study for each disease, as multiple transmission pathways exist. Another reason for using plumules is that they can be used directly in subsequent clonal processes (Hornung *et al.*, 2001). As plumules and zygotic embryos serve different purposes, these cryopreservation protocols were developed independently. The first success in using cryopreserving plumules was reported in 2008 by N'Nan *et al.* (2008) who reported that 20% of plumules cryopreserved through encapsulation dehydration showed *in vitro* shoot growth and new leaf formation after 8 months. Application to six other varieties showed *in vitro* shoot growth and new leaf formation at rates ranging from 0% to 24% (N'Nan *et al.*, 2014). Lédo *et al.* (2020) reported the successful application of droplet vitrification to plumules. After cryopreservation, however, they did not aim for direct plant regeneration but used the cryopreserved plumules for the induction of somatic embryos, which resulted in 30% of the cryopreserved tissues forming embryogenic callus; however, regeneration of the callus to plantlets was not discussed. While cryopreservation of plumules is a promising method with alternative uses, more research about the efficiency of cloning and transfer to soil is needed.

Meristems excised from *in vitro* shoot cultures are, for most plant species, the material of choice for plant cryopreservation. As regeneration from these tissues generally happens without an intervening callus phase, the chance of somaclonal variation is low. Until recently, the only material available for this method for coconut was *in vitro* shoots regenerated from germinated zygotic embryos. Wilms *et al.* (2018) reported on the successful cryopreservation of 1 mm^3 meristem shoot tips taken from the coconut variety Malayan Yellow Dwarf by applying a droplet vitrification protocol similar to that used for banana (Panis *et al.*, 2005). More than 50% of the cryopreserved shoots survived and gave rise to small *in vitro* shoots. Full plant regeneration, however, was not obtained. Since the development of the new axillary shoot multiplication protocol for coconut (Wilms *et al.*, 2021), new possibilities for cryopreservation using this shoot meristematic tissue can now be considered as an easier-to-obtain source of multiple shoot tips for cryopreservation. The first tissues that are produced, indicating axillary shoots in coconut, are proliferating shoot clumps (Wilms *et al.*, 2021). This plant material is highly meristematic and can be multiplied on a substantial scale and regenerated into plants. Thus, protocols similar to those developed for proliferating cultures of banana could be established (Panis *et al.*, 1996), with preliminary results for the cryopreservation of these proliferating shoot clumps looking very promising (Fig. 9.4) (H. Wilms and E. Rosiers, personal communication).

With the recent developments to induce somatic embryogenesis from responsive explants such as plumules (Pérez-Núñez *et al.*, 2006; Rajesh *et al.*, 2014; Sáenz *et al.*, 2018) or immature inflorescences (Cueto *et al.*, 1996), the totipotent embryogenic callus produced can also be used as an explant for cryopreservation (Nguyen *et al.*, 2015). Although research on cryopreserving coconut somatic embryos is still in its infancy, parallels can be drawn from other plants for which protocols have been successfully developed, such as its close relative the oil palm (Beulé *et al.*, 2018; Palanyandy *et al.*, 2020). Recent research has already reported recovery rates of embryogenic callus of up to 25% in coconut after cryopreservation (Welewanni *et al.*, 2020). Cryopreservation of cell suspensions derived from these somatic embryos, when eventually developed, should also be possible (Kong *et al.*, 2020).

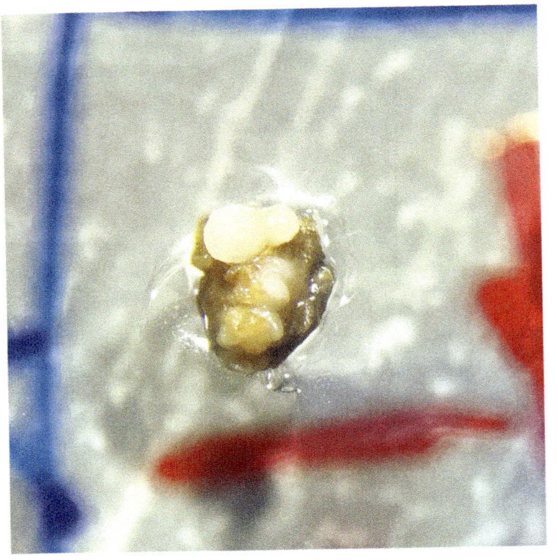

Fig. 9.4. Proliferating meristematic clump obtained via the shoot multiplication protocol after cryopreservation. Photo courtesy of H. Wilms.

Cryopreservation of pollen is another option. This should be considered separately from the previous tissues discussed, as plant regeneration from coconut pollen has not yet been described. If possible, such plants would be haploid and could be used to produce double-haploid lines, which are important in other breeding programmes. Of course, pollen cryopreservation itself is useful as such pollen is a genetic resource. Cryopreserved pollen could also be used on a short-term basis in breeding programmes (Karun *et al.*, 2014; Perera and Vidhanaarachchi, 2021). Karun *et al.* (2014) showed in a multi-year experiment that cryopreserved pollen, after desiccation, remained viable and able to fertilize, with pollen germination rates ranging from 24% to 49% depending on the accession.

9.3.4 DNA banks

A final method for preserving coconut diversity is storing its DNA (Dulloo *et al.*, 2005), although this is more a way of record keeping similar to herbarium collections (Hodkinson *et al.*, 2007) than for any other purpose, as synthetic biology is not yet advanced enough to rebuild a plant purely from its DNA (Powell, 2018). Nevertheless, it is a valuable asset in mapping populations for diversity, building trait markers for breeding objectives, and for other biotechnological-assisted breeding methods (Hodkinson *et al.*, 2007; Bourdeix and Prades, 2017). In contrast to other methods, which are often destructive to the plant or where you need to collect derivatives such as pollen or seeds, DNA sampling can be done simply by taking small quantities of leaf or root tissue, which is unlikely to affect the health of the palm (Hodkinson *et al.*, 2007).

There are multiple methods to store this DNA. It can be left within the plant sample, which can then by stored by different means such as on silica gel, flash frozen in liquid nitrogen or as lyophilized leaves (Gostel *et al.*, 2016). Alternatively, the DNA can be stored as purified DNA. Ultimately, the sequenced DNA can be stored. The *in silico* method of DNA storage makes the data it contains easily accessible, transferable over the internet and, if properly stored, able to be saved forever. Ideally, all these sequences would be grouped together in one database allowing easy access for all research. The first draft of the coconut genome was mapped in 2017 (Xiao *et al.*, 2017). As such, the biggest challenge in coconut genome research has already been overcome and can act as a reference for all future research. This has made all subsequent mapping much easier to undertake, validated by the sequencing advances made recently (Lantican *et al.*, 2019; Wang *et al.*, 2021).

9.4 Conclusion

This chapter has outlined the range of options for the banking of coconut genetic diversity. Many of these are not yet in widespread use for conserving coconut genetic resources, but a reliable approach to conservation should use multiple, complementary methods of both *in situ* and *ex situ* form. All the methods discussed have their benefits and drawbacks. *In situ* conservation allows the continued adaptation of populations to local circumstances but the samples are difficult to collect and manage. Field banks allow easier access to seednuts but are affected by the environment. *In vitro* culture makes the transport of sterile material much easier, needing less space, but needs frequent subculture. Cryopreservation allows indefinite storage, but the tissue regenerates slowly.

Given that all methods face challenges, a good conservation approach involves not only multiple locations but also multiple complementary techniques, as this is the only way to safely conserve material for the future. Those managing coconut gene banks should thus widen their focus beyond field gene banks (currently the mainstay of coconut germplasm conservation) to the implementation of the newer methods described in this chapter, as these are complementary and would help balance the risks associated with the field banks already in existence.

9.5 References

Abraham, Z., Senthilkumar, R., John, K.J., Sharma, T.V.R.S., Nair, N.V. *et al.* (2008) Collection of plant genetic resources from Andaman and Nicobar Islands. *Genetic Resources and Crop Evolution* 55, 1279–1289.

Akhtar, R. (2019) *Extreme Weather Events and Human Health: International Case Studies*. Springer Nature, Cham, Switzerland.

Assy-Bah, B. and Engelmann, F. (1992a) Cryopreservation of immature embryos of coconut (*Cocos nucifera* L.). *Cryoletters* 126, 67–74.

Assy-Bah, B. and Engelmann, F. (1992b) Cryopreservation of mature embryos of coconut (*Cocos nucifera* L.) and subsequent regeneration of plantlets. *CryoLetters* 126, 117–126.

Australian Government (2015) Pulu Keeling National Park Management Plan 2015–2025. Available at: www.dcceew.gov.au/parks-heritage/national-parks/pulu-keeling-national-park/publications/management-plan-2015-2025 (accessed 22 August 2023).

Bandupriya, H.D.D., Iroshini, W.W.M.A., Perera, S.A.C.N., Vidhanaarachchi, V.R.M., Fernando, S.C. *et al.* (2017) Genetic fidelity testing using SSR marker assay confirms trueness to type of micropropagated coconut (*Cocos nucifera* L.) plantlets derived from unfertilized ovaries. *Open Plant Science Journal* 10, 46–54.

Bandyopadhyay, R. and Frederiksen, R.A. (1999) Contemporary global movement of emerging plant diseases. *Annals of the New York Academy of Sciences* 894, 28–36.

Basheer, M. and K.M. Aarif (2013) Birds associated with the coconut palm *Cocos nucifera* in an agroecosystem in the Western Ghats region of Kerala, Southern India. *Podoces* 8, 19–21.

Batugal, P., Bourdeix, R. and Baudouin, L. (2009) Coconut breeding. In: *Breeding Plantation Tree Crops: Tropical Species*. Springer, New York, pp. 327–375.

Baudouin, L. and Lebrun, P. (2009) Coconut (*Cocos nucifera* L.) DNA studies support the hypothesis of an ancient Austronesian migration from Southeast Asia to America. *Genetic Resources and Crop Evolution* 56, 257–262.

Baudouin, L., Lebrun, P., Berger, A., Myrie, W., Been, B. *et al.* (2008) The Panama Tall and the Maypan hybrid coconut in Jamaica: did genetic contamination cause a loss of resistance to lethal yellowing. *Euphytica* 161, 353–360.

Benson, E.E. (2008) Cryopreservation theory. In: Reed, B.M. (ed.) *Plant Cryopreservation: A Practical Guide Plant Cryopreservation: A Practical Guide*. Springer, New York, pp. 15–32.

Bettoni, J.C., Mathew, L., Pathirana, R., Wiedow, C., Hunter, D.A. *et al.* (2022) Eradication of potato virus S, potato virus A, and potato virus M from infected *in vitro*-grown potato shoots using *in vitro* therapies. *Frontiers in Plant Science* 13: 878733.

Beulé, T., Ilbert, P., Adeoti, K., Durand-Gasselin, T., Dumet, D. *et al.* (2018) Recovery of oil palm (*Elaeis guineensis* Jacq.) somatic embryos cryostored for 20 years. *CryoLetters* 39, 60–66.

BGCI (Botanic Gardens Conservation International) (2022) PlantSearch online database. BGCI, Richmond, UK. Available at: https://tools.bgci.org/plant_search.php (accessed 22 August 2023).

Bourdeix, R. and Prades, A. (2017) *A Global Strategy for the Conservation and use of Coconut Genetic Resources 2018–2028*. Bioversity International, Montpellier, France.

Bourdeix, R., Guarino, L., Rao, V.R. and Baudouin, L. (2005) Status, gaps and strategy in coconut germplasm collecting. In: Batugal, P., Rao, V.R. and Oliver, J. (eds) *Coconut Genetic Resources*. International Plant Genetic Resources Institute–Regional Office for Asia, the Pacific and Oceania (IPGRI-APO), Serdang, Malaysia, pp. 44–64.

Bourdeix, R., Baudouin, L., Bambridge, T., Joly, H.I., Planes, S. *et al.* (2009) Dynamics and conservation of the coconut palm *Cocos nucifera* L. in the Pacific region: towards a new conservation approach. In: *Proceedings of the 11th Pacific Science Inter-Congress,* 2–6 March 2009, Papeete, Tahiti, French Polynesia, pp. 2–6.

Bourdeix, R., Batugal, P., Oliver, J.T. and George, M.L. (2010) *Catalogue of Conserved Coconut Germplasm*. International Plant Genetic Resources Institute – Regional Office for Asia, the Pacific and Oceania (IPGRI-APO), Serdang, Malaysia.

Bourdeix, R., Johnson, V., Baudouin, L., Tuia, V.S., Kete, T. *et al.* (2011) Polymotu: a new concept of island-based germplasm bank based on an old Polynesian practice. *Ogasawara Research* 37, 33–51.

Bourdeix, R., Johnson, V., Saena Tuia, S.V., Kapé, J. and Planes, S. (2013) Traditional conservation areas of coconut varieties and associated knowledge in Polynesian Islands (South Pacific Ocean). In: Larrue, S. (ed.) *Biodiversity and Societies in the Pacific Islands*. Presses Universitaires de Provence (PUP), Université Aix-Marseille, Aix-en-Provence, France, pp. 199–222.

Bourdeix, R., Adkins, S., Johnson, V., Perera, L. and Sisunandar. (2020) *In situ* and *ex situ* conservation of coconut genetic resources. In: Adkins, S., Foale, M., Bourdeix, R., Nguyen, Q. and Biddle, J. (eds) *Coconut Biotechnology: Towards the Sustainability of the 'Tree of Life'*. Springer, Cham, Switzerland, pp. 51–75.

Bourdeix, R. and Sourisseau, J.M. and Lin, J. (eds) (2021) *Coconut Risk Management and Mitigation Manual for the Pacific Region*. Land Resource Division, SPC.

Campbell, M.O. (2006) The sustainability of coconut palm (*Cocos nucifera* Linnaeus 1753) groves in coastal Ghana. *Journal of Coastal Research* 22, 1118–1124.

Caren, C.B., Cecilia, M.M. and Aggrey, B.N. (2019) *In vitro* regeneration of coconut (*Cocos nucifera* L) through indirect somatic embryogenesis in Kenya. *African Journal of Biotechnology* 18, 1113–1122.

Chakravarthy, A.K. and Thyagaraj, N.E. (2012) The palm squirrel in coconut plantations: ecosystem services by therophily. *Mammalia (Paris)* 76, 193–199.

Chan, E. and Elevitch, C.R. (2006) *Cocos nucifera* L. (coconut), version 2.1. In: Elevitch, C.R. (ed.) *Species Profiles for Pacific Island Agroforestry. Permanent Agriculture Resources*, Holualoa, Hawaii.

COGENT (International Coconut Genetic Resources Network) (2022) COGENT network. Available at: www.cogentco.com/en/network (accessed 29 June 2022).

Cueto, C.A., Ebert, A.W., Rillo, E.P. and Orense, O.D. (1996) *In vitro* regeneration of coconut (*Cocos nucifera* L.) from immature inflorescences of tall cultivars. *Acta Horticulturae* 447, 121–122.

Cueto, C., Rivera, R.L., Kim, H.H., Kong, H.J., Baek, H.J. *et al.* (2013) Development of cryopreservation protocols for cryobanks of coconut zygotic embryos. *Acta Horticulturae* 1039, 297–302.

Cutter, V.M. and Wilson, K.S. (1954) Effect of coconut endosperm and other growth stimulants upon the development *in vitro* of embryos of *Cocos nucifera*. *Botanical Gazette* 115, 234–240.

de Guzman, E.V. and del Rosario, D.A. (1964) The growth and development of *Cocos nucifera* L. makapuno embryo *in vitro*. *Philippine Agriculturist* 48, 82–94.

de Smet, K., du Montcel, H.T. and Swennen, R. (1995) Variability in storage potential of banana shoot cultures under medium term storage conditions. *Plant Cell, Tissue and Organ Culture* 42, 269–274.

Dhillon, B.S., Dua, R.P., Brahmi, P. and Bisht, I.S. (2004) On-farm conservation of plant genetic resources for food and agriculture. *Current Science* 87, 557–559.

Dulloo, M.E., Ramanatha Rao, V., Engelmann, F. and Engels, J.M. (2005) *Complementary Conservation of Coconuts*. In: Batugal, P., Rao, V.R. and Oliver, J. (eds) *Coconut Genetic Resources*. International Plant Genetic Resources Institute – Regional Office for Asia, the Pacific and Oceania (IPGRI-APO), Serdang, Malaysia, pp. 75–90.

Dussert, S., Chabrillange, N., Anthony, F., Engelmann, F., Recalt, C. and Hamon, S. (1997) Variability in storage response within a coffee (*Coffea* spp.) core collection under slow growth conditions. *Plant Cell Reports* 16, 344–348.

EEA (European Environment Agency) (2023) In-situ Conditions. EEA, Copenhagen. Available at: www.eea.europa.eu/help/glossary/chm-biodiversity/in-situ-conditions (accessed 21 August 2023).

El-Dawayati, M.M. (2017) *In vitro* conservation of date palm shoot-tip explants and callus cultures under minimal growth conditions. *Methods in Molecular Biology* 1638, 49–59.

Engelmann, F., Malaurie, B. and N'Nan, O. (2011) *In vitro* culture of coconut (*Cocos nucifera* L.) zygotic embryos. *Plant Embryo Culture* 710, 63–72.

Engels, J. (2003) *A Guide to Effective Management of Germplasm Collections*. International Plant Genetic Resources Institute (IPGRI), Rome.

Engels, J.M.M., Maggioni, L., Maxted, N. and Dulloo, M.E. (2008) Complementing *in situ* conservation with *ex situ* measures. In: Iriondo, J.M., Maxted, N. and Dulloo, M.E. (eds) *Conserving Plant Genetic Diversity in Protected Areas: Population Management of Crop Wild Relatives*. CAB International, Wallingford, UK, pp. 169–181.

FAOSTAT (2022) Statistical database. Food and Agricultural Organisation of the United Nations, Rome. Available at: www.fao.org/faostat/ (accessed 22 August 2023).

Foale, M., Biddle, J., Bazrafshan, A. and Adkins, S. (2020) Biology, ecology, and evolution of coconut. In: Adkins, S., Foale, M., Bourdeix, R., Nguyen, Q. and Biddle, J. (eds) *Coconut Biotechnology: Towards the Sustainability of the 'Tree of Life'*. Springer, Cham, Switzerland, pp. 17–27.

Garcia, J.N.M., Aguilar, E.A., Sangalang, J.B., Alcantara, A.J., Habito, R.C.F. *et al.* (2009) Valuation of ecosystem services of coconut types: framework and methodology development. *Journal of Environmental Science and Management* 12, 68–78.

Gostel, M.R., Kelloff, C., Wallick, K. and Funk, V.A. (2016) A workflow to preserve genome-quality tissue samples from plants in botanical gardens and arboreta. *Applications in Plant Sciences* 4: 1600039.

Gruezo, W.S. and Harries, H.C. (1984) Self sown wild type coconuts in the Philippines. *Biotropica* 16, 140–147.

Gunn, B.F., Baudouin, L. and Olsen, K.M. (2011) Independent origins of cultivated coconut (*Cocos nucifera* L.) in the Old World tropics. *PLoS One* 6: e21143.

Harries, H.C. (1978) The evolution, dissemination and classification of *Cocos nucifera* L. *Botanical Review* 44, 265–319.

Harries, H.C. (2012) Germination rate is the significant characteristic determining coconut palm diversity. *AoB Plants* 2012: pls045.

Harries, H.C. and Clement, C.R. (2014) Long-distance dispersal of the coconut palm by migration within the coral atoll ecosystem. *Annals of Botany* 113, 565–570.

Hassan, M.M. (2017) *In vitro* conservation of date palm somatic embryos using growth-retardant conditions. *Methods in Molecular Biology* 1638, 61–70.

Hebbar, K.B., Mathew, A.C. and Manikantan, M.R. and Kukkamgai. (2013) Value added products from neera. *Indian Coconut Journal* 56, 28–33.

Helliot, B., Panis, B., Poumay, Y., Swennen, R., Lepoivre, P. *et al.* (2002) Cryopreservation for the elimination of cucumber mosaic and banana streak viruses from banana (*Musa* spp.). *Plant Cell Reports* 20, 1117–1122.

Hodkinson, T.R., Waldren, S., Parnell, J.A., Kelleher, C.T., Salamin, K. *et al.* (2007) DNA banking for plant breeding, biotechnology and biodiversity evaluation. *Journal of Plant Research* 120, 17–29.

Hornung, R., Domas, R. and Lynch, P.T. (2001) Cryopreservation of plumular explants of coconut (*Cocos nucifera* L.) to support programmes for mass clonal propagation through somatic embryogenesis. *CryoLetters* 22, 211–220.

ICC (International Coconut Community) (2021) Introduction to ICG. Available at: https://coconutcommunity.org/ (accessed 22 August 2023).

IUCN (International Union for Conservation of Nature) (2022) IUCN Red List. Available at: www.iucnredlist.org/ (accessed 22 August 2023).

Jerard, B.A., Rajesh, M.K., Thomas, R.J., Niral, V. and Samsudeen, K. (2017) Island ecosystems host rich diversity in coconut (*Cocos nucifera* L.): evidence from Minicoy Island, India. *Agricultural Research* 6, 214–226.

Kalaipandian, S., Mu, Z., Kong, E.Y.Y., Biddle, J., Cave, R. *et al.* (2021) Cloning coconut via somatic embryogenesis: a review of the current status and future prospects. *Plants* 10: 2050.

Kameswara Rao, N. (2004) Plant genetic resources: advancing conservation and use through biotechnology. *African Journal of Biotechnology* 3, 136–145.

Karun, A., Sajini, K.K., Niral, V., Amarnath, C.H., Remya, P. *et al.* (2014) Coconut (*Cocos nucifera* L.) pollen cryopreservation. *CryoLetters* 35, 407–417.

Kong, E.Y., Biddle, J., Foale, M. and Adkins, S.W. (2020) Cell suspension culture: a potential *in vitro* culture method for clonal propagation of coconut plantlets *via* somatic embryogenesis. *Industrial Crops and Products* 147: 112125.

Krishna, H., Alizadeh, M., Singh, D., Singh, U., Chauhan, N. *et al.* (2016) Somaclonal variations and their applications in horticultural crops improvement. *3 Biotech* 6: 54.

Kumar, P.L., Cuervo, M., Kreuze, J.F., Muller, G., Kulkarni, G. *et al.* (2021) Phytosanitary interventions for safe global germplasm exchange and the prevention of transboundary pest spread: the role of CGIAR germplasm health units. *Plants* 10: 328.

Kumara, A.D.N.T., Chandrashekharaiah, M., Kandakoor, S.B. and Chakravarthy, A.K. (2015) Status and management of three major insect pests of coconut in the tropics and subtropics. In: *New Horizons in Insect Science: Towards Sustainable Pest Management*. Springer India, New Delhi, pp. 359–381.

Kumaunang, J., Damasco, O.P. and Protacio, C.M. (2020) Cryopreservation of 'Laguna Tall' coconut (*Cocos nucifera* Linn.) mature zygotic embryos. *Philippine Agricultural Scientist* 103, 93–101.

Labouisse, J.P. and Caillon, S. (2001) Une approche de la conservation *in situ* par l'étude d'un système semencier informel: cas du cocotier au Vanuatu (Pacifique Sud). *Oilseeds and Fats, Crops and Lipids* 8, 534–539.

Lantican, D.V., Strickler, S.R., Canama, A.O., Gardoce, R.R., Mueller, L.A. *et al.* (2019) *De novo* genome sequence assembly of dwarf coconut (*Cocos nucifera* L. 'Catigan Green Dwarf') provides insights into genomic variation between coconut types and related palm species. *G3: Genes, Genomes, Genetics* 9, 2377–2393.

Leach, B. J., Foale, M.A. and Ashburner, G.R. (2003) Some characteristics of wild and managed coconut palm populations and their environment in the Cocos (Keeling) Islands, Indian Ocean. *Genetic Resources and Crop Evolution* 50, 627–638.

Lédo, A.D.S., Santana, F.V., Oliveira, A.C.A.D., Oliveira, L.A.R.D. and Silva, A.V.C.D. (2019) Cryopreservation of Brazilian green dwarf coconut plumules by droplet-vitrification. *Ciência Rural* 50, 1–8.

Leu, A. (2021) Identifying the diversity of self-sown native coconuts (*Cocos nucifera* L.) and the introduced phenotype in Queensland, Australia. *International Journal of Plant Biology* 12: 9192.

Longley-Wood, K., Engels, M., Lafferty, K.D., McLaughlin, J.P. and Wegmann, A. (2022) Transforming Palmyra Atoll to native-tree dominance will increase net carbon storage and reduce dissolved organic carbon reef runoff. *PLoS One* 17: e0262621.

Moorhead, A. (2012) South Pacific coconut gene bank under threat. *Nature.* Available at: www.nature.com/articles/nature.2012.11792.pdf (accessed 22 August 2023).

Nair, M.K. and Karun, A. (1999) Coconut embryo culture: present status and future thrust. *CORD* 15: 34.

Nayar, N.M. (2016) Early history, lore, and economic botany. In: *The Coconut: Phylogeny, Origins, and Spread.* Elsevier Science and Technology, San Diego, California, pp. 9–24.

Nayar, N.M. (2018) Opportunistic flowering and pollinating system as a survival strategy of the coconut (*Cocos nucifera* L.), a monotypic species of the Arecaceae family. *Genetic Resources and Crop Evolution* 65, 333–342.

Nguyen, Q.T., Bandupriya, H.D., López-Villalobos, A., Sisunandar, S., Foale, M. *et al.* (2015) Tissue culture and associated biotechnological interventions for the improvement of coconut (*Cocos nucifera* L.): a review. *Planta* 242, 1059–1076.

N'Nan, O., Hocher, V., Verdeil, J.L., Konan, J.L., Ballo, K. *et al.* (2008) Cryopreservation by encapsulation–dehydration of plumules of coconut (*Cocos nucifera* L.). *CryoLetters* 29, 339–350.

N'Nan, O., Borges, M., Konan, J.L.K., Hocher, V., Verdeil, J.L. *et al.* (2012) A simple protocol for cryopreservation of zygotic embryos of ten accessions of coconut (*Cocos nucifera* L.). *In Vitro Cellular & Developmental Biology – Plant* 48, 160–166.

N'Nan, O., Bi, S.G., Tiécoura, K. and Konan, J.L.K. (2014) Use of plumules cryopreservation to save coconut germplasm in areas infected by lethal yellowing. *African Journal of Biotechnology* 13, 1702–1706.

Oropeza, C., Cordova, I., Puch-Hau, C., Castillo, R., Chan, J.L. *et al.* (2017) Detection of lethal yellowing phytoplasma in coconut plantlets obtained through *in vitro* germination of zygotic embryos from the seeds of infected palms. *Annals of Applied Biology* 171, 28–36.

Oropeza-Salín, C., Sáenz, L., Narvaez, M., Nic-Matos, G., Córdova, I. *et al.* (2020) Dealing with lethal yellowing and related diseases in coconut. In: Adkins, S., Foale, M., Bourdeix, R., Nguyen, Q. and Biddle, J. (eds) *Coconut Biotechnology: Towards the Sustainability of the 'Tree of Life'*. Springer, Cham, Switzerland, pp. 169–197.

Palanyandy, S.R., Gantait, S., Subramaniam, S. and Sinniah, U.R. (2020) Cryopreservation of oil palm (*Elaeis guineensis* Jacq.) polyembryoids via encapsulation–desiccation. *3 Biotech* 10: 9.

Panis, B. (2008) Cryopreservation of monocots. In: Reed, B.M. (ed.) *Plant Cryopreservation: A Practical Guide*. Springer, New York, pp. 241–280.

Panis, B. (2018) Sixty years of plant cryopreservation: from freezing hardy mulberry twigs to establishing reference crop collections for future generations. *Acta Horticulturae* 1234, 1–8.

Panis, B., Totte, N., van Nimmen, K., Withers, L.A. and Swennen, R. (1996) Cryopreservation of banana (*Musa* spp.) meristem cultures after preculture on sucrose. *Plant Science* 121, 95–106.

Panis, B., Piette, B. and Swennen, R. (2005) Droplet vitrification of apical meristems: a cryopreservation protocol applicable to all Musaceae. *Plant Science* 168, 45–55.

Panis, B., Nagel, M. and Van denhouwe, I. (2020) Challenges and prospects for the conservation of crop genetic resources in field genebanks, in *in vitro* collections and/or in liquid nitrogen. *Plants (Basel)* 9: 1634.

Pence, V.C. (2011) Evaluating costs for the *in vitro* propagation and preservation of endangered plants. *In Vitro Cellular & Developmental Biology – Plant* 47, 176–187.

Perera, P.I. and Vidhanaarachchi, V.R. (2021) Anther culture in coconut (*Cocos nucifera* L.). In: *Doubled Haploid Technology*. Humana, New York, pp. 167–178.

Pérez-Núñez, M.T., Chan, J.L., Saenz, L., González, T., Verdeil, J.L. *et al.* (2006) Improved somatic embryogenesis from *Cocos nucifera* (L.) plumule explants. *In Vitro Cellular & Developmental Biology – Plant* 42, 37–43.

Powell, K. (2018) How biologists are creating life-like cells from scratch. *Nature* 563, 172–175.

Pritchard, H.W. and Nadarajan, J. (2008) Cryopreservation of orthodox (desiccation tolerant) seeds. In: Reed, B.M. (ed.) *Plant Cryopreservation: A Practical Guide*. Springer, New York, pp. 485–501.

Raiesh, M.K., Nagarajan, P., Jerard, B.A., Arunachalam, V. and Dhanapal, R. (2008) Microsatellite variability of coconut accessions (*Cocos nucifera* L.) from Andaman and Nicobar Islands. *Current Science* 94, 1627–1631.

Rajesh, M.K., Radha, E., Sajini, K.K. and Karun, A. (2014) Polyamine-induced somatic embryogenesis and plantlet regeneration *in vitro* from plumular explants of dwarf cultivars of coconut (*Cocos nucifera*). *Indian Journal of Agricultural Sciences* 84, 527–530.

Rajora, O.P. and Mosseler, A. (2001) Challenges and opportunities for conservation of forest genetic resources. *Euhpytica* 118, 197–212.

Randles, J.W., Wefels, E., Hanold, D., Miller, D.C., Morin, J.P. *et al.* (1999) Detection and diagnosis of coconut foliar decay disease. In: Oropeza, C., Verdeil, J.L., Ashburner, G.R., Cardeña, R., Santamaría, J.M. (eds) *Current Advances in Coconut Biotechnology. Current Plant Science and Biotechnology in Agriculture*, Vol. 35. Springer, Dordrecht., pp. 247–258.

Research and Markets (2019) *Coconut Products Market by Type Application and Form: Global Opportunity Analysis and Industry Forecast, 2019–2026*. Research and Markets, Dublin. Available at: www.researchandmarkets.com/reports/4989541/coconut-products-market-by-type-application-and (accessed 21 August 2023).

Royal Botanic Gardens (2022) Seed Information Database (SID). Version 7.1. Royal Botanic Gardens, Kew. Available at: https://data.kew.org/sid (accessed 22 August 2023).

Sáenz, L., Chan, J.L., Narvaez, M. and Oropeza, C. (2018) Protocol for the micropropagation of coconut from plumule explants. In: Loyola-Vargas, V.M. and Ochoa-Alejo, N. (eds) *Plant Cell Culture Protocols*. Humana Press, New York, pp. 161–170.

Sajini, K.K., Karun, A., Amarnath, C.H. and Engelmann, F. (2011) Cryopreservation of coconut (*Cocos nucifera* L.) zygotic embryos by vitrification. *CryoLetters* 32, 317–328.

Salm, R. and Feder, L. (2015) Bringing the Kayapó villages of the Amazon: evaluation of a sustainable development project. *Palms* 59, 127–134.
Salm, R., Feder, L., Jardim, M.A.G., Hass, N., Jalles-Filho, E. *et al.* (2009) Conservation value of an exotic species: the case of coconuts on the Kayapo indigenous lands, south-eastern Amazonia. *Environment, Development and Sustainability* 11, 161–173.
Samosir, Y. and Adkins, S. (2014) Improving acclimatization through the photoautotrophic culture of coconut (*Cocos nucifera*) seedlings: an *in vitro* system for the efficient exchange of germplasm. *In Vitro Cellular & Developmental Biology – Plant* 50, 493–501.
Samsudeen, K. and Thamban, C. (2017) Farm conservation and utilisation of coconut diversity. *Indian Coconut Journal* 1, 16–20.
Samsudeen, K., Thamban, C., Niral, V., Augustine Jerard, B., Rajesh, M.K. *et al.* (2013) *In situ* approach for rapid characterization to aid on farm conservation of coconut germplasm – a case study of two ecotypes from West coast of India. *Journal of Plantation Crops* 41, 357–363.
Santos, G.A., Batugal, P.A., Othman, A., Baudouin, L. and Labouisse, J.P. (eds) (1996) *Manual on Standardized Research Techniques in Coconut Breeding*. International Plant Genetic Resources Institute (IPGRI), Rome.
Sharma, S.D. (2005) Cryopreservation of somatic embryos – an overview. *Indian Journal of Biotechnology* 4, 47–55.
Sisunandar, Sopade, P.A., Samosir, Y.M., Rival, A. and Adkins, S.W. (2010) Dehydration improves cryopreservation of coconut (*Cocos nucifera* L.). *Cryobiology* 61, 289–296.
Sisunandar, Novarianto, H., Mashud, N., Samosir, Y. and Adkins, S.W. (2014) Embryo maturity plays an important role for the successful cryopreservation of coconut (*Cocos nucifera*). *In Vitro Cellular & Developmental Biology – Plant* 50, 688–695.
Summerhayes, G.R. (2018) Coconuts on the move: archaeology of Western Pacific. *Journal of Pacific History* 53, 375–396.
The Phillipine Government (1994) Proclamation No. 469. Available at: www.officialgazette.gov.ph/1994/09/26/proclamation-no-469-s-1994/#:~:text=The%20destruction%20of%20the%20terrestrial,found%20therein%2C%20are%20hereby%20prohibited (accessed 22 August 2023).
Vadamalai, G., Perera, A.A.F.L.K., Hanold, D., Rezaian, M.A. and Randles, J.W. (2009) Detection of coconut cadang-cadang viroid sequences in oil and coconut palm by ribonuclease protection assay. *Annals of Applied Biology* 154, 117–125.
Verdeil, J.L., Huet, C., Grosdemange, F. and Buffard-Morel, J. (1994) Plant regeneration from cultured immature inflorescences of coconut (*Cocos nucifera* L.): evidence for somatic embryogenesis. *Plant Cell Reports* 13, 218–221.
Volis, S. and Blecher, M. (2010) Quasi *in situ*: a bridge between *ex situ* and *in situ* conservation of plants. *Biodiversity and Conservation* 19, 2441–2454.
Wang, S., Xiao, Y., Zhou, Z.W., Yuan, J., Guo, H. *et al.* (2021) High-quality reference genome sequences of two coconut cultivars provide insights into evolution of monocot chromosomes and differentiation of fiber content and plant height. *Genome Biology* 22: 304.
Weckx, S., Inzé, D. and Maene, L. (2019) Tissue culture of oil palm: finding the balance between mass propagation and somaclonal variation. *Frontiers in Plant Science* 10: 722.
Welewanni, I., Perera, C., Jayasekera, A. and Bandupriya, D. (2020) Recovery, histological observations, and genetic integrity in coconut (*Cocos nucifera* L.) embryogenic calli cryopreserved using encapsulation-dehydration procedure. *Journal of the National Science Foundation of Sri Lanka* 48, 175–186.
Wilms, H., Rhee, J.H., Rivera, R.L., Longin, K. and Panis, B. (2018) Developing coconut cryopreservation protocols and establishing cryo-genebank at RDA; a collaborative project between RDA and Bioversity International. *Acta Horticulturae* 1234, 343–348.
Wilms, H., de Bièvre, D., Longin, K., Swennen, R., Rhee, J. and Panis, B. (2021) Development of the first axillary *in vitro* shoot multiplication protocol for coconut palms. *Scientific Reports* 11: 18367.
Wilms, H., de Bièvre, D., Rosiers, E., Swennen, R., Rhee, J. and Panis, B. (2022) Tissue necrosis prevention during shoot multiplication of coconut. *Acta Horticulturae* 1339, 173–180.
Withers, L.A. and Engelmann, F. (1997) *In vitro* conservation of plant genetic resources. In: Altman, A. (ed.) *Biotechnology in Agriculture*. Marcel Dekker, New York, pp. 57–88.
Xiao, Y., Xu, P., Fan, H., Baudouin, L., Xia, W. *et al.* (2017) The genome draft of coconut (*Cocos nucifera* L.). *GigaScience* 6: gix095.
Young, H.S., Raab, T.K., McCauley, D.J., Briggs, A.A. and Dirzo, R. (2010) The coconut palm, *Cocos nucifera*, impacts forest composition and soil characteristics at Palmyra Atoll, Central Pacific. *Journal of Vegetation Science* 21, 1058–1068.

10 Coconut Biotechnology

Eveline Y.Y. Kong[1*], ZhiHua Mu[1], Vijitha R. M. Vidhanaarachchi[2], Quang Thien Nguyen[1], Sisunandar Sisunandar[3], Sundaravelpandian Kalaipandian[1] and Bart Panis[4,5]

[1]*School of Agriculture and Food Sustainability, The University of Queensland, Gatton, QLD, Australia;* [2]*Tissue Culture Division, Coconut Research Institute, Lunuwila, Sri Lanka;* [3]*Biology Education Department, Muhammadiyah University of Purwokerto, Purwokerto, Indonesia;* [4]*The Alliance of Bioversity International and CIAT, Leuven, Belgium;* [5]*Dept of Biosystems, KU Leuven, Leuven, Belgium*

Abstract

Coconut is a valuable industrial crop that has a multitude of uses and applications, such as in the production of copra, various foods and drinks, soaps and detergents, furniture and household goods, and bulk materials for manufacturing, medicines and cosmetics. A wide range of biotechnology approaches are available to bring advancements in coconut propagation, conservation, crop improvement, and tolerance to pests and diseases. Many have been accomplished with great success, while others still require further optimization prior to commercial application. This chapter provides a summary of these advancements, in particular a history of the advancements in coconut tissue culture, as well as the challenges and opportunities of this technology.

10.1 History of Coconut Biotechnology

Coconut (*Cocos nucifera* L.), a palm naturally found or grown in the lowlands of tropical and subtropical countries from East Africa to eastern Polynesia (Foale *et al.*, 2020) and the seas of the Indian and Pacific Oceans, is regarded as one of the most useful palms to humankind (Nayar, 2017). As one of the world's most important palm crops, coconut is cultivated on approximately 12 million ha of land and contributes directly to the livelihoods of many millions of farmers and their families (Nguyen *et al.*, 2015; Pham, 2016). Over the past few decades, demand for coconut planting materials has substantially increased due to several factors. Lethal pests and diseases, such as coconut rhinoceros beetle (*Oryctes rhinoceros* L.) and phytoplasma-related diseases, have applied pressure to coconut growers and local communities (Bedford, 2013; Gurr *et al.*, 2016). As well as biotic threats, abiotic stresses due to natural events and climate change have been causing long-lasting issues for coconut farming systems (Hebbar *et al.*, 2013). Furthermore, new planting materials are in great demand to help replace the growing number of senile coconut palms around the world (Salum *et al.*, 2020). To address the shortage of planting materials, breeding programmes and biotechnologies are being used to facilitate coconut production in many coconut-producing countries; however, conventional seed propagation programmes are

* Email: e.kong@uq.edu.au

©CAB International 2024. *The Coconut: Botany, Production and Uses* (eds S. Adkins *et al.*)
DOI: 10.1079/9781789249736.0010

unlikely to produce enough seedlings of the desired varieties in a timely manner (Batugal *et al.*, 2009).

Tissue culture, one type of coconut biotechnology, is likely to become an important technology to enable coconut to fulfil the ever-increasing world demand for high-quality planting material. Fernando *et al.* (2010b) has suggested that tissue culture is an important tool to speed up the process of production and multiplication of new and elite varieties and hybrids. *In vitro* culture of coconut could also be used for the short- and medium-term storage of coconut genetic resources (Lédo and Vendrame, 2021), and is also used to recover plants from cryopreservation, a long-term form of super-cold storage of germplasm (Nguyen *et al.*, 2015). The availability of efficient *in vitro* culture regeneration protocols is also indispensable to modern plant improvement techniques, such as genetic engineering and gene editing. In the future, these methods may lead to the production of improved coconut varieties that are not only high yielding but also tolerant to salinity, drought and heat stress, as well as being disease-free (Hussain *et al.*, 2012). However, research conducted in this area has been limited (Samosir, 1999; Andrade-Torres *et al.*, 2011).

The early *in vitro* coconut studies were focused on zygotic embryo culture (EC). The first study, conducted by Cutter and Wilson (1954), determined the stimulatory effect of the coconut liquid endosperm and other growth stimulants on zygotic embryo germination and growth (Fig. 10.1). A decade later, the first *in vitro* coconut plantlets were regenerated from mature Makapuno zygotic embryos (de Guzman and del Rosario, 1964). The further advancements of coconut EC have been summarized by Biddle *et al.* (2020). These studies were focused mainly on establishing the optimal growth conditions

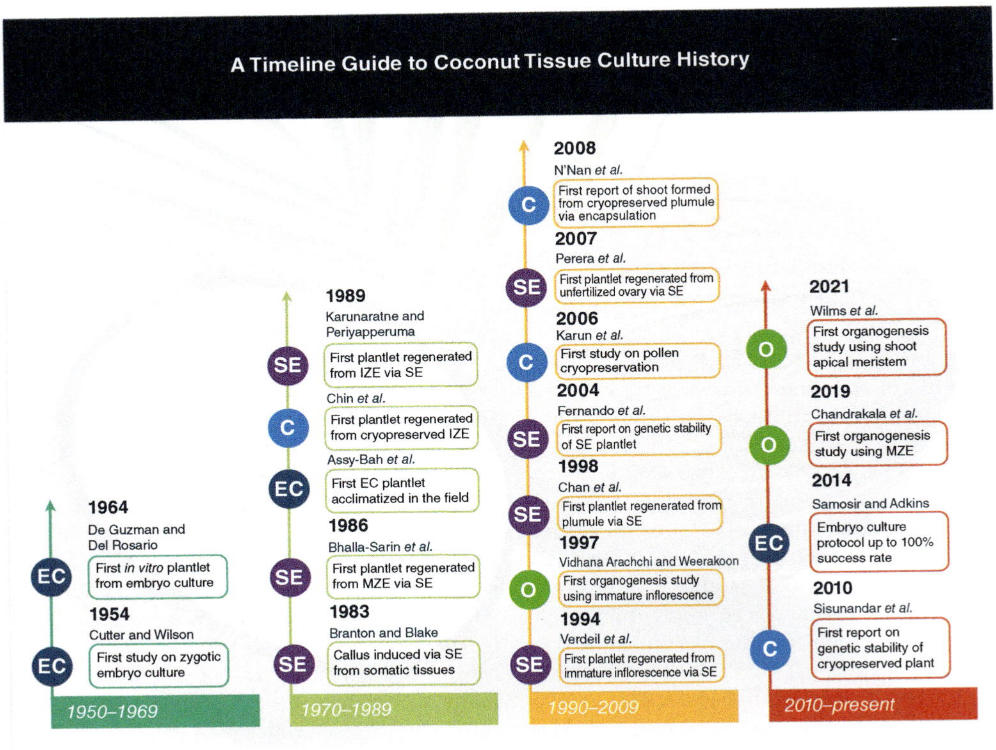

Fig. 10.1. A chronology of coconut tissue-culture studies on embryo culture (EC), somatic embryogenesis (SE), cryopreservation (C), direct shoot organogenesis (O), and the genetic fidelity of regenerated plants. IZE, immature zygotic embryo; MZE, mature zygotic embryo; SAM, shoot apical meristem.

for zygotic embryos via research on germination, photosynthesis and rooting ability, and their responses to medium additives. EC of coconut can be used for embryo rescue of coconut varieties that cannot germinate *in vivo*, for exchange and conservation of germplasm, and for micropropagation (cloning), as well as for genetic transformation (Biddle *et al.*, 2020).

Later, the primary focus of coconut biotechnological studies was in cloning technology, including somatic embryogenesis (SE), which has been well studied, and direct shoot organogenesis, which has been less well studied. Direct shoot organogenesis, in this context, refers to the direct formation of shoots without a callus phase, with shoots originating from either axillary or adventitious tissues. One of the earliest coconut SE studies was reported by Branton and Blake (1983), who induced callus from a coconut seedling's stem and from immature inflorescence tissues using the auxin 2,4-dichlorophenoxyacetic acid (2,4-D), obtaining nodular structures that could develop into embryos after reducing the media's auxin concentration. Later, the first whole plantlets were regenerated from embryogenic structures induced on callus from mature and immature zygotic embryos (Bhalla-Sarin *et al.*, 1986; Karunaratne and Periyapperuma, 1989), immature inflorescence tissues (Verdeil *et al.*, 1994), plumules (Chan *et al.*, 1998) and unfertilized ovary tissues (Perera *et al.*, 2007). The possibility of using tissue-culture techniques (i.e. SE) to regenerate genetically stable plantlets with normal growth in the field was first reported by Fernando *et al.* (2004). More recently, the genetic fidelity of coconut plantlets (regenerated from unfertilized ovaries) has been screened through simple sequence repeat (SSR) markers, demonstrating that there is potential for genetically stable coconut micropropagation using SE (Bandupriya *et al.*, 2017).

Among all the explants tried for SE, plumule tissues were found to be the most responsive, as highlighted by Pérez-Núñez *et al.* (2006). This study calculated the theoretical number of somatic embryos (98,000) that could possibly be produced from one plumule. Improvements in SE using plumule tissues have been obtained by increasing the efficiency of embryogenic callus induction, and the effectiveness of somatic embryo formation from callus and their germination and plantlet formation (Azpeitia *et al.*, 2003; Montero-Córtes *et al.*, 2010; Rajesh *et al.*, 2014). Although plumule tissues might be easier to work with, somatic tissues are a more suitable type of explant to use for SE, especially for the cross-pollinated tall coconut varieties as they ensure true-to-type clones. Hence, more recent studies have shifted towards undertaking SE using immature inflorescence tissues (Sandoval-Cancino *et al.*, 2016; Shareefa *et al.*, 2019). Moreover, the potential of using liquid cell-suspension culture for mass propagation of coconut clones is now being investigated (Kong *et al.*, 2020).

The first study on the clonal propagation of coconut using direct shoot organogenesis was reported by Vidhanaarachchi and Weerakoon (1997), where shoots could be regenerated from immature inflorescence sections directly without a callus phase, but this was at a low frequency. In later studies, complete plantlets were reported to be formed from immature inflorescence tissues, with one to 14 vegetative shoots produced per explant (Raju, 2006; Shareefa *et al.*, 2019). More recently, the first studies on the use of mature zygotic embryos (Chandrakala *et al.*, 2019) and *in vitro* seedlings (Wilms *et al.*, 2021) to regenerate coconut plantlets via direct shoot organogenesis (axillary shoots) have been conducted.

Tissue culture is essential during the recovery stage following cryopreservation. The selection of the correct recovery medium and growth conditions can significantly impact the success of a cryopreservation protocol (Nguyen *et al.*, 2015). The first *in vitro* plantlet successfully regenerated following cryopreservation was from an immature zygotic embryo (Chin *et al.*, 1989). Although Bajaj (1984) was the first to conduct a cryopreservation study using immature embryos of the variety West Coast Tall, the embryos did not regenerate into plantlets during recovery. Later, plantlets that were successfully recovered from the cryopreserved zygotic embryos of four Indonesian varieties were shown to be morphologically, genetically and epigenetically like non-cryopreserved plants (Sisunandar *et al.*, 2010). Cryopreservation of whole coconut embryos has achieved the most consistent (and successful) results to date, with a new rapid, physical dehydration protocol recently developed using a prototype apparatus, which improved the dehydration rate of embryos to 20% moisture content in 6 h. A pre-treatment of embryos with 0.4 M sucrose further reduced the drying time to just 4 h to reach a moisture content of 20% and improved the embryo germination rate (~38%) (Bazrafshan, 2022). Cryopreservation using an encapsulation dehydration technique for plumules, a typical cryopreservation method used for other species, has resulted in the recovery of limited leafy *in vitro* shoots and no field-growing plants (N'Nan *et al.*, 2008). Droplet vitrified and cryopreserved shoot apical meristems were found to survive treatment, but their subsequent growth into normal plants requires further improvement (Wilms *et al.*, 2019). A more recent study showed that shoot-tip cryopreservation using droplet vitrification with pre-treatment in sucrose had a 60% survival rate with approximately 22% forming shoots (Bazrafshan, 2022) and establishing in soil (S. Sisunandar, 2023, personal communication). Pollen recovered via *in vitro* germination was shown to retain viability and fertility after being stored in liquid nitrogen for 4 years (Karun *et al.*, 2006, 2014).

In the initial period of coconut tissue-culture research (from the 1950s to 1980s), EC was the focus for obtaining coconut plantlets. From the 1980s onwards, research on the clonal propagation of coconut via SE became an area of interest, mainly because coconut does not propagate vegetatively under natural environmental conditions. Currently, there is interest in studies on cell-suspension culture and the use of bioreactors for the potential upscaling of embryogenic callus production and improving the growth of somatic embryos, but much work still needs to be undertaken before these types of technologies can be used for coconut propagation.

In addition to a focus on coconut propagation, tissue-culture research has also concentrated on regenerating coconut palms using tissue culture from various explants for the purpose of germplasm collection, preservation and exchange. As the clonal propagation method became more popular, researchers in the 2000s began to question the genetic stability of SE clonally propagated and cryopreserved plantlets. However, to date, no genetic variation or off-types have been reported in any form of coconut tissue culture. Nevertheless, the stresses associated with SE, particularly when the same cultured material goes through several rounds of multiplication, can sometimes result in unwanted genetic change, as seen in the resultant plantlets of other species. This possibility of genetic change with SE prompted researchers in the 2000s to direct their work towards organogenesis and direct shoot multiplication, but only recently have promising results been obtained with these methods.

10.2 Coconut Micropropagation

As coconut cannot be clonally propagated by conventional approaches, one of the most important coconut biotechnologies that can be used for large scale multiplication of coconut varieties is micropropagation (Misra, 1994; Morcillo *et al.*, 2006; Nguyen *et al.*, 2015). As mentioned earlier, research on coconut micropropagation can be categorized into three types: EC, SE and direct shoot multiplication.

10.2.1 Embryo culture

Traditionally, whole coconut fruit must be transported to the planting location to start the establishment of a new palm. However, the size and weight of the coconut fruit during transport can increase the difficulty for research institutes in terms of sharing, growing and breeding coconut seedlings (Mu *et al.*, 2020). Furthermore, the endocarp (shell) and mesocarp (coir fibres) of the coconut fruit may carry insects and pathogens that raise phytosanitary concerns within the receiving laboratory. Thus, scientists have developed an alternative way to regenerate coconut palms in a space-effective and sterile way. Coconut embryos have proven to be a welcome alternative that can reduce the phytosanitary concerns and the freight costs incurred when transporting the whole fruit, especially when moving germplasm internationally. This method is known as embryo culture (EC), whereby the embryos are excised and cultured on to a specifically designed culture medium that can nurture their germination, growth and development (Biddle *et al.*, 2020). Embryo culture is now used to multiply high-value elite coconut varieties in a commercial setting, especially for varieties that cannot germinate naturally, such as Makapuno and Kopyor types (Nguyen *et al.*, 2016). As well as being useful for transportation, EC can also be used for the purpose of germplasm conservation. Cryopreservation of coconut zygotic embryos has been shown to be an effective, long-term preservation method for coconut genetic resources (Sisunandar *et al.*, 2010), and EC plays an important role in plant recovery from cryopreserved embryos.

When compared with other plant species, development of the EC protocol for coconut has been slow, probably due to coconut being a recalcitrant species (Karun, 2017), and sometimes producing poor and inconsistent growth responses, including the slow growth of *in vitro*-produced seedlings during early *ex vitro* conditions (Fernando *et al.*, 2010a). Despite these difficulties, the International Plant Genetic Resources Institute (Frison *et al.*, 1993), the Australian Centre for International Agricultural Research (ACIAR) (Ashburner *et al.*, 1995; Samosir *et al.*, 2008), the Food and Agriculture Organization of the United Nations (Engelmann and Batugal, 2002) and Bioversity International (Cueto *et al.*, 2012) have developed and improved several practical protocols for EC.

For successful EC, several stepwise techniques must be followed (Fig. 10.2). In the early days of the coconut EC technology, a range of culture media were tested, and the Y3 basal medium created by Eeuwens (1976) proved to be superior (Biddle *et al.*, 2020). Along with Y3 medium, a specific sucrose concentration (25–60 g L^{-1}, as the main energy source and osmotically active compound) and activated charcoal (1.0–2.5 g L^{-1}, to limit tissue necrosis) are generally required for optimal embryo growth (Nguyen *et al.*, 2015). Several plant growth regulators have been employed to accelerate the germination and growth of embryos. Plant growth regulators can promote various stages of development, from germination to acclimatization stage. For instance, gibberellic acid has been employed to optimize the germination of embryos (Magdalita *et al.*, 2015), while indole-3-butyric acid or naphthalene acetic acid has commonly been applied during acclimatization to strengthen root initiation (Ashburner *et al.*, 1993; Rillo, 1998).

At 5–6 months after initiation (including monthly subcultures), *in vitro* plantlets are ready for transplantation into soil, a process known as acclimatization. This is to ensure the adaptability and survival of *in vitro* plants in the field (Preece and Sutter, 1991). Acclimatization of *in vitro* coconut plantlets remains one of the main bottlenecks of all types of coconut tissue culture

Fig. 10.2. Embryo culture (EC) of Kopyor Brown Dwarf coconut. (A) Isolated embryo. (B) Shoot emergence and elongation from embryos. (C) EC seedlings with opened leaves and roots. (D) Acclimatization of EC seedlings. (E) Field-grown mature Kopyor coconut palm propagated using the EC technique.

(Mu *et al.*, 2020). During acclimatization of coconut plantlets, the lighting system, gaseous conditions, medium composition, relative humidity, base substrate and carbon sources are extremely important to obtain a high success rate and shorten the acclimatization period (Kalaipandian *et al.*, 2021). There are successful methods for acclimatization, but they can be slow and rely heavily on operator expertise. There is room for improvement in all of these methods.

In summary, EC is an important coconut tissue-culture method that requires further research, particularly so that rapid and effective acclimatization can be achieved.

10.2.2 Somatic embryogenesis

SE in coconut has been investigated for more than 40 years, starting with a study by Eeuwens and Blake (1977). In recent years, clonal propagation using SE has been substantially improved (Pérez-Núñez *et al.*, 2006; Sandoval-Cancino *et al.*, 2016). This tissue-culture pathway, by which multiple clones can be generated from a single explant, is a promising solution to the issue of quality coconut seedling shortages (Kalaipandian *et al.*, 2021). SE technology has been recommended for elite coconut varieties that are highly sought after (Nguyen *et al.*, 2016). With the possibility of upscaling via a 'biofactory' and rapid acclimatization, this technology is believed to offer a pragmatic solution to the needs of farming communities (Samosir and Adkins, 2014; Sáenz-Carbonell *et al.*, 2020).

SE is already being used for the mass propagation of various plant species, including woody plants and palms (Tautorus *et al.*, 1991; Gaj, 2004; Solís-Ramos *et al.*, 2012). The mass multiplication of coconut palms through SE has been developed for different explants, including the plumules (shoot meristems excised from a zygotic embryo), unfertilized ovaries and immature inflorescences. These explants are cultured on to a 2,4-D-containing tissue-culture medium, thus forming embryogenic callus clumps (Fig. 10.3).

The plumule-derived callus is multiplied and secondary SE callus is obtained (Pérez-Núñez *et al.*, 2006; Sáenz-Carbonell *et al.*, 2013), and this has been scaled up to multiply dwarf varieties (Sandoval-Cancino *et al.*, 2016). However, mass propagation of pure tall and hybrid varieties using plumule explants is only viable if the plumules are obtained from fruit collected from selective crossing. This is because plumule tissue is not naturally true to the mother type in these cross-pollinating types. Fortunately, recent advances in coconut SE have been made to effectively initiate an embryogenic callus from an immature inflorescence, which has a somatic origin (Sandoval-Cancino *et al.*, 2016). Although this type of explant is a little more difficult to deal with, research shows that it holds enormous potential for producing true-to-type clones.

The unfertilized ovary has also been identified as a potential source (Perera *et al.*, 2007, 2009), as this tissue can produce true-to-type palms, which is not possible using their embryos due to segregation (Vidhanaarachchi *et al.*, 2013). The genetic fidelity of the clonal plants generated from unfertilized ovaries was confirmed by Bandupriya *et al.* (2017). However, the SE protocol using unfertilized ovaries still needs improvements to be commercially viable.

Scale-up of SE

Secondary SE and subsequent multiplication of embryogenic callus, together with the regeneration capacity of these embryogenic structures, allows an increase in the rate of mass propagation (Pérez-Núñez *et al.*, 2006). A semi-commercial model, or biofactory, has been established in association with the Centro de Investigación Científica de Yucatán in Mexico (Sáenz-Carbonell *et al.*, 2020). This biofactory approach provides a model platform for scaling up of SE in coconut; however, there have been several technical and legislative factors that have prevented it being reproduced in other laboratories. Recent advances in smart farming, with technologies such as a vertical farming system and artificial intelligence, might help improve the effectiveness of the existing biofactory model. It is also worth noting that such large intellectual and capital investments will probably need to come from many stakeholders. Recently, a public–private partnership was established to facilitate the adoption of SE technology in coconut (Kalaipandian *et al.*, 2021).

Limitations of coconut SE

Research achievements on coconut SE leading to the mass multiplication of coconut plantlets has

Fig. 10.3. Somatic embryogenesis of high-yielding CRIC 65 hybrid coconut (a cross between Sri Lankan Tall and Sri Lankan Green Dwarf) using ovary tissues. (A) Immature inflorescences (circled: –4 developmental stage). (B) A female flower (at centre) in rachis. (C) Ovary tissue. (D) An embryogenic callus. (E) A somatic embryo. (F) A germinating somatic embryo. (G) Shoot and root development. (H) Plantlets ready for acclimatization. (I) Clonal plants ready for field planting. (J) A clonal palm regenerated from somatic embryogenesis.

progressed slowly compared with other crops. This is due to reasons such as the heterogeneous response of the different coconut explants used, the slow growth of its tissues *in vitro* and the undefined culture conditions created by the incorporation of activated charcoal into the culture medium (Fernando *et al.*, 2010b). According to Ebert and Taylor (1990), *in vitro*-cultured tissues of coconut exhibit marked differences in their growth responses due to the variation in 2,4-D concentration used in the culture medium as a result of the presence of activated charcoal in the same medium. In addition, the production of non-embryogenic callus and the formation of incomplete somatic embryos limits the success of SE and the conversion of somatic embryos into clonal plants. The plumule explant tissue has been identified as the most responsive tissue, yet the production of embryogenic callus is often low (40–60%), due in part to the development of non-embryogenic callus (Sáenz-Carbonell *et al.*, 2013). Formation of aberrant somatic embryos without a shoot meristem or with fused cotyledonary leaves has also been reported by Sáenz *et al.* (2006) and Jayaraj *et al.* (2015).

The SE-derived plantlets are often fragile during the deflasking phase, and this implies that *ex vitro* acclimatization is of foremost importance to improve the whole SE process. Research has shown that acclimatization approaches (i.e. an atmospheric enrichment with CO_2 in a mini growth chamber) can improve plantlet production, as well as speed up the process (Samosir and Adkins, 2014; Sisunandar *et al.*, 2018). The use of an appropriate potting substrate is often dependent on what is available to the practitioners. Excellent substrates include rice husk charcoal, coco peat and vermiculite (Sisunandar *et al.*, 2018). However, a good balance between moisture retention and drainage should be attained, together with sufficient availability of macro- and micronutrients.

The genetic stability of the SE-derived plantlets needs to be confirmed prior to planting, as field determination may take many years to establish (Kalaipandian *et al.*, 2021). Unwanted abnormalities could result in fruit losses; therefore, pre-screening must be undertaken rigorously to minimize the chances of planting out off-types. This concern is often raised by the industry partners, pushing back full commitment to the scale-up of SE technology in coconut. In addition, an assessment of field performance of these plantlets is highly recommended for productivity assurance.

10.2.3 Direct shoot organogenesis

The differentiation of a single cell or a cluster of cells to form shoots or roots is termed organogenesis (Su, 2002). *In vitro* organogenesis can theoretically come from one of three origins: axillary, adventitious or *de novo*. An axillary origin involves pre-formed meristematic cells, while an adventitious origin does not, and *de novo* origin refers to those that arise from an unorganized proliferating cell mass such as a callus (Phillips and Garda, 2019). Direct shoot organogenesis, which involves the shoot formation directly from a tissue without a callus intervention, regardless of whether it is of axillary or adventitious origin, has the potential to be a clonal propagation pathway for coconut (Kong *et al.*, 2021).

Vegetative propagation of coconut was originally only thought to be achievable through SE (Tisserat, 1991). However, there are some exceptions of coconut palms producing multiple growing shoots (Kong *et al.*, 2021), sparking a rethink of the potential of coconut explants producing multiple shoots by tissue culture. Such shoot multiplication systems could avoid putative somaclonal variation. Direct shoot organogenesis that develops from pre-formed meristematic cells has a lower chance of accumulating spontaneous mutations (Phillips and Garda, 2019). Hence, cloning plantlets through direct shoot organogenesis may be a more secure pathway compared with SE, as the risk of somaclonal variation is likely to be lower (Abahmane, 2011). So far, little research has been undertaken on propagating coconut by direct shoot organogenesis (Vidhanaarachchi and Weerakoon, 1997; Raju, 2006; Chandrakala *et al.*, 2019; Shareefa *et al.*, 2019; Wilms *et al.*, 2021). To date, only explants with pre-existing meristems (e.g. immature inflorescence sections, zygotic embryos and young shoots developed from zygotic embryos) have been used successfully.

A recent and very promising development is the establishment of an axillary shoot multiplication system starting from *in vitro* seedlings derived from embryos (Wilms *et al.*, 2021). For this, 3-month-old *in vitro* plants are spliced and these halves are grown on a thidiazuron-containing medium (Fig. 10.4). The resulting 'cauliflower-like' meristem clumps can be multiplied en masse and/or regenerated into plants. This method comes with many advantages compared with the existing *in vitro* multiplication protocols. It is user friendly, efficient (20–30% of the explants respond) and appears to be variety independent (applicable to all varieties studied to date), and the plants are likely to be less prone to somaclonal variation. Although the protocol is promising, the mass clonal propagation of coconut by direct shoot organogenesis still requires much work to improve, optimize and genetically assess this pathway before it can be practically and commercially implemented.

10.3 Biotechnological Approaches to Conserve Coconut Germplasm

To date, coconut germplasm conservation has relied on classical *in situ* conservation and *ex situ* field gene banks. Alternative approaches using biotechnological techniques including *in vitro* slow-growth conservation and cryopreservation are still in the pilot phase (reviewed by Bourdeix *et al.*, 2020). An overview of the biotechnological techniques such as *in vitro* culture and cryopreservation that can be applied to the conservation of coconut germplasm can be found in Chapter 9 (this volume).

Fig. 10.4. Direct shoot multiplication using Malayan Yellow Dwarf coconut seedlings. (A) Halved shoots derived from zygotic embryos cultured on medium containing thidiazuron. (B) Proliferating meristem clumps after 6 weeks. (C) Shoot clusters formed on thidiazuron-free medium. (D) Regenerated shoots. Photos courtesy of H. Wilms (KU Leuven, Belgium).

If applied successfully, these biotechnological approaches would be able to sustainably maintain germplasm for many years and, importantly, ensure that no genetic change occurs during this time of storage (Bourdeix *et al.*, 2020). It is known that certain conservation approaches can allow environmental stresses to affect plant cells, with the cells modifying their genome as a response to these stresses, allowing a phenotypic change or a 'silent' mutation to occur (McClintock, 1984; Phillips *et al.*, 1994). It has been shown that various changes at the phenotypic, cytological, biochemical and molecular levels can be induced by tissue culture (McClintock, 1984; Phillips *et al.*, 1994; Rani and Raina, 2000). Until now, no genetic instabilities have been observed within the coconut plants produced through tissue culture, at either the phenotypic or molecular level (Bandupriya *et al.*, 2017; Shareefa *et al.*, 2019). Similarly, no genetic instabilities have been observed in coconut plants recovered from short- to medium-term conservation (Sisunandar *et al.*, 2012).

In coconut, the genetic fidelity of the plants recovered from cryopreservation has been studied in a number of ways (Sisunandar *et al.*, 2010). No significant differences were found between plants recovered from cryopreservation and from non-cryopreserved plants, based on morphological observations made on four coconut varieties. Similarly, at the cytological level, no significant changes were observed for plants recovered from cryopreservation. Similarity was seen in chromosome number, the total short and long arm lengths, and the centromeric index. A genetic fidelity assessment made using microsatellite markers generated from ten SSR primers showed no significant difference in allele frequency between non-cryopreserved and cryopreserved materials. Moreover, epigenetic assessment, by measuring global DNA methylation rates, also showed similarity between plants recovered from cryopreserved and non-cryopreserved embryos.

10.4 Genomics for Crop Improvement, Pest and Disease Tolerance

Whole-genome sequencing of both tall and dwarf coconut types has created the possibility of a better understanding of the role of genes and is likely to lead to the development of improved varieties (Xiao *et al.*, 2017; Lantican *et al.*, 2019). Genome-wide identification of molecular markers and annotation of genes will be an invaluable resource for crop improvement against various biotic and abiotic stresses (Lantican *et al.*, 2019; Rajesh *et al.*, 2020). The Hainan Tall variety was the first coconut variety to be sequenced, and was predicted to have around 28,039 protein-coding genes in its genome, representing 74% of genome annotation for the predicted genes (Xiao *et al.*, 2017). In the Catigan Green Dwarf variety, whole-genome sequencing predicted about 34,958 genes associated with various traits including biotic and abiotic stress resistance and oil content. Furthermore, about 7139 SSR markers were designed for marker-assisted breeding and 58,503 variants were identified by comparing Hainan Tall and Catigan Green Dwarf genome sequences (Lantican *et al.*, 2019). Another whole-genome sequencing study of Chowghat Green Dwarf predicted that around 13,707 genes were estimated to code for 11,181 proteins. Of these, 6828 coding genes were confirmed by studying the RNA-sequencing data of various coconut tissues (Rajesh *et al.*, 2020). These genomic studies have led to the identification of genes for various traits in coconut.

Wang *et al.* (2021) recently performed DNA sequencing, transcriptome and metabolome analysis, and demonstrated that the fibre synthesis genes (e.g. lignin and cellulose genes) were present at a higher copy number and therefore more highly expressed in a tall variety, compared with a dwarf variety, which might explain the higher fibre content of tall types. Genome-wide association studies of tall and dwarf varieties have shown a single-nucleotide change in the promoter region of the gibberellin 20-oxidase gene. This might explain the differences in coconut plant height of the two types (Wang *et al.*, 2021). Genetic mapping of a biparental mapping population using the method of genotyping by sequencing has revealed that coconut could have gone through a substantial invasion by transposable elements in the last million years, and a fluctuation in sea levels could be one of the reasons for this. A time-course RNA-sequencing profile experiment for tall and dwarf varieties under salt stress yielded significant results to support this hypothesis (Yang *et al.*, 2021).

Flowering time is one of the most complicated traits in plants. As coconut takes several years to maturity, it is important to understand the molecular mechanisms for the control of flowering time to manipulate this characteristic in coconut. Transcriptome sequencing of leaves at the seedling and reproductive stages of tall and dwarf coconuts revealed the differential expression of key genes in the photoperiod pathway, including suppressor of overexpression of constans (*SOC1*), flowering locus T (*FT*) and Apetala 1 (*AP1*). Alternative splicing revealed that the transcripts of the *FT* gene were different in tall and dwarf coconut, where a shorter alternative splice variant, with a 6 bp deletion, was found in dwarf types but was absent in most tall types (Xia *et al.*, 2020).

Molecular markers are very powerful and could dramatically increase the efficiency of germplasm conservation, genotyping and marker-assisted breeding for desirable traits. Wu *et al.* (2019) developed approximately 84 SSR markers with a selectively amplified microsatellite technique with the utilization of a recently sequenced whole genome of coconut and confirmed more than 90% had transferability within the palm family. The development of more molecular markers will be helpful for germplasm conservation and evaluation, and for constructing genetic linkage maps with good coverage in every chromosome and for the designing of breeding programmes for producing superior varieties of coconut (Wu *et al.*, 2019). A genome-wide genotyping-by-sequencing study developed more than 19,000 single-nucleotide polymorphisms (SNPs) from a panel with 112 accessions from the Atlantic and Pacific gene pools. Around 10,338 and 4606 SNPs pertained to Atlantic and Pacific germplasm, respectively, and 3432 SNPs were found to differentiate these two gene pools (Muñoz-Pérez *et al.*, 2022).

Bogia coconut syndrome (BCS), first found in Papua New Guinea, is a type of lethal yellowing disease caused by phytoplasma. Phytoplasmas pose a substantial threat to the coconut industry worldwide. BCS-associated phytoplasmas were collected from different regions in Papua New Guinea and their 16S rRNA gene sequences analysed. This revealed a high identity among them represented by strain BCS-Bo^{R} (Miyazaki *et al.*, 2018). The molecular study of these organisms could help to identify alternative hosts and help prevent spreading. In another study of coconut foliar decay disease, caused by coconut foliar decay virus, DNA sequencing showed a difference in this virus from those of the family *Geminiviridae* or *Nanoviridae*, which suggests a new genus and family (Gronenborn *et al.*, 2018). Dual RNA sequencing of both the infecting oomycete (*Phytophthora palmivora*) and the infected host tissues (coconut leaflets) provided data to compare virulence of a fungus and the host response mechanism (Gangaraj and Rajesh, 2020).

Coconut rhinoceros beetle (CRB) is currently one of the most dangerous pests of coconut worldwide. A new invasive haplotype of CRB found in Guam (CRB-G) was shown to be insensitive to *Oryctes rhinoceros* nudivirus (OrNV), a previously effective biocontrol agent. A draft genome study of CRB-G identified a total of 21,999 putative SSR markers and will be an invaluable resource for studying the population genetics, dispersal patterns, phylogenetics and species behaviour (Rajesh *et al.*, 2021). In addition, the OrNV genome was sequenced and annotated, helping to elucidate the mechanisms that compromised the efficacy of OrNV as a biocontrol agent against CRB (Etebari *et al.*, 2020).

10.5 Conclusion

Compared with other important horticultural crops, coconut biotechnological research and its applications are in their infancy. The first tissue-culture technique applied to coconut, for zygotic EC, was developed 60 years ago and proved to be the ideal tool to propagate elite Makapuno-type varieties. SE in coconut was first reported about 40 years ago and was established with the main aim of clonally propagating elite coconut palms on a large scale. However, this technique has not been developed to a commercial scale. More recently, there have been efforts to develop a clonal multiplication system based on axillary shoot multiplication or regeneration of plants from immature inflorescences. These might provide welcome alternatives but are still subject to further improvement before they can be applied commercially. It is a similar story for the biotechnologically-driven conservation techniques, such as *in vitro* conservation and cryopreservation. Promising results have been reported, particularly for the cryopreservation of whole embryos, but their widespread application has not yet been implemented in coconut gene banks. Any work on genetic engineering or gene editing of coconut is likely to be far in the future due to the absence of a suitable single-cell regeneration system such as embryogenic cell-suspension culture. In addition, some of these tools may not match the organic, clean and green image that coconut currently has.

This chapter has demonstrated that, to keep pace with the developments of other important horticultural crops and to develop varieties that are resistant to biotic and abiotic stresses, thus improving the livelihoods of (often poor, small-scale) farmers, substantially more biotechnological coconut research is needed. This can only be realized by investing more resources and enhanced collaboration among the different research groups working on coconut biotechnological techniques and conservation.

10.6 References

Abahmane, L. (2011) Date palm micropropagation via organogenesis. In: Jain, S.M., Al-Khayri, J.M. and Johnson, D.V. (eds) *Date Palm Biotechnology*. Springer, Dordrecht, Netherlands, pp. 69–90.

Andrade-Torres, A., Oropeza, C., Sáenz, L., González-Estrada, T., Ramírez-Benítez, J. E. *et al.* (2011) Transient genetic transformation of embryogenic callus of *Cocos nucifera*. *Biologia* 66, 790–800.

Ashburner, G.R., Thompson, W.K. and Burch, J.M. (1993) Effect of α-naphthaleneacetic acid and sucrose levels on the development of cultured embryos of coconut. *Plant Cell, Tissue and Organ Culture* 35, 157–163.

Ashburner, G.R., Faure, M.G., Tomlinson, D.R. and Thompson, W.K. (1995) *A Guide to the Zygotic Embryo Culture of Coconut Palms* (Cocos nucifera L.). ACIAR Technical Reports 36. Australian Centre for International Agricultural research (ACIAR), Bruce, ACT, Australia.

Azpeitia, A., Chan, J.L., Sáenz, L. and Oropeza, C. (2003) Effect of 22(S),23(S)-homobrassinolide on somatic embryogenesis in plumule explants of *Cocos nucifera* (L.) cultured *in vitro*. *Journal of Horticultural Science and Biotechnology* 78, 591–596.

Bajaj, Y.P.S. (1984) Induction of growth in frozen embryos of coconut and ovules of citrus. *Current Science* 53, 1215–1216.

Bandupriya, H.D.D., Iroshini, W.W.M.A., Perera, S.A.C.N., Vidhanaarachchi, V.R.M., Fernando, S.C. *et al.* (2017) Genetic fidelity testing using SSR marker assay confirms trueness to type of micropropagated coconut (*Cocos nucifera* L.) plantlets derived from unfertilized ovaries. *Open Plant Science Journal* 10, 46–54.

Batugal, P., Bourdeix, R. and Baudouin, L. (2009) Coconut breeding. In: Jain, S.M. and Priyadarshan, P.M. (eds) *Breeding Plantation Tree Crops: Tropical Species*. Springer, New York, pp. 327–375.

Bazrafshan, A. (2022) *Ex situ* conservation of coconut (*Cocos nucifera* L.) germplasm using cryopreservation. PhD thesis, The University of Queensland, Brisbane, Australia.

Bedford, G. (2013) Long-term reduction in damage by rhinoceros beetle *Oryctes rhinoceros* (L.) (Coleoptera: Scarabaeidae: Dynastinae) to coconut palms at *Oryctes* Nudivirus release sites on Viti Levu, *Fiji. African Journal of Agricultural Research* 8, 6422–6425.

Bhalla-Sarin, N., Bagga, S., Sopory, S.K. and Guha-Mukherjee, S. (1986) Induction and differentiation of callus from embryos of *Cocos nucifera* L. by IAA-conjugates. *Plant Cell Reports* 5, 322–324.

Biddle, J., Nguyen, Q., Mu, Z., Foale, M. and Adkins, S. (2020) Germplasm reestablishment and seedling production: embryo culture. In: Adkins, S., Foale, M., Bourdeix, R., Nguyen, Q. and Biddle, J. (eds) *Coconut Biotechnology: Towards the Sustainability of the 'Tree of Life'*. Springer, Cham, Switzerland, pp. 199–225.

Bourdeix, R., Adkins, S., Johnson, V., Perera, L. and Sisunandar. (2020) *In situ* and *ex situ* conservation of coconut genetic resources. In: Adkins, S., Foale, M., Bourdeix, R., Nguyen, Q. and Biddle, J. (eds) *Coconut Biotechnology: Towards the Sustainability of the 'Tree of Life'*. Springer, Cham, Switzerland, pp. 51–75.

Branton, R.L. and Blake, J. (1983) Development of organized structures in callus derived from explants of *Cocos nucifera* L. *Annals of Botany* 52, 673–678.

Chan, J.L., Saénz, L., Talavera, C., Hornung, R., Robert, M. *et al.* (1998) Regeneration of coconut (*Cocos nucifera* L.) from plumule explants through somatic embryogenesis. *Plant Cell Reports* 17, 515–521.

Chandrakala, D., Renuka, R. and Sushmitha, D.S. (2019) Influence of 2,4-D and TDZ on direct organogenesis in coconut var. East Coast Tall. *International Journal of Chemical Studies* 7, 4111–4115.
Chin, H.F., Krishnapillay, B. and Hor, Y.L. (1989) A note on the cryopreservation of embryos from young coconuts (*Cocos nucifera* var. Mawa). *Pertanika* 12, 183–186.
Cueto, C.A., Johnson, V.B., Engelmann, F., Kembu, A., Konan, J.L. *et al.* (2012) Technical guidelines for the safe movement and duplication of coconut (*Cocos nucifera* L.) germplasm using embryo culture transfer protocols. COGENT; Bioversity International, Montpellier, France.
Cutter, V.M.J. and Wilson, K.S. (1954) Effect of coconut endosperm and other growth stimulants upon the development *in vitro* of embryos of *Cocos nucifera*. *Botanical Gazette* 115, 234–240.
de Guzman, E.V. and del Rosario, D.A. (1964) The growth and development of *Cocos nucifera* L. makapuno embryo *in vitro*. *Philippine Agriculturist* 48, 82–94.
Ebert, A. and Taylor, H.F. (1990) Assessment of the changes of 2,4-dichlorophenoxyacetic acid concentrations in plant tissue culture media in the presence of activated charcoal. *Plant Cell, Tissue and Organ Culture* 20, 165–172.
Eeuwens, C.J. (1976) Mineral requirements for growth and callus initiation of tissue explants excised from mature coconut palms (*Cocos nucifera* L.) and cultured *in vitro*. *Physiologia Plantarum* 36, 23–28.
Eeuwens, C.J. and Blake, J. (1977) Culture of coconut and date palm tissue with a view to vegetative propagation. *Acta Horticulturae* 78, 277–286.
Engelmann, F. and Batugal, P. (2002) Background on the development and implementation of the coconut embryo *in vitro* culture project. In: Engelmann, F., Batugal, P. and Oliver, J. (eds) *Coconut Embryo In Vitro Culture: Part 2*. International Plant Genetic Resources Institute – Regional Office for Asia, the Pacific and Oceania (IPGRI-APO), Serdang, Malaysia, pp. 1–4.
Etebari, K., Filipović, I., Rašić, G., Devine, G.J., Tsatsia, H. *et al.* (2020) Complete genome sequence of *Oryctes rhinoceros* nudivirus isolated from the coconut rhinoceros beetle in Solomon Islands. *Virus Research* 278: 197864.
Fernando, S.C., Weerakoon, L.K., Perera, P.I.P., Bandupriya, H.D.D., Ambagala, I.E. *et al.* (2004) Genetic fidelity and *ex vitro* performance of tissue-cultured coconut plants. In: *Proceedings of the International Conference of the Coconut Research Institute of Sri Lanka – Part II*, 8–11 September, Colombo, Sri Lanka. Coconut Research Institute, Lunuwila, Sri Lanka.
Fernando, S.C., Santha, E.S. and Hewarathna, D.J.A. (2010a) Activated coconut shell charcoal as a component of tissue culture media of *Cocos nucifera* L. *Journal of The National Science Foundation of Sri Lanka* 38, 181–185.
Fernando, S.C., Vidhanaarachchi, V.R.M., Weerakoon, L.K. and Santha, E.S. (2010b) What makes clonal propagation of coconut difficult? *Asia-Pacific Journal of Molecular Biology and Biotechnology* 18, 163–165.
Foale, M., Biddle, J., Bazrafshan, A. and Adkins, S. (2020) Biology, ecology, and evolution of coconut. In: Adkins, S., Foale, M., Bourdeix, R., Nguyen, Q. and Biddle, J. (eds) *Coconut Biotechnology: Towards the Sustainability of the 'Tree of Life'*. Springer, Cham, Switzerland, pp. 17–27.
Frison, E., Putter, C.A.J. and Diekmann, M. (eds) (1993) *FAO/IBPGR Technical Guidelines for the Safe Movement of Coconut Germplasm*. Food and Agriculture Organization of the United Nations, Rome/International Board for Plant Genetic Resources, Rome.
Gaj, M.D. (2004) Factors influencing somatic embryogenesis induction and plant regeneration with particular reference to *Arabidopsis thaliana* (L.) Heynh. *Plant Growth Regulation* 43, 27–47.
Gangaraj, K.P. and Rajesh, M.K. (2020) Dataset of dual RNA-sequencing of *Phytophthora palmivora* infecting coconut (*Cocos nucifera* L.). *Data in Brief* 30: 105455.
Gronenborn, B., Randles, J.W., Knierim, D., Barrière, Q., Vetten, H.J. *et al.* (2018) Analysis of DNAs associated with coconut foliar decay disease implicates a unique single-stranded DNA virus representing a new taxon. *Scientific Reports* 8: 5698.
Gurr, G.M., Johnson, A.C., Ash, G.J., Wilson, B.A.L., Ero, M.M. *et al.* (2016) Coconut lethal yellowing diseases: a phytoplasma threat to palms of global economic and social significance. *Frontiers in Plant Science* 7: 1521.
Hebbar, K.B., Balasimha, D. and Thomas, G.V. (2013) Plantation crops response to climate change: coconut perspective. In: Singh, H.C.P., Rao, N.K.S. and Shivashankara, K.S. (eds) *Climate-resilient Horticulture: Adaptation and Mitigation Strategies*. Springer, New Delhi, pp. 177–187.
Hussain, A., Qarshi, I.A., Nazir, H. and Ullah, I. (2012) Plant tissue culture: current status and opportunities. In: Leva, A. and Rinaldi, L. (eds) *Recent Advances in Plant In Vitro Culture*. InTech Open, London, pp. 1–28.

Jayaraj, K.L., Bhavyashree, U., Fayas, T.P., Sajini, K.K., Rajesh, M.K. *et al.* (2015) Histological studies of cellular differentiation during somatic embryogenesis of coconut plumule-derived calli. *Journal of Plantation Crops* 43, 196–203.

Kalaipandian, S., Mu, Z., Kong, E.Y.Y., Biddle, J., Cave, R. *et al.* (2021) Cloning coconut via somatic embryogenesis: a review of the current status and future prospects. *Plants* 10: 2050.

Karun, A. (2017) Coconut tissue culture: the Indian initiatives, experiences and achievements. *CORD* 33: 11.

Karun, A., Sajini, K.K., Nair, M., Kumaran, P.M. and Samsudheen, K. (2006) Cryopreservation of coconut (*Cocos nucifera* L.) pollen. *Journal of Plantation Crops* 34, 568–571.

Karun, A., Sjini, K.K., Niral, V., Amarnth, C.H., Remya, P. *et al.* (2014) Coconut (*Cocos nucifera* L.) pollen cryopreservation. *CryoLetters* 35, 407–417.

Karunaratne, S. and Periyapperuma, K. (1989) Culture of immature embryos of coconut, *Cocos nucifera* L.: callus proliferation and somatic embryogenesis. *Plant Science* 62, 247–253.

Kong, E.Y.Y., Biddle, J., Foale, M. and Adkins, S.W. (2020) Cell suspension culture: a potential *in vitro* culture method for clonal propagation of coconut plantlets via somatic embryogenesis. *Industrial Crops and Products* 147, 112–125.

Kong, E.Y.Y., Biddle, J., Foale, M., Panis, B. and Adkins, S.W. (2021) The potential to propagate coconut clones through direct shoot organogenesis: a review. *Scientia Horticulturae* 289: 110400.

Lantican, D.V., Strickler, S.R., Canama, A.O., Gardoce, R.R., Mueller, L.A. *et al.* (2019) *De novo* genome sequence assembly of dwarf coconut (*Cocos nucifera* L. 'Catigan Green Dwarf') provides insights into genomic variation between coconut types and related palm species. *G3 Genes, Genomes, Genetics* 9, 2377–2393.

Lédo, A.S. and Vendrame, W.A. (2021) Coconut micropropagation and cryopreservation. *Horticultural Reviews* 48, 307–337.

Magdalita, P.M., Damasco, O.P., Samosir, Y.M.S., Mashud, N., Novariento, H. *et al.* (2015) An enhanced embryo culture methodology for coconut (*Cocos nucifera* L.). *International Journal of Innovation Sciences and Research* 4, 485–493.

McClintock, B. (1984) The significance of responses of the genome to challenge. *Science* 226, 792–801.

Misra, S. (1994) Conifer zygotic embryogenesis, somatic embryogenesis, and seed germination: biochemical and molecular advances. *Seed Science Research* 4, 357–384.

Miyazaki, A., Shigaki, T., Koinuma, H., Iwabuchi, N., Rauka, G.B. *et al.* (2018) '*Candidatus* phytoplasma noviguineense', a novel taxon associated with Bogia coconut syndrome and banana wilt disease on the island of New Guinea. *International Journal of Systematic and Evolutionary Microbiology* 68, 170–175.

Montero-Córtes, M., Sáenz, L., Córdova, I., Quiroz, A., Verdeil, J.L. *et al.* (2010) GA_3 stimulates the formation and germination of somatic embryos and the expression of a *KNOTTED*-like homeobox gene of *Cocos nucifera* (L.). *Plant Cell Reports* 29, 1049–1059.

Morcillo, F., Gagneur, C., Adam, H., Richaud, F., Singh, R. *et al.* (2006) Somaclonal variation in micropropagated oil palm. Characterization of two novel genes with enhanced expression in epigenetically abnormal cell lines and in response to auxin. *Tree Physiology* 26, 585–594.

Mu, Z., Guo, X., Biddle, J., Foale, M., Li, Z. *et al.* (2020) A newly designed chamber for the acclimatization of coconut plantlets coming from *in vitro*. *In Vitro Cellular and Developmental Biology – Animal* 56: 44.

Muñoz-Pérez, J.M., Cañas, G.P., López, L. and Arias, T. (2022) Genome-wide diversity analysis to infer population structure and linkage disequilibrium among Colombian coconut germplasm. *Scientific Reports* 12: 2958.

Nayar, N.M. (2017) The coconut in the world. In: *The Coconut: Phylogeny, Origins, and Spread*. Elsevier Science and Technology, San Diego, California, pp. 1–8.

Nguyen, Q.T., Bandupriya, H.D.D., López-Villalobos, A., Sisunandar, S., Foale, M. *et al.* (2015) Tissue culture and associated biotechnological interventions for the improvement of coconut (*Cocos nucifera* L.): a review. *Planta* 242, 1059–1076.

Nguyen, Q.T., Bandupriya, H.D.D., Foale, M. and Adkins, S.W. (2016) Biology, propagation and utilization of elite coconut varieties (makapuno and aromatics). *Plant Physiology and Biochemistry* 109, 579–589.

N'Nan, O., Hocher, V., Verdeil, J.L., Konan, J.L., Ballo, K. *et al.* (2008) Cryopreservation by encapsulation–dehydration of plumules of coconut (*Cocos nucifera* L.). *CryoLetters* 29, 339–350.

Perera, P.I.P., Hocher, V., Verdeil, J.L., Doulbeau, S., Yakandawala, D.M.D. *et al.* (2007) Unfertilized ovary: a novel explant for coconut (*Cocos nucifera* L.) somatic embryogenesis. *Plant Cell Reports* 26, 21–28.

Perera, P.I.P., Vidhanaarachchi, V.R.M., Gunathilake, T.R., Yakandawala, D.M.D., Hocher, V. *et al.* (2009) Effect of plant growth regulators on ovary culture of coconut (*Cocos nucifera* L.). *Plant Cell, Tissue and Organ Culture* 99, 73–81.

Pérez-Núñez, M.T., Chan, J.L., Sáenz, L., González, T., Verdeil, J.L. *et al.* (2006) Improved somatic embryogenesis from *Cocos nucifera* (L.) plumule explants. *In Vitro Cellular and Developmental Biology – Plant* 42, 37–43.

Pham, L.J. (2016) Coconut (*Cocos nucifera*). In: McKeon, T.A., Hayes, D.G., Hildebrand, D.F. and Weselake, R.J. (eds) *Industrial Oil Crops*. AOCS Press, Urbana, Illinois, pp. 231–242.

Phillips, G.C. and Garda, M. (2019) Plant tissue culture media and practices: an overview. *In Vitro Cellular and Developmental Biology – Plant* 55, 242–257.

Phillips, R.L., Kaeppler, S.M. and Olhoft, P. (1994) Genetic instability of plant tissue cultures: breakdown of normal controls. *Proceedings of the National Academy of Sciences U S A* 91, 5222–5226.

Preece, J.E. and Sutter, E.G. (1991) Acclimatization of micropropagated plants to the greenhouse and field. In: Debergh, P.C. and Zimmerman, R.H. (eds) *Micropropagation: Technology and Application*. Springer, Dordrecht, Netherlands, pp. 71–93.

Rajesh, M.K., Radha, E., Sajini, K.K. and Karun, A. (2014) Polyamine-induced somatic embryogenesis and plantlet regeneration *in vitro* from plumular explants of dwarf cultivars of coconut (*Cocos nucifera*). *Indian Journal of Agricultural Sciences* 84, 527–530.

Rajesh, M.K., Chowdappa, P., Behera, S.K., Kasaragod, S., Gangaraj, K.P. *et al.* (2020) Assembly and annotation of the nuclear and organellar genomes of a dwarf coconut (Chowghat Green Dwarf) possessing enhanced disease resistance. *OMICS* 24, 726–742.

Rajesh, M.K., Antony, G., Arvind, K., Godwin, J., Gangaraj, K.P. *et al.* (2021) Draft genome sequence, annotation and SSR mining data of *Oryctes rhinoceros* Linn. (Coleoptera: Scarabaeidae), the coconut rhinoceros beetle. *Data in Brief* 38: 107424.

Raju, C.R. (2006) Direct *in vitro* shoot development from immature rachilla explants of coconut (*Cocos nucifera* L.). *CORD* 22, 51–58.

Rani, V. and Raina, S.N. (2000) Genetic fidelity of organized meristem-derived micropropagated plants: a critical reappraisal. *In Vitro Cellular and Developmental Biology – Plant* 36, 319–330.

Rillo, E.P. (1998) PCA's embryo culture technique in the mass production of Makapuno coconuts. In: Batugal, P.A. and Engelmann, F. (eds) *Coconut Embryo In Vitro Culture: Part I. Proceedings of the First Workshop on Embryo Culture*, Banao, Guinobatan, Albay, Philippines 27–31 October 1997. International Plant Genetic Resources Institute (IPGRI), Rome, pp. 69–78.

Sáenz, L., Azpeitia, A., Chuc-Armendariz, B., Chan, J.L., Verdeil, J.L. *et al.* (2006) Morphological and histological changes during somatic embryo formation from coconut plumule explants. *In Vitro Cellular and Developmental Biology – Plant* 42, 19–25.

Sáenz-Carbonell, L., Montero-Cortes, M., Pérez-Nuñez, T., Azpeitia-Morales, A., Andrade-Torres, A. *et al.* (2013) Coconut (*Cocos nucifera* L.) somatic embryogenesis and related gene expression. In: Aslam, J., Srivastava, P.S. and Sharma, M.P. (eds) *Somatic Embryogenesis and Gene Expression*. Narosa Publishing House, New Delhi, pp. 172–187.

Sáenz-Carbonell, L., Nguyen, Q., López-Villalobos, A. and Oropeza-Salín, C. (2020) Coconut micropropagation for worldwide replanting needs. In: Adkins, S., Foale, M., Bourdeix, R., Nguyen, Q. and Biddle, J. (eds) *Coconut Biotechnology: Towards the Sustainability of the 'Tree of Life'*. Springer, Cham, Switzerland, pp. 227–240.

Salum, U., Foale, M., Biddle, J., Bazrafshan, A. and Adkins, S. (2020) Towards the sustainability of the 'Tree of Life': an introduction. In: Adkins, S., Foale, M., Bourdeix, R., Nguyen, Q. and Biddle, J. (eds) *Coconut Biotechnology: Towards the Sustainability of the 'Tree of Life'*. Springer, Cham, Switzerland, pp. 1–15.

Samosir, Y.M.S. (1999) Optimisation of somatic embryogenesis in coconut (*Cocos nucifera* L.). PhD thesis, The University of Queensland, Queensland, Australia.

Samosir, Y.M.S. and Adkins, S. (2014) Improving acclimatization through the photoautotrophic culture of coconut (*Cocos nucifera*) seedlings: an *in vitro* system for the efficient exchange of germplasm. *In Vitro Cellular and Developmental Biology – Plant* 50, 493–501.

Samosir, Y., Rival, A. and Adkins, S.W. (2008) The survival and the genetic fidelity of coconut embryos after cryopreservation. In: *Sixth International Symposium on In Vitro Culture and Horticultural Breeding*, 24–28 August 2008, Brisbane, Australia, p. 43.

Sandoval-Cancino, G., Sáenz, L., Chan, J.L. and Oropeza, C. (2016) Improved formation of embryogenic callus from coconut immature inflorescence explants. *In Vitro Cellular and Developmental Biology – Plant* 52, 367–378.

Shareefa, M., Thomas, R.J., Sreelekshmi, J.S., Rajesh, M.K. and Karun, A (2019) *In vitro* regeneration of coconut plantlets from immature inflorescence. *Current Science* 117, 813–820.

Sisunandar, Rival, A., Turquay, P., Samosir, Y. and Adkins, S.W. (2010) Cryopreservation of coconut (*Cocos nucifera* L.) zygotic embryos does not induce morphological, cytological or molecular changes in recovered seedlings. *Planta* 232, 435–447.

Sisunandar, Sopade, P.A., Samosir, Y.M.S., Rival, A. and Adkins, S.W. (2012) Conservation of coconut (*Cocos nucifera* L.) germplasm at sub-zero temperature. *CryoLetters* 33, 465–475.

Sisunandar, Husin, A., Julianto, T., Yuniaty, A., Rival, A. *et al.* (2018) *Ex vitro* rooting using a mini growth chamber increases root induction and accelerates acclimatization of Kopyor coconut (*Cocos nucifera* L.) embryo culture-derived seedlings. *In Vitro Cellular and Developmental Biology – Plant* 54, 508–517.

Solís-Ramos, L.Y., Andrade-Torres, A., Sáenz Carbonell, L.A., Oropeza Salín, C.M. and Castaño de la Serna, E. (2012) *Somatic Embryogenesis in Recalcitrant Plants*. IntechOpen, Rijeka, Croatia, pp. 597–618.

Su, W.W. (2002) Cell culture and regeneration of plant tissues. In: Khachatourians, G.G., McHughen, A., Scorza, R., Nip, W.-K. and Hui, Y.H. (eds) *Transgenic Plants and Crops*. Marcel Dekker, New York, pp. 151–167.

Tautorus, T., Fowke, L. and Dunstan, D. (1991) Somatic embryogenesis in conifers. *Canadian Journal of Botany* 69, 1873–1899.

Tisserat, B. (1991) Clonal propagation of palms. In: Lindsey, K. (ed) *Plant Tissue Culture Manual*. Springer, Dordrecht, Netherlands, pp. 1–14.

Verdeil, J.-L., Huet, C., Grosdemange, F. and Buffard-Morel, J. (1994) Plant regeneration from cultured immature inflorescences of coconut (*Cocos nucifera* L.): evidence for somatic embryogenesis. *Plant Cell Reports* 13, 218–221.

Vidhana Arachchi, V.R.M. and Weerakoon, L.K. (1997) Callus induction and direct shoot formation in *in vitro* cultured immature inflorescence tissues of coconut. *COCOS* 12, 39–43.

Vidhana Arachchi, V., Fernando, S.C., Perera, P. and Weerakoon, L.K. (2013) Application of un-fertilized ovary culture to identify elite mother palms of *Cocos nucifera* L. with regenerative potential. *Journal of the National Science Foundation of Sri Lanka* 41, 29–34.

Wang, S., Xiao, Y., Zhou, Z.-W., Yuan, J., Guo, H. *et al.* (2021) High-quality reference genome sequences of two coconut cultivars provide insights into evolution of monocot chromosomes and differentiation of fiber content and plant height. *Genome Biology* 22: 304.

Wilms, H., Rhee, J.H., Rivera, R.L., Longin, K. and Panis, B. (2019) Developing coconut cryopreservation protocols and establishing a cryo-genebank at RDA; a collaborative project between RDA and Bioversity International. *Acta Horticulturae* 1234, 343–348.

Wilms, H., de Bièvre, D., Longin, K., Swennen, R., Rhee, J. *et al.* (2021) Development of the first axillary *in vitro* shoot multiplication protocol for coconut palms. *Scientific Reports* 11: 18367.

Wu, Y., Yang, Y., Qadri, R., Iqbal, A., Li, J. *et al.* (2019) Development of SSR markers for coconut (*Cocos nucifera* L.) by selectively amplified microsatellite (SAM) and its applications. *Tropical Plant Biology* 12, 32–43.

Xia, W., Liu, R., Zhang, J., Mason, A.S., Li, Z. *et al.* (2020) Alternative splicing of flowering time gene *FT* is associated with halving of time to flowering in coconut. *Scientific Reports* 10: 11640.

Xiao, Y., Xu, P., Fan, H., Baudouin, L., Xia, W. *et al.* (2017) The genome draft of coconut (*Cocos nucifera*). *GigaScience* 6, 1–11.

Yang, Y., Bocs, S., Fan, H., Armero, A., Baudouin, L. *et al.* (2021) Coconut genome assembly enables evolutionary analysis of palms and highlights signaling pathways involved in salt tolerance. *Communications Biology* 4: 105.

11 Coconut Oil: Nutrition and Health

Fabian M. Dayrit*
Ateneo de Manila University, Loyola Heights, Quezon City, Manila, Philippines

Abstract
This chapter provides an overview of coconut oil as it is one of the most multi-functional vegetable oils, providing an efficient energy source for mitochondria, as well as therapeutic properties and antimicrobial activity. Virgin coconut oil has been shown in a clinical study to be effective in improving mild cases of coronavirus disease 2019 (COVID-19). There is positive evidence for its benefits for Alzheimer's disease. Unfortunately, coconut oil has been erroneously identified with animal fats and margarines. Given its numerous health benefits and the absence of evidence regarding the link between coconut oil and heart disease, it is clear from the information presented in this chapter that the nutritional advice against coconut oil needs to be revisited.

11.1 Introduction

The coconut palm (*Cocos nucifera* L.) grows throughout the tropical and subtropic regions of the world within about 25°N and 25°S of the equator, mainly within a few kilometres of the coastline. Numerous varieties exist and continue to be developed (Batugal *et al.*, 2005; Chan and Elevitch, 2006). The entire coconut palm may be the most useful and multi-functional of all plants (Sangamithra *et al.*, 2013). The coconut fruit can be processed into both food and non-food products (Fig. 11.1). The food products from the coconut fruit are obtained mainly from fresh coconut kernel, dried kernel (copra) and coconut water. The oil obtained from copra requires chemical refining and can be used for both food and non-food applications. One of the products from the fresh coconut kernel is virgin coconut oil (VCO), which has both food and non-food applications. The non-food applications of VCO are often intended for external use, such as cosmetics and massage oils.

The global market value of food products from the coconut fruit was estimated to be about US$4 billion in 2021 (Mordor Intelligence, 2022); however, because the coconut is grown mainly by subsistence farmers who use the coconut fruit for their own food consumption and for local commerce only, it is difficult to measure its full contribution to the economy.

During its productive life of about 60–70 years, the coconut palm may yield 70–100 fruit year^{-1} with minimal management, and up to 150 fruits in well-managed and fertilized areas (van Damm, 2002). The coconut fruit may be consumed at around 6 months' maturity, when the coconut water is at its highest volume and sweetness (Nathanael, 1966; Prades *et al.*, 2012), or at 12 months, when the coconut oil content is at its

* Email: fdayrit@ateneo.edu

DOI: 10.1079/9781789249736.0011

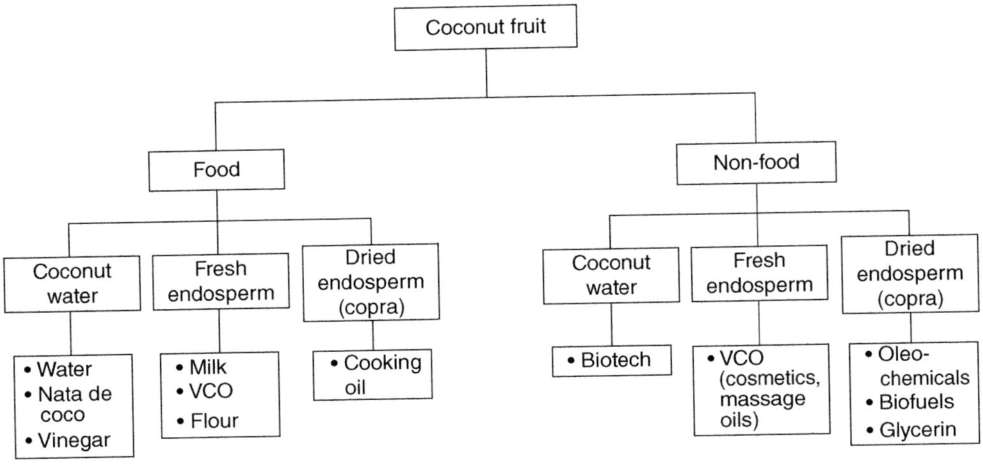

Fig. 11.1. The coconut fruit is a multi-product resource with both food and non-food uses. VCO, virgin coconut oil.

highest (Nathanael, 1966). The endosperm, the firm white kernel of the fruit, starts to develop at approximately 5 months and peaks at around 12 months. The weight of the kernel and its oil content both increase as the fruit matures (Fig. 11.2; Nathanael, 1966). From these data, the amount of oil per mature fruit is estimated at about 125 g.

The fatty acid composition of the endosperm likewise changes with age. At 6 months, the predominant fatty acid is oleic acid (an 18-carbon fatty acid with one double bond, C18:1). The amount of oleic acid rapidly drops and at 9 months the predominant fatty acid is lauric acid (C12:0) (Fig. 11.3). The fatty acid composition of coconut oil is generally measured in the 12-month-old fruit. The fatty acid composition of the oil varies depending on variety, genotype and phenotype, fertilization and irrigation (Laureles *et al.*, 2002; Naresh Kumar, 2011).

11.2 The Different Types of Coconut Oil

11.2.1 Refined, bleached and deodorized coconut oil

The first major commercial product from the coconut fruit was copra. The practice of drying the coconut kernel to produce copra was first introduced in the 1800s by British traders to supply the European luxury soap market. The copra industry rapidly expanded between the world wars in South-east Asia, in particular Indonesia and the Philippines, due to the high European and American demand for coconut oil as a versatile material for soap and edible products, such as confectionery and margarine, which was made from hydrogenated coconut oil (Cerepak, 2019; Asba *et al.*, 2020).

Farmers traditionally make copra by drying the fresh coconut kernel in the sun or by heating it by burning coconut husk or charcoal. Such drying reduces the moisture content of the kernel from 40% in the fresh kernel to about 6% in copra. However, during farm drying, these handling conditions are not enough to prevent microbial growth. The oil obtained from copra requires washing with an acid and a base (chemical refining), bleaching to remove colour and then heating to a high temperature to remove volatile compounds (deodorization). This oil, which is called refined, bleached and deodorized coconut oil, is used for frying or may be further processed into other food products, such as partially or fully hydrogenated coconut oil for use in margarines, shortenings and creamers (Jeyarani *et al.*, 2009). The commercial production of VCO is a more recent development, which started in the 2000s.

11.2.2 Virgin coconut oil

One of the high-value products from coconut is VCO. Codex Alimentarius (2015) defines virgin oils as those that are obtained 'without altering the

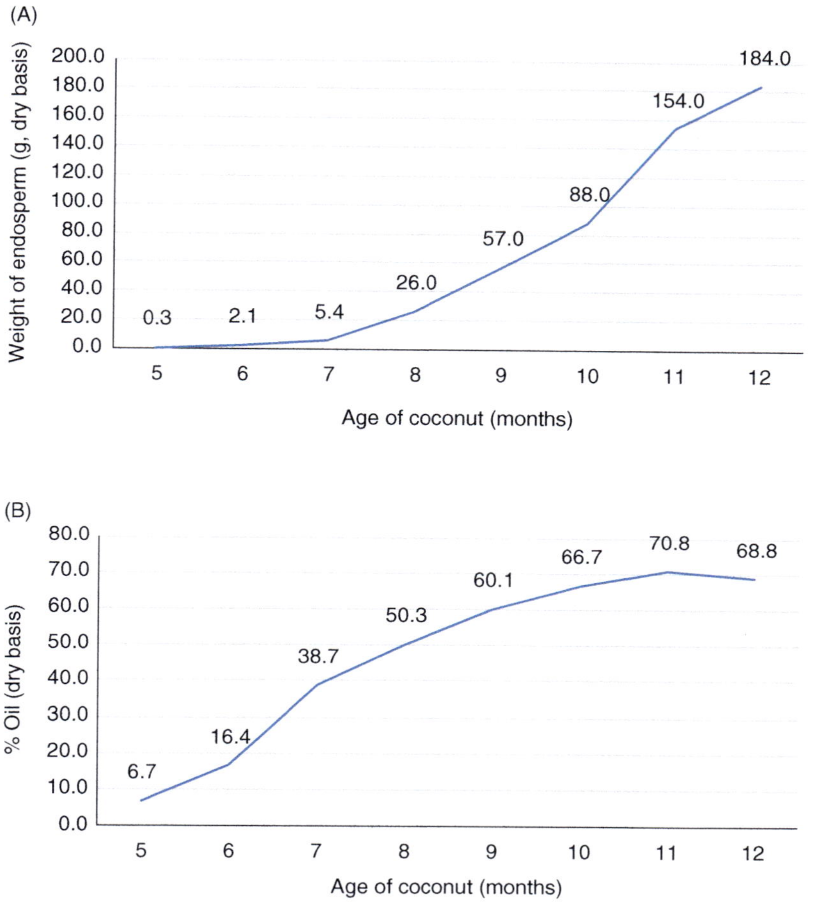

Fig. 11.2. The endosperm starts to form at 5 months after fertilization. The weight of (A) the endosperm and (B) its oil content both increase as the fruit matures. Data from Nathanael, 1966.

nature of the oil, by mechanical procedures, e.g. expelling or pressing, and the application of heat only. They may have been purified by washing with water, settling, filtering, and centrifuging only'.

VCO can be made in two ways, through dry or wet processes (Fig. 11.4). The dry method refers to mechanical expelling of the oil from the kernel by applying high pressure, while the wet method refers to the preparation of a milky emulsion by the addition of water to the ground kernel, followed by separation of the oil from the aqueous mixture. This can be achieved using a centrifuge, or by natural separation or fermentation, which entails the action of enzymes from the kernel itself and from microorganisms present in the local environment. The products from the dry method are an oil and a residue, which can be processed into coco flour. The wet method produces a milk, from which the oil can be obtained, along with additional products, which may include coco flour. Neither of these processes use chemical processing or high temperatures.

11.3 Composition and Physico-chemical Properties of Coconut Oil

Fats and oils are composed of triglyceride molecules to which the various constituent fatty acids are attached. Coconut oil has a distinctive

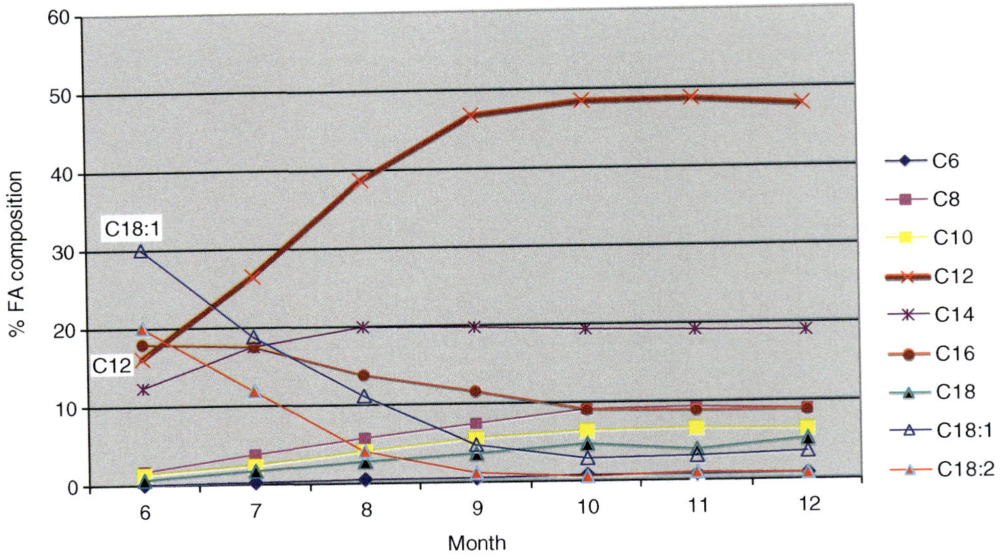

Fig. 11.3. The fatty acid (FA) composition of the coconut endosperm changes as the fruit matures. Note that at 6 months, oleic acid (C18:1) makes up about 30% of the fatty acids, while lauric acid (C12) makes up only about 16%. The fatty acid composition where lauric acid makes up about 50% usually refers to the 12-month-old fruit. From Resurreccion and Banzon, 1979.

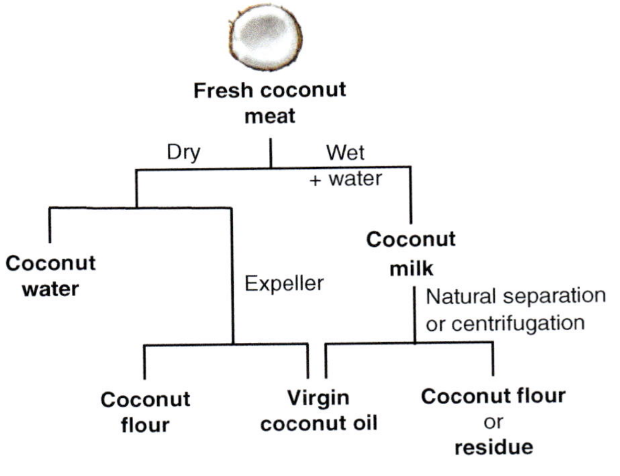

Fig. 11.4. Overview of pathways for preparing virgin coconut oil from fresh coconut kernel.

fatty acid composition compared with other vegetable oils: it consists of about 64% medium-chain fatty acids (C6:0–C12:0), about 28% long-chain saturated fatty acids (C14:0–C18:0) and about 8% unsaturated fatty acids (C18:1 and C18:2). The fatty acid profiles of crude coconut oil, refined, bleached and deodorized coconut oil, and VCO are similar. The following section will discuss the significance of this composition with regard to its health properties.

11.3.1 Fatty acid composition

The fatty acid compositions of coconut oil and other common vegetable oils are distinctly different (Table 11.1). In general, fatty acids are grouped into four classes: medium-chain saturated fatty acids (MCFAs; C6:0–C12:0), long-chain saturated fatty acids (LCFAs; C14:0–C18:0), monounsaturated fatty acids (MUFAs; C18:1) and polyunsaturated fatty acids (PUFAs; C18:2 and C18:3). From their fatty acid profiles, coconut oil is classified as a medium-chain saturated oil, while maize (*Zea mays* L.) oil and soybean (*Glycine max* (L.) Merr.) oil are PUFA oils, olive (*Olea europaea* L.) oil is a MUFA oil and palm (*Elaeis guineensis* Jacq.) oil is mixed.

Biologically, oleic acid is produced endogenously in the human liver, whereas linoleic acid and linolenic acid are not. For this reason, linoleic acid and linolenic acid are classified as 'essential fatty acids', which means that these fatty acids need to be obtained from the diet. Linoleic acid and linolenic acid are the major dietary sources of ω-6 and ω-3 fats, respectively. The World Health Organization gave the following recommended minimum and maximum daily intakes for these fatty acids (% of total energy): linoleic acid (C18:2) 2.5–9.0% and linolenic acid (C18:3) 0.5–2.0% (WHO, 2008). This also gives a recommended ratio of ω-6:ω-3 in the diet of 5:1. Excessive consumption of PUFAs and a high ω-6:ω-3 ratio are unhealthy and are a suspected cause of obesity (Harris *et al.*, 2009; Simopoulos and di Nicolantonio, 2016).

11.3.2 Chemical properties

Vegetable oils and animal fats are made up of linear fatty acids, which may be saturated, monounsaturated or polyunsaturated (Fig. 11.5A). Knowledge of their chemical stability in the presence of oxygen, light and heat is needed to understand the chemical changes that occur during frying. In general, the stabilities follow the trend: saturated fatty acids > MUFAs > PUFAs. The unsaturated fatty acids (MUFAs and PUFAs) are grouped separately because of their different chemical and metabolic characteristics. Chemically, MUFA and PUFA have different stabilities, with PUFA being more unstable, particularly in the diallylic position between the two double bonds, which is found in linoleic acid, C18:2 (Fig. 11.5B), and linolenic acid, C18:3.

Unsaturated compounds, such as PUFAs, are susceptible to reaction with oxygen radicals, and to photochemical and thermal degradation. High levels of monounsaturation (C18:1) and polyunsaturation (C18:2 and C18:3) make such vegetable oils more prone to oxidative rancidity. When oils that are predominantly PUFAs are used for frying, numerous degradation products are produced, such as *trans*-fats, aldehydes, peroxides and free radicals. These compounds induce detrimental health effects, including inflammation, cytotoxicity and genotoxicity (Fig. 11.5; Moumtaz

Table 11.1. Comparison of % fatty acid compositions of selected vegetable oils. Data from Codex Alimentarius, 2013, 2015.

Classification	Fatty acid	Coconut oil[1]	Maize oil[a]	Olive oil[b]	Palm oil[a]	Soybean oil[a]
MCFA, saturated	C6:0, caproic	0.4	ND	ND	ND	ND
	C8:0, caprylic	7.0	ND	ND	ND	ND
	C10:0, capric	6.3	ND	ND	ND	ND
	C12:0, lauric	49.0	0.2	ND	0.3	0.1
LCFA, saturated	C14:0, myristic	18.9	0.2	0.1	1.0	0.1
	C16:0, palmitic	7.0	12.6	13.8	40.8	10.8
	C18:0, stearic	2.0	1.7	2.8	4.3	3.7
MUFA	C18:1, oleic	7.5	31.1	69.0	41.6	23.5
PUFA	C18:2, linoleic	1.8	49.8	12.3	11.8	53.5
	C18:3, linolenic	0.1	1.0	1.5	0.3	7.8

LCFA, long-chain saturated fatty acids; MCFA, medium-chain saturated fatty acids; MUFA, monounsaturated fatty acids; PUFA, polyunsaturated fatty acids; ND, not detectable.
Sources: [a]Codex Alimentarius (2015); [b]Codex Alimentarius (2013).

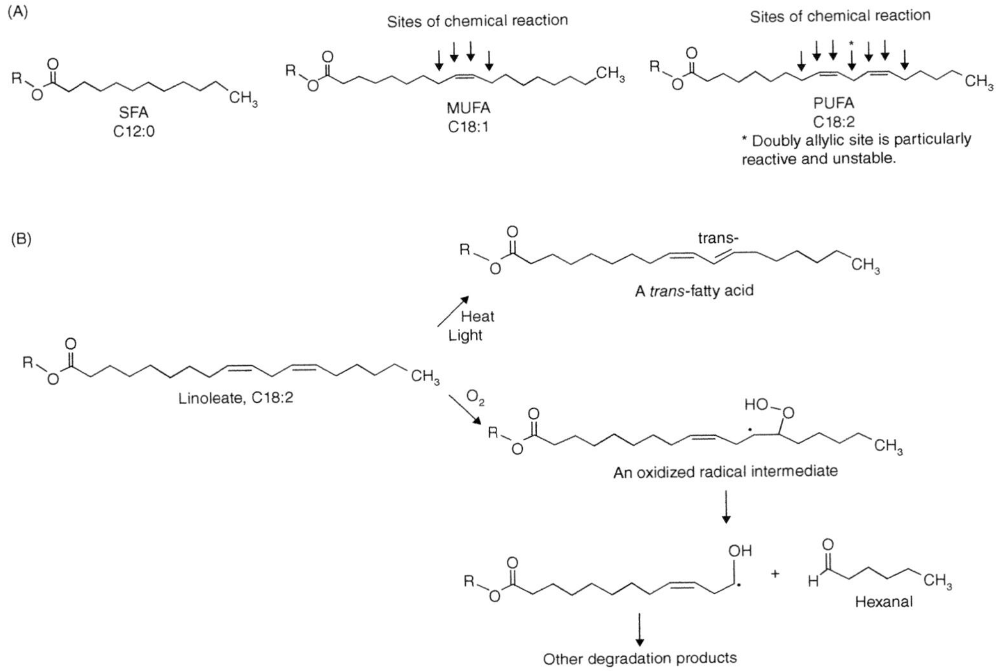

Fig. 11.5. Chemical structures of the main classes of fatty acids found in coconut oil: saturated (SFA), monounsaturated (MUFA) and polyunsaturated (PUFA) fatty acids. (A) The carbon positions where reactions are most likely to occur are indicated by arrows. The trend of chemical stability against oxygen, light and heat is as follows: SFAs > MUFAs > PUFAs. SFAs are generally stable at a frying temperature of 200°C.(B) Unsaturated fatty acids, in particular PUFAs, are susceptible to degradation due to oxidation, heat and light. All three conditions are present during frying. Shown here are some of the reactions that may occur in linoleic acid (C18:2), which lead to the formation of *trans*-fats, peroxides, free radicals and aldehydes, among others. Hexanal is a potential carcinogen.

et al., 2019; Yee and Tiu, 2019). During frying, where the temperature can exceed 200°C for extended periods of time, the likelihood that these degradation compounds will be produced in PUFA oils is high. Among the fats present in the various oils (Table 11.1), the most susceptible to degradation are maize oil and soybean oil. Frying with PUFA oils produces high levels of reactive oxygen species, polymeric compounds and other breakdown products (Choe and Min, 2007). These compounds may lead to chronic inflammatory conditions, which can result in obesity, diabetes, atherosclerosis, heart disease and cancer (Kim *et al.*, 2008). In contrast, as coconut oil is low in unsaturated fatty acids, it is stable against oxidation and is suitable for frying, even at 250°C (Pramitha and Karta, 2021).

11.3.3 Physical properties

A common source of misunderstanding about coconut oil is due to its melting point. Coconut oil, which melts at about 25°C, is liquid at room temperature in the tropics. However, unlike PUFA oils, and like lard and margarine, coconut oil is a solid at room temperature in temperate countries. Thus, coconut oil is often grouped together with lard, butter and margarine, a convenient but erroneous conflation. For example, the 2015 Dietary Guidelines for Americans grouped coconut oil, lard and margarine under the category of 'solid fats' (USDA and DHHS, 2015). These may be solid at typical temperatures in the USA, but their fatty acid compositions are very different (Table 11.2).

Table 11.2. What is 'saturated fat'? Comparison of the fatty acid profiles of coconut oil and some common animal fats: lard, tallow, chicken fat, butter and milk. Cheese has a very similar fatty acid profile to the milk from which it is produced.

	Median values (% of total fatty acids)					
		Animal fat				
Fatty acid	Coconut oil[a]	Lard, pork fat[b]	Tallow, beef fat[b]	Chicken fat[b]	Butter[b]	Bovine milk fat[c]
C4:0	–	–	–	–	2.6	4.4
C6:0	0.4	–	–	–	1.6	2.4
C8:0	7.3	–	–	–	0.9	1.4
C10:0	6.5	0.1	–	–	2.0	2.7
C12:0	49.2	0.2	0.9	0.1	2.3	3.3
C14:0	18.9	1.3	3.7	0.9	8.2	10.9
C16:0	8.9	23.8	24.9	21.6	21.3	30.6
C18:0	3.0	13.5	18.9	6.0	9.8	12.2
C20:0	0.1	–	–	–	–	0.2
C16:1	0.0	2.7	4.2	5.7	1.8	1.0
C18:1	7.5	41.2	36.0	37.3	20.4	22.8
C20:1	0.1	1.0	0.3	1.1	–	–
C18:2	1.8	10.2	3.1	19.5	1.8	1.6
C18:3	0.1	1.0	0.6	1.0	1.2	0.7
Total saturated fat	**94**	**39**	**48**	**29**	**46**	**64**
MCFA (saturated)	63	0	1	0	7	10
LCFA (saturated)	31	39	48	29	39	54
Total unsaturated fat	**9**	**56**	**44**	**65**	**25**	**26**
MUFA	8	45	41	44	22	24
PUFA	2	11	4	21	3	2
Cholesterol (mg 100 g^{-1})	<0.3	95	109	85	219	15[d]

LCFA, long-chain saturated fatty acids; MCFA, medium-chain saturated fatty acids; MUFA, monounsaturated fatty acids; PUFA, polyunsaturated fatty acids.
Sources: [a]Codex Alimentarius (2015); [b]Reeves and Weihrauch (1979); [c]Mansson (2008); [d]USDA (2017).

11.4 Saturated Fat, Coconut Oil and Heart Disease

11.4.1 What is saturated fat?

One of the most enduring dogmas of modern medicine and nutrition is the claim that 'saturated fats cause heart disease'. This was first proclaimed in 1961 by the American Heart Association's (AHA) Ad Hoc Committee on Dietary Fat and Atherosclerosis, and was reaffirmed in 2017 by an AHA Presidential Advisory, which stated that: 'Taking into consideration the totality of the scientific evidence, satisfying rigorous criteria for causality, we conclude strongly that lowering intake of saturated fat and replacing it with unsaturated fats, especially polyunsaturated fats, will lower the incidence of CVD [cardiovascular disease]' (Sacks *et al.*, 2017). This dogma rests on the definition of what a 'saturated fat' is.

In 1961, the AHA Ad Hoc Committee on Dietary Fat and Atherosclerosis gave examples of food items that were classified as 'saturated fat' (Page *et al.*, 1961). This included whole milk, cream, butter, cheese, coconut oil, chocolate fat and meat (they did not specify the type of fat). When a comparison of the fatty acid profiles (Table 11.2) of coconut oil and some common animal fats (e.g. lard, tallow, chicken fat, butter and cow's milk) is made, dairy cheese and cream have very similar profiles to whole cow's milk. Clearly, only coconut oil can be considered a true saturated fat and the statement of the AHA Ad Hoc Committee that coconut oil and animal fats can be categorized in the same group is incorrect. Except for milk, animal fats are not saturated. Furthermore, animal fats have much higher levels of cholesterol than coconut oil. Thus, there is no basis to claim that coconut oil and animal fats are the same, or that animal fats are saturated fats.

Although the AHA Committee noted the presence of hydrogenated food products in the diet, such as margarine and shortening, it failed to recognize the health risks posed by these products, even though it was already known that hydrogenated products contained up to 42% *trans*-fat, and that *trans*-fat had been detected in significant amounts in the heart, aortic tissue and atheroma of subjects who had died of atherosclerosis (Johnston *et al.*, 1957). In 1997, it was conservatively estimated that 30,000 deaths year^{-1} in the USA could be attributed to the consumption of *trans*-fats (Ascherio *et al.*, 1994). As the warning against *trans*-fats was not officially recognized by dietary authorities until the 2000s, most research undertaken before this time did not account for the effects of *trans*-fat in the diet.

11.4.2 Animal fat and *trans*-fat in the western diet

The correct definition of terms is very important in solving any problem. The erroneous definition of 'saturated fat' can only give erroneous conclusions. The seminal Seven Countries Study of Ancel Keys (1986) committed such an error where he famously concluded that 'death rates were related positively to average percentage of dietary energy from saturated fatty acids'. However, a later analysis of the food consumed in the seven countries revealed that the predictors of heart disease were butter, lard, margarine and meat (Menotti *et al.*, 1999). This error was repeated in the Cochrane meta-analysis looking at the saturated fat intake and incidence of cardiovascular disease (Hooper *et al.*, 2015). Here, the authors selected 15 randomized controlled trials (RCTs) conducted between 1965 and 2006 with a combined total of approximately 59,000 participants, which used a variety of interventions to reduce 'saturated fat' intake. The authors concluded that there was moderate-quality evidence that heart disease can be reduced by reducing 'saturated fat' intake and replacing it with polyunsaturated fats. However, none of the 15 RCT studies may have been dealing with saturated fat. Consider the following: (i) all of the studies were undertaken in western countries, namely the USA, Canada, certain European countries and New Zealand, that consume western diets; (ii) studies on the fats consumed in these countries show that they contained mainly animal fats and margarine (*trans*-fat) (Michels and Sacks, 1995; Jensen, 2012); and (iii) 'coconut' and 'coconut oil' were not mentioned anywhere in the meta-analysis paper, nor in any of the 15 RCT studies. In other words, it was unlikely that there was any coconut oil included in any of these studies.

The erroneous identification of animal fat as a saturated fat and the failure to account for the presence of *trans*-fats have led to erroneous conclusions. As most dietary and epidemiological studies done in the west have involved animal fats and *trans*-fats, the conclusion that saturated fat causes heart disease is incorrect. This error means that corrections should be made to dietary recommendations that have used these papers as justification for advice. In particular, the World Health Organization (WHO) should correct its warning against the consumption of coconut oil in its Healthy Diet Advisory and its Replace Trans Fat programme (WHO, 2015). The various national dietary guidelines should likewise review their recommendations.

11.4.3 Coconut oil is not linked to heart disease in the Tropics

If coconut oil causes heart disease, the evidence should be clear in countries that consume large amounts of coconut oil regularly in their diet. However, this has not been the case. Kaunitz (1972) noted that 'evidence from epidemiological studies of arteriosclerosis in populations consuming large amounts of coconut oil are cited to show that coconut oil in a natural diet is not disadvantageous and may even be of advantage'. This conclusion was supported by Prior *et al.* (1981), who studied the impact of a high consumption of dietary coconut on heart disease among inhabitants of the South Pacific Islands. In their traditional diet, the inhabitants of Pukapuka and Tokelau Islands were consuming 34% and 63%, respectively, of their total daily energy as coconut. As coconut oil accounts for about 65% of the endosperm, their total coconut oil intake can be estimated to be 22% and 40%, respectively, of their total calories. Yet Prior *et al.* (1981) reported that 'vascular disease is uncommon in both populations and there is no evidence of the high saturated fat intake having a harmful effect in these populations'.

In a nationwide diet survey conducted in the Philippines in 1987, the data showed that in one of the regions (Bicol Province), where coconut (grated and milk) amounted to about 18% of their daily energy intake, one of the lowest incidences of acute myocardial infarction and atherosclerosis was found (Florentino and Aguinaldo, 1987). The WHO reported that Pacific islanders who abandoned their traditional coconut-based diet in favour of imported western processed foods were 2.2 times more likely to be obese and 2.4 times more likely to be diabetic (WHO, 2003). Obesity and diabetes are risk factors for heart disease.

Two studies conducted in India showed that coconut oil is not linked to coronary heart disease (CHD). Kumar (1997) investigated this link in 32 CHD patients and matched healthy controls in a Kerala hospital. The results showed no link between coconut or coconut oil in the causation of CHD. In 2016, a 2-year study comparing coconut oil with sunflower oil showed no link with CHD (Vijayakumar *et al.*, 2016).

In 2016, Eyres *et al.* (2016) selected and analysed 21 research papers on coconut oil, involving eight clinical trials and 13 observational studies. They concluded that 'observational evidence suggests that consumption of coconut flesh (i.e. kernel) or squeezed coconut (i.e. VCO), in the context of traditional dietary patterns, does not lead to adverse cardiovascular outcomes'.

The effects of butter, VCO and extra virgin olive oil (EVOO) on various health parameters were compared in a 4-week feeding study conducted among British volunteers in Cambridge, UK. The results showed that both VCO and EVOO lowered low-density lipoprotein (LDL) and glucose levels, while butter raised LDL levels. In addition, VCO raised high-density lipoprotein (HDL) levels to a greater extent than EVOO or butter (Khaw *et al.*, 2018). Interestingly, VCO also lowered the level of C-reactive protein (a measure of inflammation). This same result was noted in a 2021 study conducted in the Philippines on the use of VCO against mild cases of coronavirus 2019 (COVID-19) (see below).

However, despite all the evidence, detractors continue to link coconut oil to heart disease. In an editorial in the AHA journal *Circulation*, Sacks (2020) criticized the evidence that supported coconut oil and focused only on negative studies that used surrogate LDL measurements as proof that coconut oil causes heart disease. However, LDL has been found to be an unreliable predictor of heart disease. LDLs can be subdivided into three or four subgroups and not all of the subgroups are linked to atherosclerosis (Ivanova *et al.*, 2017; Hua and Malinski, 2019). The use of undifferentiated LDL as the sole biomarker for heart disease has been questioned (Astrup *et al.*, 2019). Thus, the claim that coconut oil causes heart disease based solely on undifferentiated LDL data is no longer tenable. Furthermore, Sacks claimed, without evidence and contrary to well-known chemistry, that PUFA oils are better frying oils than coconut oils (Moumtaz *et al.*, 2019; Yee and Tiu, 2019). As stated above, the degradation products from the heating of PUFA oils have been shown to form *trans*-fats, aldehydes and peroxides (Fig. 11.5). These are known to produce oxidized LDL and cholesterol, which are known to cause atherosclerosis (Stapran*s et al.*, 2000; Holvoet, 2004; Meisinger *et al.*, 2005; Parthasarathy *et al.*, 2010).

11.5 Healthy Properties of Coconut Oil

Coconut oil is unique among the major vegetable oils in that it contains a high proportion of MCFA (about 65%). What is different about MCFAs is that these fatty acids are not produced by the body and are not stored in significant amounts. In contrast, LCFAs are stored in adipose tissues.

11.5.1 Medium-chain triglycerides, MCFAs and LCFAs

The term 'medium-chain triglyceride' (MCT) was first applied to a commercial product produced in the 1960s from caprylic acid (C8:0) and capric acid (C10:0), which were waste by-products from coconut oil. This product, called 'MCT oil', was found to have use as a dietary lipid (Harkins and Sarett, 1968). Thus, the term 'medium-chain' was originally associated with C8:0 and C10:0 fatty acids only. Research on the biochemistry of fatty acid metabolism has resulted in a more scientifically valid grouping of MCFAs, which was made up of caproic acid (C6:0) to lauric acid (C12:0) (Bergen *et al.*, 1966; Bach and Babayan, 1982). This classification was based on several metabolic criteria. MCFAs (i.e. C6:0–C12:0) can cross the mitochondrial membrane without the assistance of carnitine, unlike the LCFAs (C14 and longer) (Wade, 1991). MCFAs are metabolized more rapidly for energy and are only minimally present in adipose fat compared with LCFAs. MCFAs are rapidly oxidized in the liver mitochondria, producing large concentrations of acetyl-coenzyme A, which drives the production of ketone bodies (β-hydroxybutyrate and acetyl acetate); this then results in a thermogenic response, with less ectopic fat compared with LCFAs (McCarty and di Nicolantonio, 2016).

The classification of lauric acid (C12:0) as a medium-chain or long-chain fatty acid has been a subject of debate, when compared with myristic acid (C14:0). The metabolism of MCFAs (C6:0–C12:0) can be affected by various conditions, such as the composition of the diet, the fasting state and sex. Lymphatic transport and absorption are affected by the amount and type of fat present (Sukhotnik *et al.*, 2001; Takeuchi *et al.*, 2006). Upon entering the gastrointestinal tract, MCFAs are absorbed more rapidly than LFCAs and are transported more efficiently into the portal blood directly to the liver. Absorption of fats via the portal vein has been determined as high for MCFAs (C6:0 > C8:0 > C10:0 > C12:0), while LCFAs are incorporated into chylomicrons and transported through the lymph (Papamandjaris *et al.*, 1998).

Although the general behaviour of C12:0 classifies it as a medium chain, there are situations where it is described as an 'intermediate' fatty acid. Schon *et al.* (1955) conducted metabolic studies on lauric acid esters and noted that C12:0 has properties of intermediate-chain-length fatty acids. Under challenging nutritional conditions, humans utilize whatever fatty acids are available. In a study on Nigerian infants, it was found that C10:0–C14:0 fatty acids are readily absorbed by the infant's small intestine and metabolized directly by mitochondria without the use of carnitine. The authors referred to C10:0–C14:0 fatty acids as 'intermediate' (Knox *et al.*, 2000). Variations in the metabolism of C12:0 can be observed, depending on the feeding conditions. A high consumption of coconut oil (45% of calories) in rats (*Rattus norvegicus domestica* (Berkenhout, 1769)) for 6 weeks resulted in storage of C12:0-containing triglycerides in adipose tissues (Gordis, 1965). This is consistent with the observation that, under constant feeding of coconut oil, C12:0 can be detected in the bloodstream for at least 4 h after feeding (Nonaka *et al.*, 2016). Thus, the distribution of C12:0 in the body after ingestion appears to depend on feeding conditions. However, what does not change is that C12:0 does not require carnitine to cross the mitochondrial membrane, while longer-chain fatty acids require carnitine (Wade, 1991).

Because lauric acid comprises about half of the fatty acids in coconut oil, it is the primary focus of attention. However, the properties of lauric acid alone and coconut oil should not be assumed to be the same. As noted above, some properties may be partially shared by closely related fatty acids, C10:0–C14:0. Unless experiments verify otherwise, the lack of information on C10:0 and C14:0 should not be taken as absence of activity. The physiological effects of VCO should consider the activities of all its components and metabolites together.

11.5.2 Metabolic disorders associated with the mitochondria

The provision of efficient and clean sources of metabolic energy is important for health. The formation of excessive amounts of reactive oxygen species can damage the mitochondria (Dorighello *et al.*, 2016) and lead to the formation of white adipose tissue, which increases inflammatory conditions that are risk factors for type 2 diabetes and heart disease (Franssens *et al.*, 2015). Such conditions have been attributed to the western diet, which is characterized by excessive amounts of ω-6 PUFAs with an unhealthy ω-6:ω-3 ratio of 20:1. This has been identified as one of the causes of the increasing incidence of Alzheimer's disease, type 2 diabetes, obesity and cancer in countries with this diet (Kopp, 2019).

A study of rat cardiac mitochondria showed that feeding with coconut oil gave the lowest concentration of oxidized proteins and peroxidized lipids when compared with fish oil and olive oil. The authors concluded that the coconut oil diet offered strong protection against oxidative stress in heart mitochondria (Lemieux *et al.*, 2011). This supports the claim that coconut oil protects the heart against oxidative stress.

Provided that ω-6 and ω-3 PUFAs are available in appropriate amounts and at the correct ratio (3–11% at a 5:1 ratio), coconut oil, which is made up of about 65% MCFAs, 27% saturated LCFAs and 8% unsaturated fats, can provide a healthy source of dietary fat. The optimum amount of dietary MCFAs, which should be considered separately from saturated LCFAs, should be determined in future studies.

11.5.3 Antibacterial and antiviral properties

The seminal research by Kabara in the 1970s and 1980s identified lauric acid and its monoglyceride derivative, monolaurin (ML), as potent antibacterial compounds present in coconut oil (Kabara *et al.*, 1972; Kabara, 1984). It was shown that C10:0 and C12:0 and their respective monoglyceride derivatives act by disrupting the lipid membranes of Gram-positive bacteria. The antiviral activities of C12:0 and ML were first noted by Sands *et al.* (1979). Hierholzer and Kabara (1982) showed that ML reduced the infectivity of 14 human RNA and DNA enveloped viruses in cell culture by disintegrating the virus envelope. Thormar *et al.* (1987) confirmed the ability of C12:0 and ML to inactivate viruses by disintegration of the cell membrane.

There are at least three mechanisms that may explain the antimicrobial activity of C10:0, C12:0 and their monoglycerides: (i) these compounds can disrupt the membrane of Gram-positive bacteria and lipid-coated viruses by physico-chemical processes; (ii) they can interfere with cellular processes of the host, such as signal transduction and transcription; and (iii) they stabilize human cell membranes. These multiple mechanisms may explain why bacteria have been unable to evolve resistance against the action of these compounds. Over 100 patents have now been filed worldwide on the use of ML in food, cosmetic and animal feed applications (Dayrit, 2015).

Despite their safety and affordability, the pharmaceutical industry and medical community have largely ignored their potential as antimicrobial agents. However, the rise of antibiotic resistance provides opportunities to explore coconut oil and its fatty acids and monoglycerides as affordable treatments against infectious diseases (Churchward *et al.*, 2018).

11.5.4 Anti-inflammatory and immunomodulatory properties

While the previous discussion has shown that coconut oil does not cause inflammation, is it an anti-inflammatory? Inflammation is the response of the human immune system to damage caused by physical, chemical or biological agents, such as pathogens, toxic compounds, heat or ionizing irradiation. Inflammation is an important defence and repair mechanism that is vital to health. However, because an excessive response can also damage cells, modulation of the immune response is equally important. VCO has been used traditionally as a treatment for various skin conditions. Visible evidence of the anti-inflammatory properties of VCO can be seen in its effect on the skin. Clinical studies have shown that topical application of VCO inhibits inflammatory cytokines, including tumour necrosis factor-α and interleukin–6, –5 and –8 (Varma *et al.*, 2019).

Proper modulation of the immune system depends on interactions between the sympathetic noradrenergic nervous system and the immune cells of the lymphoid organs (thymus and lymph nodes). The use of VCO has been shown to modulate immune responses by upregulating neuroprotective factors while suppressing inflammatory cytokines through cellular signalling pathways (Karrunanithi *et al.*, 2019).

11.5.5 The use of VCO against COVID-19

In the first clinical study on the use of VCO against COVID-19, Agdeppa *et al.* (2021) reported that feeding VCO to subjects with mild cases of COVID-9 resulted in a quicker relief of symptoms and inflammation, by observing a more rapid and larger drop in C-reactive protein levels than in the control group. This was a 28-day randomized, double-blind, controlled feeding study. The results of this study are consistent with the antiviral and immunomodulatory effects of coconut oil and the possibility that VCO can be adopted as a functional food that may enable faster recovery from COVID-19, as well as post-COVID-19 effects. It shows that VCO can be a safe, affordable and readily available functional food against COVID-19.

11.5.6 The potential of VCO against Alzheimer's disease

The brain requires a continuous supply of energy, most of which is normally provided by glucose. However, the metabolism of glucose in the brain can be impaired by a form of insulin resistance that results in Alzheimer's disease and which has molecular and biochemical features that are shared with diabetes mellitus. This condition has been called 'type 3 diabetes' (de la Monte and Wands, 2008). One of the therapeutic strategies for countering this condition is by using ketone-based interventions. 'Ketones' refers to a group of liver metabolites, namely β-hydroxybutyrate and acetylacetate (Cunnane *et al.*, 2020). Ketones serve as the brain's alternative fuel and can become the primary fuel under conditions of limited glucose availability, such as fasting. Preliminary evidence indicates that the generation of ketones in the liver from medium-chain fatty acids, in particular caprylic acid (C8:0) and capric acid (C10:0), may improve cognitive performance in Alzheimer's disease (Poff *et al.*, 2021). As VCO contains about 13% of these two fatty acids, it can potentially contribute to ketone production.

A clinical trial of 44 patients with Alzheimer's disease was conducted for 21 days, wherein half of the patients took 40 mL of coconut oil daily. The patients were evaluated through a cognitive test, the Mini-Mental State Examination, to determine possible changes. It was observed that patients who received coconut oil showed statistically significant cognitive improvement after completion of the intervention (de la Rubia Ortí *et al.*, 2017). Another study assessed brain ketone uptake in mild to moderate Alzheimer's disease, comparing the ingestion of 30 g day^{-1} of caprylic acid and capric acid. The results showed that the brain glucose deficit in Alzheimer's disease can be compensated by both types of MCFA (Croteau *et al.*, 2018). The longest duration Alzheimer's disease/ketone study used caprylic acid in a 6-month randomized, double-blind, placebo-controlled, crossover study, with an additional 6-month open-label extension. Twenty subjects, with an average age 72.6 years, consumed approximately 25 g day^{-1} (about 1.8 tablespoons). Eighty per cent of the subjects showed stabilization or an improvement in cognition, and even better responses were obtained with the extension period (Juby *et al.*, 2022).

These studies support the potential of VCO, in particular caprylic (C8:0) and capric (C10:0) acids, in the management of Alzheimer's disease. There is much interest in determining whether VCO can prevent the development of this disease.

11.6 Conclusion

The coconut is one of the most useful fruits in the world, yielding food and drink and other useful non-food materials. VCO has been shown to be one of the most multi-functional among the vegetable oils, providing a clean energy source for mitochondria, as well as antimicrobial and anti-inflammatory activity. Inflammation is a major cause of many diseases.

As VCO is antimicrobial, it can help address conditions caused by harmful microorganisms. There is sufficient evidence that supports the healthy properties of VCO and its role as a functional food and a healthy frying vegetable oil.

There is no evidence that links VCO or coconut oil to heart disease. Studies on the effect of VCO on lipid profiles show that VCO raises both the LDL and HDL levels, resulting in a healthy LDL:HDL ratio. Research on VCO has shown favourable effects in some studies in lowering blood glucose and obesity.

Unfortunately, coconut oil continues to be the most misunderstood vegetable oil. It has been erroneously identified with animal fats, even though their fatty acid and cholesterol profiles are very different. Recently, coconut oil has been grouped with lard and margarine because these are 'solid fats', a classification that has no scientific basis. The Dietary Guidelines for Americans have been steadfast in their advice since its first edition in 1980 to avoid coconut oil and saturated fat. Heeding this advice, coconut oil consumption fell from 10% in 1980 to 4% in 2010 (IndexMundi, 2016). However, during this same period, heart disease remained as the number one cause of death in the USA (CDC, 2015), which suggests that there may be no link between coconut oil or saturated fat and heart disease (Mozaffarian *et al.*, 2015).

Given the numerous health benefits that have been discovered from its consumption and the absence of evidence regarding the link between coconut oil and heart disease, the nutritional advice against coconut oil needs to be revisited.

11.7 References

Agdeppa, I.A., Nacis, J.S., Capanzana, M.V., Dayrit, F.M. and Tanda, K.V. (2021) Virgin coconut oil is effective in lowering C-reactive protein levels among suspect and probable cases of COVID-19. *Journal of Functional Foods* 83: 104557.

Asba, A.R., Evita, A.L., Rahman, F. and Nur, N. (2020) The globalizing of copra and coconut oil industry of Makassar before the second world war. *IOP Conference Series: Earth and Environmental Science* 575: 012094.

Ascherio, A., Hennekens, C.H., Buring, J.E., Master, C., Stampfer, M.J. *et al.* (1994) Trans-fatty acids intake and risk of myocardial infarction. *Circulation* 89, 94–101.

Astrup, A., Bertram, H.C.S., Bonjour, J.-P., de Groot, L.C., de Oliveira Otto, M.C. *et al.* (2019) WHO draft guidelines on dietary saturated and trans fatty acids: Time for a new approach? *British Medical Journal* 366, l4137.

Bach, A. and Babayan, V.K. (1982) Medium-chain triglycerides: an update. *American Journal of Clinical Nutrition* 36, 950–962.

Batugal, P., Ramanatha Rao, V. and Oliver, J. (2005) *Coconut Genetic Resources*. International Plant Genetic Resources Institute – Regional Office for Asia, the Pacific and Oceania (IPGRI-APO), Serdang, Malaysia.

Bergen, S.S. Jr, Hashim, S.A. and van Itallie, T.B. (1966) Hyperketonemia induced in man by medium-chain triglyceride. *Diabetes* 15, 723–725.

CDC (Centers for Disease Control and Prevention) (2015) *Leading Causes of Death and Numbers of Deaths, by Age: United States, 1980 and 2014*. CDC, Atlanta, Georgia. Available at: www.cdc.gov/nchs/data/hus/2015/020.pdf (accessed 30 August 2023).

Cerepak, P. (2019) Coconuts: Catalysts of conflict. Available at: https://edgeeffects.net/coconuts-conflict/ (accessed 30 August 2023).

Chan, E. and Elevitch, C.R. (2006) *Cocos nucifera* (coconut), version 2.1. In: Elevitch, C.R. (ed.) *Species Profiles for Pacific Island Agroforestry*. Permanent Agriculture Resources, Holualoa, Hawaii. Available at: https://raskisimani.files.wordpress.com/2013/01/cocos-nucifera-coconut.pdf (accessed 30 August 2023).

Choe, E. and Min, D.B. (2007) Chemistry of deep-fat frying oils. *Journal of Food Science* 72, R77–R86.

Churchward, C.P., Alany, R.G. and Snyder, L.A.S. (2018) Alternative antimicrobials: the properties of fatty acids and monoglycerides. *Critical Reviews in Microbiology* 44, 561–570.

Codex Alimentarius (2013) *Standard for Olive Oils and Olive Pomace Oils. Codex Stan 33–1981, Amendment: 2013*. Food and Agricultural Organization of the United Nations, Rome.

Codex Alimentarius (2015) *Standard for Named Vegetable Oils. Codex Stan 210–1999. Amendment: 2015*. Food and Agriculture Organization of the United Nations, Rome.

Croteau, E., Castellano, C.-A., Richard, M.A., Fortier, M., Nugent, S. *et al.* (2018) Ketogenic medium chain triglycerides increase brain energy metabolism in Alzheimer's disease. *Journal of Alzheimer's Disease* 64, 551–561.

Cunnane, S.C., Trushina, E., Morland, C., Prigione, A., Casadesus, G. *et al.* (2020) Brain energy rescue: an emerging therapeutic concept for neurodegenerative disorders of ageing. *Nature Reviews Drug Discovery* 19, 609–633.

Dayrit, F.M. (2015) The properties of lauric acid and their significance in coconut oil. *Journal of the American Oil Chemists Society* 92, 1–15.

de la Monte, S.M. and Wands, J.R. (2008) Alzheimer's disease is type 3 diabetes – evidence reviewed. *Journal of Diabetes Science and Technology* 2, 1101–1113.

de la Rubia Ortí, J.E., Bueno Cayo, A.M., Rochina, M.J. and Yang, H.I. (2017) How does coconut oil affect cognitive performance in Alzheimer patients? *Nutricion Hospitalaria* 34, 352–356.

Dorighello, G.G., Paim, B.A., Kiihl, S.F., Ferreira, M.S., Catharino, R.R. *et al.* (2016) Correlation between mitochondrial reactive oxygen and severity of atherosclerosis. *Oxidative Medicine and Cellular Longevity* 2016: 7843685.

Eyres, L., Eyres, M.F., Chisholm, A. and Brown, R.C. (2016) Coconut oil consumption and cardiovascular risk factors in humans. *Nutrition Reviews* 74, 267–280.

Florentino, R.F. and Aguinaldo, A.R. (1987) Diet and cardiovascular disease in the Philippines. *Philippine Journal of Coconut Studies* 13, 56–70.

Franssens, B.T., van der Graaf, Y., Kappelle, L.J., Westerink, J., de Borst, G.J. *et al.* (2015) Body weight, metabolic dysfunction, and risk of type 2 diabetes in patients at high risk for cardiovascular events or with manifest cardiovascular disease: a cohort study. *Diabetes Care* 38, 1945–1951.

Gordis, E. (1965) The long-term stability of triglyceride molecules in adipose tissue. *Journal of Clinical Investigation* 44, 1978–1985.

Harkins, R.W. and Sarett, H.P. (1968) Nutritional evaluation of medium-chain triglycerides in the rat. *Journal of the American Oil Chemists Society* 45, 26–30.

Harris, W.S., Mozaffarian, D., Rimm, E., Kris-Etherton, P., Rudel, L.L. *et al.* (2009) Omega-6 fatty acids and risk for cardiovascular disease. *Circulation* 119, 902–907.

Hierholzer, J.C. and Kabara, J.J. (1982) *In vitro* effects of monolaurin compounds on enveloped RNA and DNA viruses. *Journal of Food Safety* 4, 1–12.

Holvoet, P. (2004) Oxidised LDL and coronary heart disease. *Acta Cardiologica* 59, 479–484.

Hooper, L., Martin, N., Abdelhamid, A. and Davey Smith, G. (2015) Reduction in saturated fat intake for cardiovascular disease. *Cochrane Database of Systematic Reviews* 10: CD011737.

Hua, J. and Malinski, T. (2019) Variable effects of LDL subclasses of cholesterol on endothelial nitric oxide/peroxynitrite balance – the risks and clinical implications for cardiovascular disease. *International Journal of Nanomedicine* 14: 8973.

IndexMundi (2016) United States Coconut Oil Domestic Consumption by Year. IndexMundi, Charlotte, North Carolina. Available at: www.indexmundi.com/agriculture/?country=us&commodity=coconut-oil&graph=domestic-consumption (accessed 30 August 2023).

Ivanova, E.A., Myasoedova, V.A., Melnichenko, A.A., Grechko, A.V. and Orekhov, A.N. (2017) Small dense low-density lipoprotein as biomarker for atherosclerotic diseases. *Oxidative Medicine and Cellular Longevity* 2017: 1273042.

Jensen, T. (2012) The consumption of fats in Denmark 1900–2000. *Anthropology of Food* S7. Available at: http://journals.openedition.org/aof/7100 (accessed 30 August 2023).

Jeyarani, T., Khan, M.I. and Khatoon, S. (2009) *Trans*-free plastic shortenings from coconut stearin and palm stearin blends. *Food Chemistry* 114, 270–275.

Johnston, P.V., Johnson, O.C. and Kummerow, F.A. (1957) Occurrence of *trans* fatty acids in human tissue. *Science* 126, 698–699.

Juby, A.G., Blackburn, T.E. and Mager, D.R. (2022) Use of medium chain triglyceride (MCT) oil in subjects with Alzheimer's disease: a randomized, double-blind, placebo-controlled, crossover study, with an open-label extension. *Alzheimer's Dementia* 8: e12259.

Kabara, J.J. (1984) Antimicrobial agents derived from fatty acids. *Journal of the American Oil Chemists Society* 61, 397–403.

Kabara, J.J., Swieczkowski, D.M., Conley, A.J. and Truant, J.P. (1972) Fatty acids and derivatives as antimicrobial agents. *Antimicrobial Agents and Chemotherapy* 2, 23–28.

Karrunanithi, S., Ravichandran, K.A., Hima, L., Pratap, U.P., Vasantharekha, R. *et al.* (2019) Virgin coconut oil enhances neuroprotective and anti-inflammatory factors in the thymus and mesenteric lymph nodes of rats. *Clinical and Experimental Neuroimmunology* 11, 65–72.

Kaunitz, H. (1972) Nutritional properties of coconut oil. *Ceylon Coconut Quarterly* 23, 85–91.

Keys, A., Menotti, A., Karvonen, M.J., Aravanis, C., Blackburn, H. *et al.* (1986) The diet and 15-year death rate in the Seven Countries Study. *American Journal of Epidemiology* 124, 903–915.

Khaw, K.-T., Sharp, S.J., Finikarides, L., Afzal, I., Lentjes, M. *et al.* (2018) Randomised trial of coconut oil, olive oil or butter on blood lipids and other cardiovascular risk factors in healthy men and women. *British Medical Journal Open* 8: e020167.

Kim, J., Wei, Y. and Sowers, J.R. (2008) Role of mitochondrial dysfunction in insulin resistance. *Circulation Research* 102, 401–414.

Knox, E., Vander Jagt, D.J., Shatima, D., Huang, Y.S. and Chuang, L.T. (2000) Nutritional status and intermediate chain-length fatty acids influence the conservation of essential fatty acids in the milk of northern Nigerian women. *Prostaglandins, Leukotrienes and Essential Fatty Acids* 63, 195–202.

Kopp, W. (2019) How western diet and lifestyle drive the pandemic of obesity and civilization diseases. *Diabetes, Metabolic Syndrome and Obesity: Targets and Therapy* 12, 2221–2236.

Kumar, P.D. (1997) The role of coconut and coconut oil in coronary heart disease in Kerala, South India. *Tropical Doctor* 27, 215–217.

Laureles, L.R., Rodriguez, F.M., Reaño, C.E., Santos, G.A., Laurena, A.C. *et al.* (2002) Variability in fatty acid and triacylglycerol composition of the oil of coconut (*Cocos nucifera* L.) hybrids and their parentals. *Journal of Agricultural and Food Chemistry* 50, 1581–1586.

Lemieux, H., Bulteau, A.L., Friguet, B., Tardif, J.C. and Blier, P.U. (2011) Dietary fatty acids and oxidative stress in the heart mitochondria. *Mitochondrion* 11, 97–103.

Mansson, H.L. (2008) Fatty acids in bovine milk fat. *Food and Nutrition Research* 52: 1821.

McCarty, M.F. and di Nicolantonio, J.J. (2016) Lauric acid-rich medium-chain triglycerides can substitute for other oils in cooking applications and may have limited pathogenicity. *Open Heart* 3: e000467.

Meisinger, C., Baumert, J., Khuseyinova, N., Loewel, H. and Koenig, W. (2005) Plasma oxidised low-density lipoprotein, a strong predictor for acute coronary heart disease events in apparently healthy, middle-aged men from the general population. *Circulation* 112, 651–657.

Menotti, A., Kromhout, D., Blackburn, H., Fidanza, F., Buzina, R. *et al.* (1999) Food intake patterns and 25-year mortality from coronary heart disease: cross-cultural correlations in the Seven Countries Study. *European Journal of Epidemiology* 15, 507–515.

Michels, K. and Sacks, F. (1995) Trans fatty acids in European margarines. *New England Journal of Medicine* 332, 541–542.

Mordor Intelligence (2022) Coconut market size & share analysis – Growth trends & forecasts (2023–2028) Available at: www.mordorintelligence.com/industry-reports/coconut-products-market (accessed 30 August 2023).

Moumtaz, S., Percival, B.C., Parmar, D., Grootveld K.L., Jansson, P. *et al.* (2019) Toxic aldehyde generation in and food uptake from culinary oils during frying practices: peroxidative resistance of a monounsaturate-rich algae oil. *Scientific Reports* 9: 4125.

Mozaffarian, D., Benjamin, E.J., Go, A.S., Arnett, D.K., Blaha, M.J. *et al.* (2015) Heart disease and stroke statistics – 2015 update. A report from the American Heart Association. *Circulation* 131, e29–e322.

Naresh Kumar, S. (2011) Variability in coconut (*Cocos nucifera* L.) germplasm and hybrids for fatty acid profile of oil. *Journal of Agricultural and Food Chemistry* 59, 13050–13058.

Nathanael, W.R.N. (1966) Moisture and other quality factors of copra. *Ceylon Coconut Quarterly* 17, 1–41.
Nonaka, Y., Takagi, T., Inai, M., Nishimura, S., Urashima, S. *et al.* (2016). Lauric acid stimulates ketone body production in the KT-5 astrocyte cell line. *Journal of Oleo Science* 65, 693–699.
Page, I.H., Allen, E.V., Chamberlain, F.L., Keys, A., Stamler, J. *et al.* (1961) Dietary fat and its relation to heart attacks and strokes. *Circulation* 23, 133–136.
Papamandjaris, A.A., MacDougall, D.E. and Jones, P.J. (1998) Medium chain fatty acid metabolism and energy expenditure: obesity treatment implications. *Life Science* 62, 1203–1215.
Parthasarathy, S., Raghavamenon, A., Garelnabi, M.O. and Santanam, N. (2010) Oxidised low-density lipoprotein. *Methods in Molecular Biology* 610, 403–417.
Poff, A.M., Moss, S., Soliven, M. and d'Agostino, D.P. (2021) Ketone supplementation: meeting the needs of the brain in an energy crisis. *Frontiers in Nutrition* 8: 783659.
Prades, A., Dornier, M., Diop, N. and Pain, J.P. (2012) Coconut water uses, composition and properties: a review. *Fruits* 67, 87–107.
Pramitha, D.A.I. and Karta, I.W. (2021) Analysis of fatty acids in virgin coconut oil frying at various temperatures. *Jurnal Sains dan Teknologi* 10, 104–111.
Prior, I.A., Davidson, F., Salmond, C.E. and Czochanska, Z. (1981) Cholesterol, coconuts, and diet on Polynesian atolls: a natural experiment: the Pukapuka and Tokelau Island studies. *American Journal of Clinical Nutrition* 34, 1552–1561.
Resurreccion, A.N.P. and Banzon, J.A. (1979) Fatty acid composition of the oil from progressively maturing bunches of coconut. *Philippine Journal of Coconut Studies* 4, 1–16.
Sacks, F.M. (2020) Coconut oil and heart health, fact or fiction? *Circulation* 141, 815–817.
Sacks, F.M., Lichtenstein, A.H., Wu, J.H., Appel, L.J., Creager, M.A. *et al.* (2017) Dietary fats and cardiovascular disease. A Presidential Advisory from the American Heart Association. *Circulation* 136, e1–e23.
Sands, J.A., Landin, P., Auperin, D. and Reinhardt, A. (1979) Enveloped virus inactivation by fatty acid derivatives. *Antimicrobial Agents and Chemotherapy* 15, 27–31.
Sangamithra, A., Swamy, G.J., Sorna, P.R., Chandrasekar, V., Sasikala, S. *et al.* (2013) Coconut: an extensive review on value added products. *Indian Food Industry Magazine* 32, 29–36.
Schon, H., Gey, F., Strecker, F.J. and Weitzel, G. (1955) Metabolic studies with fatty acids of intermediate chain length. III. Feeding experiments with lauric acid esters. *Hoppe-Seyler's Zeitschrift fur Physiologische* 301, 143–155.
Simopoulos, A.P. and di Nicolantonio, J.J. (2016) The importance of a balanced ω-6 to ω-3 ratio in the prevention and management of obesity. *Open Heart* 3: e000385.
Staprans, I., Pan, X.-M., Rapp, J.H., Grunfeld, C. and Feingold, K.R. (2000) Oxidised cholesterol in the diet accelerates the development of atherosclerosis in LDL receptor- and apolipoprotein E-deficient mice. *Arteriosclerosis Thrombosis and Vascular Biology* 20, 708–714.
Sukhotnik, I., Semih Gork, A., Chen, M., Drongowski, R.A., Coran, A.G. *et al.* (2001) Effect of low-fat diet on lipid absorption and fatty-acid transport following bowel resection. *Pediatric Surgery International* 17, 259–264.
Takeuchi, H., Funabashi, J., Ooyama, K., Aoyama, T. and Sugano, M. (2006) Lymphatic transport of a physical mixture of medium- and long-chain TAG in rats. *Lipids* 41, 63–66.
Thormar, H., Isaacs, C.E., Brown, H.R., Barshatzky, M.R. and Pessolano, T. (1987) Inactivation of enveloped viruses and killing of cells by fatty acids and monoglycerides. *Antimicrobial Agents and Chemotherapy* 31, 27–31.
USDA (US Department of Agriculture) (2017) 100% Cow's Milk. USDA, Washington, DC. Available at: https://fdc.nal.usda.gov/fdc-app.html#/food-details/457325/nutrients (accessed 30 August 2023).
USDA and DHHS (US Department of Agriculture and US Department of Health and Human Services) (2015) *Dietary Guidelines for Americans 2015*. USDA/DHHS, Washington, DC.
van Dam, J.E.G. (2002) Coir processing technologies: improvement of drying, softening, bleaching and dyeing coir fibre/yarn and printing coir floor coverings. Food and Agriculture Organization of the United Nations, Rome. Available at: https://edepot.wur.nl/195898 (accessed 30 August 2023).
Varma, S.R., Sivaprakasam, T.O., Arumugam, I., Dilip, N. and Raghuraman, M. *et al.* (2019) *In vitro* anti-inflammatory and skin protective properties of virgin coconut oil. *Journal of Traditional and Complementary Medicine* 9, 5–14.
Vijayakumar, M., Vasudevan, D.M., Sundaram, K.R., Krishnan, S., Vaidyanathan, K. *et al.* (2016) A randomized study of coconut oil versus sunflower oil on cardiovascular risk factors in patients with stable coronary heart disease. *Indian Heart Journal* 68, 498–506.
Wade, L.A. (1991) A role for carnitine in medium-chain fatty acid metabolism? *Nutrition Reviews* 49, 243–245.

WHO (World Health Organization) (2003) *Diet, Food Supply and Obesity in the Pacific*. WHO Regional Office for the Western Pacific, Manila, Phillipines. Available at: https://apps.who.int/iris/handle/10665/206945 (accessed 30 August 2023).

WHO (World Health Organization) (2008) *Interim Summary of Conclusions and Dietary Recommendations on Total Fat & Fatty Acids*. Joint FAO/WHO Expert Consultation on Fats and Fatty Acids in Human Nutrition, 10–14 November 2008, World Health Organization, Geneva.

WHO (2015) Healthy diet fact sheet no. 394. Available at: www.who.int/publications/m/item/healthy-diet-factsheet394 (accessed 30 August 2023).

Yee, T.P. and Tiu, T.K. (2019) Effects of oxidised oils on inflammation-related cancer risk. *Journal of Oil Palm Research* 31, 1–13.

12 Traditional Coconut Products

Julianne M. Biddle[1,2]*, Keith Chapman[3], Ravi Pandiselvam[4] and Anandu Chandra Khanashyam[5]

[1]*The Australian Centre for International Agricultural Research, Canberra, ACT, Australia;* [2]*School of Agriculture and Food Sustainability, The University of Queensland, Gatton, QLD, Australia;* [3]*InterAgConsult, Buderim, QLD, Australia;* [4]*Indian Council of Agricultural Research – Central Plantation Crops Research Institute, Kasaragod, Kerala;* [5]*Department of Food Science and Technology, Kasetsart University, Chatuchak, Bangkok, Thailand*

Abstract

Coconut (*Cocos nucifera* L.) has always been a part of daily life and culture. Coconut has traditionally been grown largely by smallholder farmers in tropical coastal regions. These farmers tend to produce a limited harvest on small areas of land that they may or may not own. This could be fewer than 20 palms, often in a mixed farming system, along with other types of agriculture. All parts of the coconut plant have a use, and there is seemingly no limit to what parts of the palm and fruit can be used for. From food to fibre to fuel, it is an unimaginably versatile species. This chapter cannot cover all coconut products; rather, it focuses on traditional products, those that are important at the household and community level for millions of smallholder farmers worldwide. Many commercial coconut products have origins at the household level. A select group of these products are also described.

12.1 Introduction

Coconut has traditionally been a smallholder crop worldwide that has sustained itself by being valuable for the production of an extremely wide range of value-added household products. Figures 12.1 and 12.2 show the many products that can be produced from the coconut fruit and other parts of the palm (Foale, 2003). However, over the last 30 years, there has generally been less interest in coconut production, so palms are ageing and declining in annual fruit production in many locations. In the context of declining productivity, together with unattractive prices for coconut bulk commodities and disease and pest pressures, smallholders have had to resort to more value-added farm products to preserve incomes and support replanting of ageing palms. Without these traditional and new coconut product options, coconut production is simply not sustainable, as it is extremely difficult for smallholders to make a decent living from the crop as an unprocessed or bulk commodity.

Fortunately, in the early 2000s, the world began to recognize the health benefits of coconut, and governments, along with international agencies, such as the Food and Agriculture Organization of the United Nations (FAO) and the International Coconut Community (ICC; formerly the Asia-Pacific Coconut Community), and development funders began to focus on a coconut industry revival and on research and

* Email: Julianne.biddle@uq.edu.au or julianne.biddle@aciar.gov.au

 The Coconut: Botany, Production and Uses (eds S. Adkins *et al.*)
DOI: 10.1079/9781789249736.0012

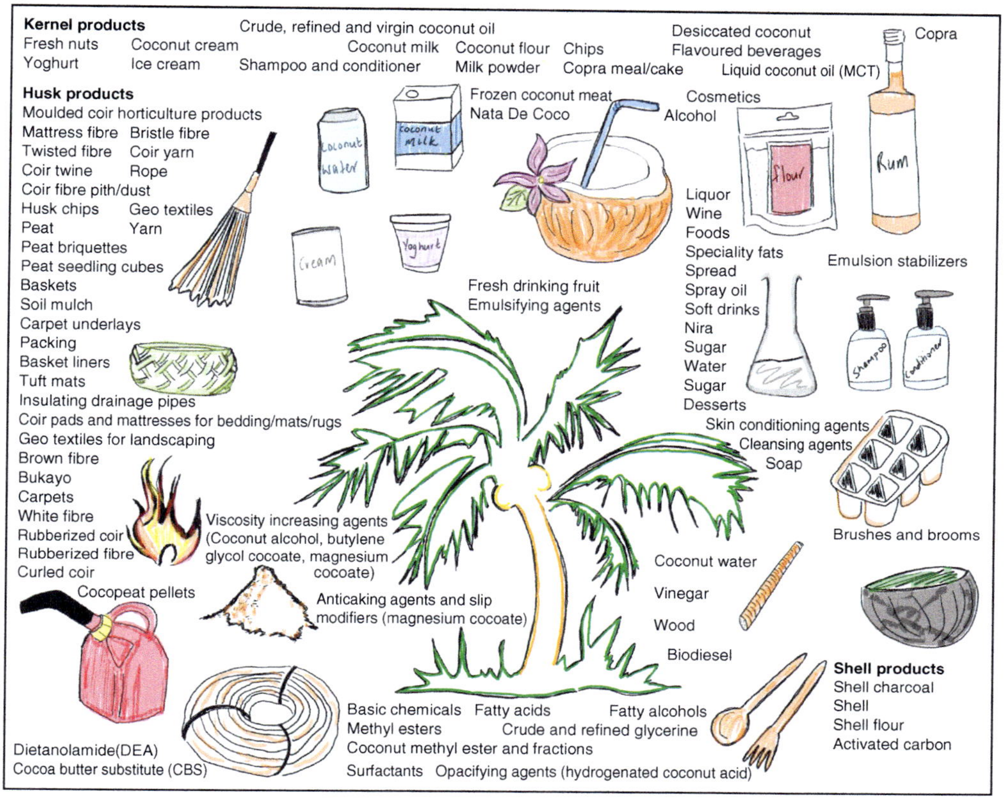

Fig. 12.1. The extremely diverse array of products that can be produced from the coconut palm.

development. As demand began to rise for the health-related nutraceutical products of coconut, this demand began to benefit smallholders and has sparked a renewal of the coconut industry in many parts of the world. In the space of the last 10 years or so, the International Conference on Coconut Oil and the Conservation and Development of Coconut Oil of Thailand Forum and interventions by governments, development donors, and the FAO and ICC have helped thousands of smallholders back to a firmer status with improved incomes. Thus began the re-emergence of an old industry that a few years prior had looked like it was fading into a sunset industry. The year 2015 saw the initiation of the Bangkok Declaration and a draft of the First International Strategic Action Plan for Coconut Oil developed at the International Conference on Coconut Oil. These actions outlined a framework for the worldwide research and development required to support the revival of the coconut industry and its smallholder partners.

The plan focused on three key components, namely: (i) processing technology and certification of coconut oil; (ii) utilization of coconut oil; and (iii) marketing of coconut oil. The plan outlined the outcomes and the actions needed to meet the goals of the three components, with the key goal being that the 'Asia Pacific Region be recognized as a reliable supplier of high-quality certified coconut oil in the global market'. The plan came at a time when demand for coconut products worldwide was increasing rapidly and actions needed to be taken to increase product supplies, so there was a focus on addressing issues such as replanting and improving productivity while meeting the need for improved product quality and product diversity.

Although coconut has been used as an important resource, with industrial use for an extremely wide range of products from explosives to soap, this chapter provides a snapshot of the diversification of products and options that can give smallholder coconut farmers improved

Fig. 12.2. Images of coconut products. (A, B) Fresh coconut-based desserts. (C–G) A wide range of commercial coconut products, including fresh drinking coconuts for sale at a café (D) and being enjoyed by the first author of this chapter (E, F) at a street market in Thailand.

incomes and livelihoods, and thus cement their continued engagement in the industry. Although some coconut-growing countries are far more engaged in the production and sale of coconut value-added products (e.g. Thailand, Indonesia, India and Pakistan), and while other countries continue to focus largely on coconut as a bulk commodity (e.g. mostly those of the Pacific Region), all coconut-growing countries have a myriad of coconut uses linked to culture and daily life. With this in mind, this chapter is not intended to be the definitive list of all products that can be produced from coconut but focuses largely on a summary of individual products produced at the household level with a brief overview of how they are made. Some that can be scaled to larger commercial processing are also covered. The chapter also discusses, to a lesser extent, factory-scale produced products, with some deeper exploration of examples related to coconut wood value-added products.

12.2 Daily Life

Coconut has been part of life and culture from pre-history to the present. In prehistoric times, coconut water, the liquid portion of young coconut fruit used as a life-sustaining drink, was the most important part of the plant. It was used to provide drinking water for long sea voyages and when available ground water was undrinkable. It is possible that it provided this service long before structured agriculture began. More recent history has seen a focus more on coconut as a food, as well as for non-food uses (Harries, 2000; Foale, 2003).

Today, coconut is a key ingredient in many foods consumed daily in coconut-growing countries, from home-cooked foods to drinks, and from meat dishes to sweets. Days can start with the collection or purchase of coconut fruit, followed by opening and processing into milk for home use. These skills, like grating coconut kernel, can be some of the first skills taught to children in their family home. Others learn to make items from coconut leaves, such as roofing materials and rope from coconut fibres. Children have been known to uproot mature coconut seedlings by the leaves while walking to school, bang them against a nearby tree to split the husk and shell, and remove and eat the coconut haustorium (coconut apple), the germinating part of the seed (Ganeshram, 2017).

Coconut is often a part of important days, such as coconut oil for the blessing of newborn children, coconut donations for religious ceremonies and foods for special occasions. In Egypt, for example, sobia is a drink enjoyed during Iftar and a coconut-based drink is used for fast-breaking during Ramadan. Similarly, in Senegal, they have coconut-rice pudding for Iftar. Egypt, Lebanon and Syria enjoy a semolina and coconut cake called hareesh or basbousa, which is favoured for many special days. Ghoriba, a small coconut biscuit, is popular with chai in Morocco (Ganeshram, 2017). Beyond consumption, coconut oil is valued as a hair and skin treatment, as a lubricant, to prevent rust, as a fuel, and for many commercial processed goods in the food, pharmaceutical, cosmetic, engineering and manufacturing industries (Divya *et al.*, 2022).

12.3 Copra

Copra, or dried coconut flesh, is the foundation for many coconut products, with the primary use being to produce crude coconut oil. Copra is produced from the endosperm (white meat) of the coconut fruit. Processing is generally undertaken close to where the coconut is grown, and the copra is produced by removing the coconut shell, the breaking up the endosperm and drying with heat, over fire (smoke), in the sun, using a kiln or hot air dryer, or more recently using solar drying systems (Divya *et al.*, 2022), as practised in the Solomon Islands (Fig. 12.3). Crude coconut oil is then extracted from this dried coconut flesh (copra). If this is done by mechanical means, the residue is generally called copra cake. If oil is then further extracted using a solvent, what remains is generally called copra meal. Copra meal has value as an animal feed, often used for dairy cattle, although the presence of aflatoxin can restrict its use, so control methods are needed to ensure quality.

Fig. 12.3. A traditional method of copra production in the Pacific region. The photo shows a fire drying system for copra production in the Solomon Islands in 2017.

12.4 Traditional Products Derived from Oil

12.4.1 Coconut oil

In many coconut-growing countries, much of the fat consumed in traditional diets comes from coconut. Coconut oil is commonly produced in the home by heating the milk until a clear oil separates (Ganeshram, 2017). Commercial production involves crushing and pressing copra (dried coconut kernel) to extract the oil (Boateng *et al.*, 2016). This copra-derived (crude) coconut oil has a long history, and commercial production has occurred in mills for over 100 years. It is one of the oldest oilseed-crushing industries (Ganeshram, 2017). It can be undertaken at a community level up to large modern ventures with automation, high-pressure expellers and solvent extraction.

The coconut oil produced can range from clear and colourless to yellow or brown in colour, with a strong coconut aroma. It is then often further refined, deodorized and bleached, using a standard vegetable oil method (Ganeshram, 2017). Coconut oil has well-established edible and inedible uses. It is most used as a cooking oil and a low-cost component of processed foods. When it is hydrogenated, the fat can be used in cooking and as a margarine- or butter-style product. Other uses include baby foods, and a flavour enhancer for a seemingly endless range of baked and processed foods. Non-edible uses include a base ingredient for soaps, cosmetics, detergents and other cleansing agents, plasticizers and synthetic resins, and fuel (biodiesel).

Coconut oil is widely used in the manufacture of cosmetics such as hair and skin products, with virgin coconut oil (VCO) having a growing use for these types of products due to its hypoallergenic properties and enhanced quality compared with crude coconut oil (see Section 12.4.3).

12.4.2 Fuel

Given current climate pressures and a push to reduce the use of fossil fuels, there is renewed interest in coconut oil and coconut oil-derived coco methyl ester as a diesel fuel replacement or as a fuel enhancer (Hossain *et al.*, 2012). To date, the Philippines, Thailand, Vanuatu and other countries in the Pacific region have used coconut oil as a replacement for diesel or as a fuel component (Bawalan and Chapman, 2006). It is anticipated that interest will continue to grow. The Philippines currently has a law that mandates the addition of biodiesel (starting from 1% and gradually increasing to a maximum of 5%) to diesel fuels for new motor vehicles being sold in the country.

12.4.3 Virgin coconut oil

VCO is a high-value coconut product extracted from fresh and mature coconut kernel by either natural or mechanical means, with or without undergoing thermal treatments. When producing VCO, the oil does not undergo chemical processing, such as bleaching and deodourizing, and can be directly used for human consumption (Bawalan and Chapman, 2006). VCO is rich in medium-chain fatty acids and has cardioprotective, anti-inflammatory, anti-stress, antimicrobial, analgesic and antipyretic properties (Dumancas *et al.*, 2016). Due to its recognized health benefits and increasing consumer trends towards edible oils that are natural and untreated with chemicals, VCO is gaining importance as a functional food and as a nutritional supplement.

VCO can be extracted from coconut kernels using a variety of processes, including solvent extraction, dry extraction and wet extraction (Agarwal and Bosco, 2017). Solvent extraction has several disadvantages over the other two methods of extraction such as a high safety risk, environmental hazards and high energy consumption. For wet extraction methods, VCO can be extracted from coconut milk with or without the application of a heat treatment. For the extraction method where heat is applied, the coconut milk is usually heated to 100–120°C for 60 min. The heat treatment destabilizes the coconut milk proteins and emulsions, resulting in the separation of VCO. In the extraction process where heat is not applied, VCO is separated from coconut milk with the aid of other separation techniques such as supercritical fluid extraction, aqueous extraction, enzymatic extraction and fermentation. In the case of dry processes, the coconut is dried to a moisture content of 3–4% and VCO is later extracted using a high-pressure expeller.

Although generally only produced on a large commercial scale, technology has been developed for home, microscale (<1000 fruit day^{-1}) and village-scale (1000–5000 fruit day^{-1}) VCO extraction (Bawalan and Chapman, 2006), as practised in locations such as the Solomon Islands (Fig. 12.4).

12.4.4 Soap

Coconut oil has good lathering properties, so it can be used in the making of soap and other household cleaning products. Coconut oil is combined with water and caustic soda to make soap. Caustic soda reacts with the fatty acids in coconut oil, and converts them into sodium salts (soap). This process requires precision to ensure that the finished product does not have excessive alkalinity, as this could burn or react with skin.

Using a cold method, soap is made by mixing one part sodium hydroxide (NaOH) to two parts water, with constant stirring to dissolve the NaOH flakes. Care needs to be taken to avoid breathing in the fumes, as the mixture can become hot. Six parts of coconut oil is added slowly while stirring in one direction until it thickens. This can take up to 2 h. Colours and fragrances can then be added, and the mixture poured into moulds to set (Fig. 12.5). The soap can be cut after 24 h but needs to be aged for at least 1 week prior to packaging or use (Ghosh, 2015).

Fig. 12.4. Village-level virgin coconut oil (VCO) production in the Solomon Islands. The images show VCO being produced and stored at a farmer cooperative in the Solomon Islands. (A) A small portable unit that can be used to expel oil. (B, C) The contents of the warehouse, including oil storage containers and purification equipment.

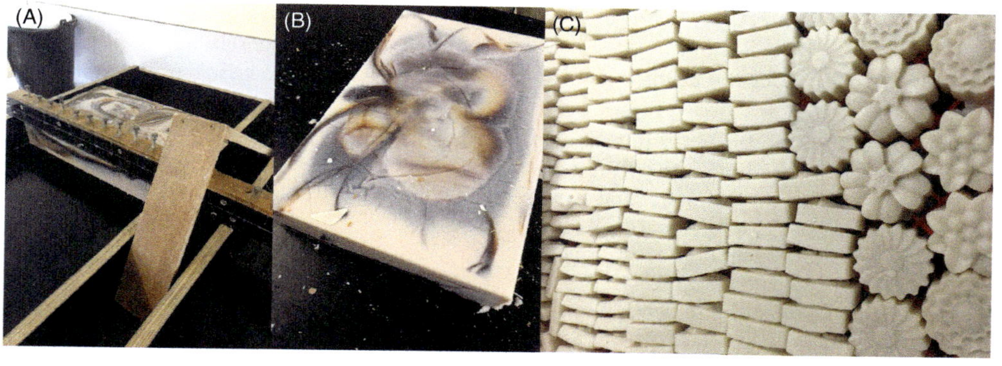

Fig. 12.5. Coconut soap production. (A) A rectangular coconut soap mould with wires for cutting. (B) A finished marble soap. (C) White coconut soap bars stacked ready for sale.

12.4.5 Hair oil

The use of hair oils extends back to time immemorial (Mysore and Arghya, 2022). In many regions, it is thought that long-term use can reduce hair loss and greying, increase shine and add volume. Hair is generally oiled by combing from the scalp to the tips weekly prior to shampooing. Coconut oil is traditionally used as a hair oil in southern India, often in coastal regions, and in South American countries, Malaysia, Indonesia and Sri Lanka. Positive beliefs regarding hair oil have led the cosmetic industry to make many commercially available oil-based hair care products. Evidence is still being collected regarding the actual benefits, but it is thought that coconut oil has the following characteristics when applied to the scalp: (i) soap-like characteristics (saponification), valuable in shampoos; (ii) antibacterial and antifungal properties; (iii) emollient properties that can be used to protect hair from drying; (iv) lubrication of hair strands, to prevent tangles and to smooth and enhance appearance; and (v) protection from protein loss by penetrating the hair shaft (Mysore and Arghya, 2022).

12.4.6 Oil pulling

Oil pulling is a widely used traditional method to improve oral health that may date back as far as 5000 years. In India, oil pulling, or Snaihik Gandoosh, is a folk remedy used to strengthen teeth, freshen breath, prevent cavities, and treat bleeding gums and mouth and lip dryness. Oil pulling involves the energetic movement of oil in the mouth, much like the use of modern commercial mouthwashes. It is also thought to heal around 30 disorders, including diabetes, asthma and migraine. Studies have examined the health benefits and there is growing evidence that it is as good as chemical oral health treatments such as toothpaste and mouthwash (Zope and Varma, 2020).

12.5 Products Produced from the Fruit

12.5.1 Coconut chips

Coconut chips are a crispy, ready-to-eat snack that can be prepared from mature coconut kernel. Today, most of the crunchy and crispy snacks available in the market are oil fried. However, present-day customers are more concerned about their health, and this has compelled snack food companies to develop low-fat items with a similar flavour and texture as the traditional fried snacks. Therefore, alternative methods for producing coconut chips have been developed where convection and osmotic drying are used to reduce the moisture content of fresh coconut slices. As the coconut kernel is rich in nutrients such as carbohydrates, minerals and healthy fat, coconut chips can be a healthier and more nutritious alternative to other fried foods like potato chips.

To produce coconut chips, mature coconut fruit of 8–9 months of age are selected. After removal of the husk, shell and testa, the kernel is cut into slices of less than 0.75 mm thickness followed by blanching and osmotic dehydration (Madhavan *et al.*, 2010). The coconut chips can then be packed and marketed in laminated aluminium pouches, and can be stored for up to 6 months.

12.5.2 Flour

Coconut flour is a screened food-quality product produced after drying and extracting oil (usually VCO) or milk from the white coconut kernel (Divya *et al.*, 2022). The kernel can be characterized based on the fibre, fat and protein composition. Coconut flour can be produced by two methods:

1. The dry method. This process involves drying the ground coconut kernel, partnered with oil extraction and pulverization of the meal. The dry method results in a high-protein flour (33%), which works well as a wheat flour substitute. This method results in a high recovery of good-quality oil.
2. The wet method. This process can be achieved at a small to medium scale. The milk is extracted from the kernel, followed by drying of the residue and then grinding to make coconut flour. When using the wet process, approximately 52% of the total oil from the endosperm is recovered. The residue can still contain a high proportion of oil, and products can be sold as VCO and flour with a high fibre content. This fibre is valuable as a functional food useful in lowering the glycaemic index and cholesterol levels in the consumer (Divya *et al.*, 2022).

12.5.3 Desiccated coconut

First manufactured in the 1880s, desiccated coconut is the dried and disintegrated kernel of mature coconut (Ganeshram, 2017). It is available in different grades, sizes and textures, from powders to shredded forms. Grated coconut is commonly produced in coconut-growing countries and consumed fresh (not dried) in home cooking. When produced and dried commercially, it is generally utilized in the confectionery, cake decorating, pastry, chocolate and biscuit industries. It is rich in dietary fibre and saturated fatty acids and contains no cholesterol.

The traditional grating of coconut kernel for home use is relatively straightforward, but in a commercial context, there are numerous steps, with many remaining manual. The major operations involved in the commercial production and processing of desiccated coconut are dehusking, deshelling, blanching/pasteurization, disintegration, drying, grading and packing. In general, 1000 fruit can yield about 100–140 kg of desiccated coconut. Before dehusking, the mature fruit are usually stored for about a month to facilitate water absorption. This aids in the easy separation of the coconut kernel from the shell. Pasteurization of the shredded coconut is usually accomplished by dipping the kernel in boiling water (>95°C) for 1–2 min. The coconut kernel then undergoes a size reduction where the kernel is cut or shredded into different shapes and sizes (a meal, long threads or slices) depending on the desired final product. These are later dried to a moisture content of 2–3% before being graded and packed.

12.5.4 Milk

Coconut milk, known as 'gata' in the Philippines and 'santan' in Indonesia and Malaysia, is the liquid gained from grating of the white solid coconut kernel, sometimes with the addition of water. When it sits, the cream comes to a layer on the top and is then used for community-level oil extraction through heating (Ganeshram, 2017).

Coconut milk is not to be confused with coconut water (see Section 12.5.7), which is the coconut liquid endosperm, and is usually served immediately as a thirst-quenching drink. Coconut milk is generally composed of around 50% water, plus fat and protein, in contrast to coconut water, which is made up of approximately 94% water (Prades *et al.*, 2002). Coconut milk is commonly used for traditional home cooking in coconut-growing countries. Coconut milk is commonly produced at the village scale, as seen in the Solomon Islands (Fig. 12.6).

12.5.5 Coconut jam

Often called coconut caramel, coconut jam is a home-made spread enjoyed in the Pacific region, particularly Samoa. It is a cooked product made from coconut milk, brown sugar and glucose. Common uses include as a spread for bread and for biscuits, dessert topping, a cake component, a fruit enhancer and a marinade (Ghosh, 2015). It has a very high sugar content of approximately

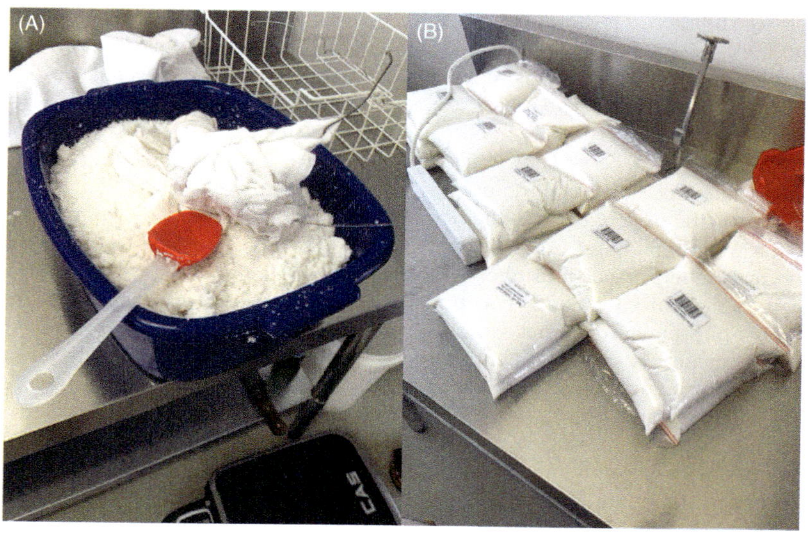

Fig. 12.6. Fresh coconut cream production in a commercial kitchen. Coconut milk is often produced at the household level in the Pacific region for home use, but this figure shows some steps in the production of coconut milk in a small commercial kitchen as part of a farmer cooperative in the Solomon Islands. The milk is for local sale, largely to local restaurants. (A) The waste left after the milk is produced. (B) The milk packaged in bags ready for fresh use or freezing for storage.

75%. It is produced from mature, undamaged fruit. The kernel from this fresh fruit (with no sign of any decay) is grated immediately after opening. Milk is extracted from the grated kernel by adding water at a ratio of 0.5 parts water to one part coconut (for a manual pressing method). No added water is needed for mechanical processing. The milk is then strained and weighed. For every 1 kg of coconut milk, 178 g of brown sugar and 88 g of glucose are needed. The sugar and glucose are initially added to half of the total amount of coconut milk and the milk is boiled slowly to dissolve it. While stirring, the mixture is further maintained at around 80°C for 10 min. The final half of the milk is added once the mixture thickens and is then boiled for another 35 min until the temperature gets to 100°C. During this process, the mixture is constantly stirred to ensure it does not burn. When a small portion of the mixture forms a firm ball in cold water, the jam is ready. It is then strained and poured into jars while hot (Ghosh, 2015).

12.5.6 Other dairy alternatives

An emerging area of interest is the option of using coconut milk as an alternative for traditional cow or goat milk-based products such as yogurt, cheese, butter and ice cream. These are largely manufactured products and are not produced as much in home kitchens. They tend to require considerable processing and the use of additives to achieve the desired texture and flavours to match traditional dairy products. For example, the production of coconut cheese is heavily reliant on the concentration of coconut protein or the use of protein from other sources. Coconut protein is generally only found in small amounts in coconut fruit derivatives and is difficult to extract and recover (Sringam, 1997). Nevertheless, methods have been developed to produce coconut cheese-like and tofu-like products, and research in the manufacture of these types of products continues, driven by strong consumer demand.

Coconut-based dairy alternatives are becoming increasingly popular in high-income countries. In Australia, numerous businesses have been established. The story of one such example, a company called COYO making coconut yogurt, is described below.

The company COYO was founded by Henry and Sandra Gosling. Henry grew up in Fiji and gained a deep understanding of coconut. He spent his youth wandering coconut plantations of Savu Savu, Fiji, and was often responsible for collecting and preparing coconut for his family. This instilled in him an understanding of the value of coconut for nourishment and culture. In June 2009, then living in Australia, Henry began looking for a new way to share the benefits of coconut. At that point he realized that he could make a unique product, yogurt. At that stage, no one was producing yogurt commercially from coconut. His aim was to produce real yogurt, containing probiotic cultures, made from coconut milk instead of the more traditional method from cow, goat or sheep milk. Thus, Henry began to explore how to produce a coconut yogurt that had the favoured characteristics of dairy yogurt – a smooth and creamy texture and exceptional taste, while maintaining the coconut and probiotic health benefits.

Henry began by handcrafting trial batches. It took 6 months of testing every day. Both Henry and Sandra were motivated by the desire to offer a superior product to people living with intolerances to dairy, lactose and gluten. In November 2009, they sampled the latest batch and found that they had discovered the ideal texture and taste. They started small, with a local health food store in Australia, and expanded to larger commercial batches. Since then, they have expanded their range of products, from natural yogurt to other flavours like vanilla, mango, berry and chocolate, and to frozen yogurt and ice cream. Their products are made from 100% natural, sustainably grown, certified organic ingredients. They use no fillers, preservatives or additives. The company is now truly international, with a footprint in other countries, including New Zealand and countries in the UK.

12.5.7 Coconut water

Not to be confused with coconut milk, coconut water (coconut liquid endosperm), often known as coconut juice is a delicious, refreshing, ancient tropical drink collected from the core of coconut fruit (Prithviraj *et al.*, 2021). Its popularity has been steadily growing on the global market in recent years, and it is currently used as a natural drink, medicine, microbial and plant growth medium and biocatalyst (Prades *et al.*, 2002). The tender coconut water (from 7–8-month-old fruit) is a rich source of ascorbic acid, vitamin B (nicotinic acid, riboflavin and biotin) and minerals such as sodium, magnesium and potassium. Coconut water is an isotonic solution with moderate acidity and a balanced sugar level, which make it an excellent rehydration and sports drink.

It is likely that the utility of coconut water, and the fruit as a natural transport mechanism, played a large part in the domestication of coconut. It would have been of importance to peoples in coastal regions as it offered nourishing, safe drinking water in locations where fresh water may have been scarce. It is thought that it was vital during sea voyages, particularly the movement between islands that occurred in the Pacific region (Clement *et al.*, 2013).

When fruit are at peak coconut water content, with no air in the cavity (no noise when shaken), they need to be cut from the palm. This is when the fruit reaches maximum size and the kernel is still a thin jelly. They must be lowered to the ground gently. If they are dropped, they can get damaged and leak. The immature fruit has a short shelf life, only lasting a matter of days. Modern cold-chain transport can extend the possible storage time.

One of the major factors that determines the quality of coconut water is its maturity (Pandiselvam *et al.*, 2019, 2020). As the coconut fruit matures, the volume of liquid endosperm inside the fruit starts to reduce, and the kernel starts to change from a jelly-like substance to a hard, thick, white coconut kernel. Significant changes also happen to the sugar content of the liquid during maturation. The invert sugar concentration of the coconut water increases during the early development stages, reaching a maximum at 220 days. After that, the sugar content tends to decrease, making mature coconut water a less popular beverage.

Coconut water is traditionally consumed directly from the fruit immediately after the fruit is picked or after a short period of refrigeration. It also has commercial uses as a packaged beverage, for other food products, and as the foundation liquid for plant tissue culture due to its chemical composition. Research has demonstrated its value for oral rehydration, and possible protection from myocardial infarction and hypertension (Jangchud *et al.*, 2007; Choeisunthon and Tongchitpakdee, 2011; Gordon and Jackson, 2017). Images of the processes involved in coconut water production in Thailand are shown in Fig. 12.7.

Snowball tender coconut, packaged coconut water and nata de coco are the major value-added products from coconut water produced commercially. Snowball tender coconuts are made from 7–8-month-old fruit by removing the husk, shell and testa. After processing, the snowball tender coconut looks like a white ball and contains coconut water, which can be consumed by piercing a straw through the tender coconut meat. The Central Plantation Crops Research Institute, Kerala, successfully developed the technology to process snowball tender coconuts (Manikantan *et al.*, 2019). Nata de coco is produced from coconut water with the action of a certain cellulose-forming bacterium (*Acetobacter* spp.). It is a gelatinous substance and can be used as a dessert when preserved in sugar syrup (Ghosh, 2015).

12.5.8 Biochar

Interest in biochar is growing as it is eco-friendly and has properties valuable for carbon sequestration, soil quality improvement and bioremediation. It is also involved in the production of biocomposites and bioenergy (Neogi *et al.*, 2022). Biochar can be produced from a wide range of waste materials, such as lignocellulosic biomass, including from coconut-shell waste, sewage and manure (Castilla-Caballero *et al.*, 2020).

Coconut-shell biochar is a carbon-rich product produced by heating the shells in the partial or full absence of oxygen, otherwise known as pyrolysis. Pyrolysis of biomass to produce biochar is generally conducted at or above 500°C to ensure a changed chemical composition and to break down biopolymers. Coconut biochar is valued as it is highly porous and is generally alkaline. These properties make it ideal to improve soil characteristics including the organic matter content, water-holding properties,

Fig. 12.7. Images of steps in coconut water production in Thailand. (A) Fruit ready to begin processing. (B, D) Manual dehusking, with the fruit stored in baskets. (C, F) Washing of fruit. (E) Bulk coconut water storage containers.

microorganism composition, and pH of the soil. It has demonstrated benefits for plant productivity and a reduction in nitrogen leeching from the soil (Zhao *et al.*, 2019). In the current context of agriculture input prices increasing globally, the prevention of fertilizer leaching from soil and the possibility of reduced fertilizer use make biochar an attractive additive for many farming systems, which can be produced at the village scale (Fig. 12.8).

12.5.9 Activated charcoal

Activated carbon (charcoal) is often grouped into wood-based, coal-based, coconut and other fruit shell-based carbon. Of all the types, coconut-activated carbon generally fetches the highest price. It is a product that is commercially produced, and activated coconut-shell carbon has generally been used in gas purification and the recovery of solvents, but more recently, uses have widened to water and air purification. Activated carbon from coconut is often considered to be an all-round absorber of a higher quality than other sources like carbon from coal, due to a more consistent particle size, (Ghosh, 2015).

Environmental protections and reforms to the coal industry, combined with emissions considerations, are likely to result in a decline in coal-based activated carbon and increases in coconut and other renewable activated carbons, although production methods for coconut still require refinement to reduce the environmental impacts of production.

12.6 Products from Other Parts of the Palm

12.6.1 Heart of palm

Easily comparable with green leafy vegetables, the coconut palm heart (large bud), gained from a cut palm with the stalks cut away, can be eaten fresh as a salad ingredient but is also able to be preserved by cooking, pickling or canning (Harries, 2000). Heart of palm is a commercially produced product from other palm species, but this is not the case for coconut, probably due to the reluctance to cut down coconut palms even when they are old (senile) and unproductive. This may reflect their cultural significance and value as a food source that may remain and

Fig. 12.8. Coconut copra and biochar production in the Solomon Islands. (A, B) A complete unit, with a large metal tray for producing copra above and a cavity to produce coconut biochar below. Coconut shells are loaded into the base through the door shown in (B) and set alight, while coconut meat is dried using heat on the top tray. Once cool, biochar can be removed from the door in (B).

support the community after conflict or extreme weather events like cyclones and tidal waves. During the civil war in Liberia, coconut heart of palm was only consumed as a last resort by a starving population (Ganeshram, 2017).

12.6.2 Woven goods

Coconut leaves and midribs have a range of uses including brooms and handicrafts (Ghosh, 2015). Coconut palm leaves are traditionally used for woven products, ranging from swiftly woven leaf baskets for rubbish collection to intricate baskets made for special occasions, hats and other ornamental items (Fig. 12.9).

In Vietnam, gift giving is a large part of special occasions, for example Mother's Day and the Tet Festival. These gifts are generally food hampers contained in hand-woven baskets. These baskets are traditionally made from a bamboo frame covered with woven coconut leaf midribs. They are made into a range of shapes and sizes, for example the lang hoa cong dua (a flower-arrangement basket), gio cong dua (a basket with a handle) and ro cong dua (a rectangular-shaped basket). Many of the premium midrib baskets are produced in the Hung Phong community in Ben Tre Province. In 2006, approximately 30% of households in that region were participating in basket production or sale (Keizer and Le Thuy, 2006).

12.6.3 Sugar

Coconut sugar (also known as jaggery) is a product obtained from toddy (phloem sap), a liquid that is collected by tapping the unopened coconut spadix (Hebbar *et al.*, 2022). It is traditionally used in many cuisines in countries such as India, Indonesia and the Philippines. To collect the toddy, the juice collector climbs the palm and beats uniformly over the surface of the spathe for about 3 weeks. This ruptures the cells and increases the sap flow to the spadix. Then, using a sickle, a cut of about 4–5 cm is made on the unopened inflorescence and the sap is collected using a container (Fig. 12.10). The toddy contains around 15 g of sugar in every 100 g of sap. The unfermented toddy is heated to 118–120°C with constant stirring until the sugar crystallizes (Fig. 12.10). The solid mass obtained is generally

Fig. 12.9. Beautiful woven objects from coconut leaves. (A) A hand-woven cricket. (B, C) A hand-woven basket.

Fig. 12.10. Images of some of the steps in coconut sugar production in Thailand. (A–C) Sap being collected high in a coconut palm canopy with a ladder (A), used for regular gathering of the sap in containers (B, C). (D) The collected liquid is reduced over a flame to produce coconut sugar.

light to dark brown in colour and is often called coconut jaggery (Pravitha *et al.*, 2021). This is later poured into moulds for hardening. Coconut jaggery contains around 68.4% sugar, 10.9% water and 2.2% ash. For the preparation of white sugar, 2% lime is initially added to the toddy to eliminate albuminous impurities. The limed toddy is then carbonated twice, each time being filtered to eliminate excess lime. This is later concentrated to 75% sugar content using an evaporator. After crystallization is completed, the crystalline sugar is separated by centrifugation.

12.6.4 Toddy

Traditional non-alcoholic and alcoholic beverages are produced from the watery phloem sap, also known as toddy. Collection is the same as for sugar production (see Section 12.6.3). In a process probably dating back to before the 17th century, the sap is collected from the unopened inflorescence tissues of many other palms (Ganeshram, 2017). Like many other fruit juices, the toddy collected can be fermented or distilled into wine and other products. Coconut water has properties that mean it can also be fermented, but this activity is less common.

Given that toddy is sweet and the collection containers are generally rinsed but not thoroughly cleaned, combined with the tropical conditions ideal for coconut growth, this means that alcohol production is almost inevitable. Toddy is generally collected morning and evening, and the morning collection can have about 3% alcohol and 10% sugar, while fermentation for over 30 h results in an 8% alcohol wine, commonly known as arrack. When put through a double-distillation process, it forms the foundation for spirits such as rum and gin. Many regions have their own type with unique flavours added. Like other fruit wines, toddy can be used as the foundation for vinegar. The process takes 6 months and involves maturation in closed casks, often with the addition of spices and caramel colouring (Ganeshram, 2017).

12.6.5 Coconut wood

Coconut wood has been widely used by tropical communities for building construction, furniture and household items (Keanu, 2020). With shortages in locally produced timber and expensive transport for imported timbers, there is a keen interest for coconut to be used more (Suprobo and Santosa, 2017; Keanu, 2020). This has led to the development of enhanced methods for coconut wood processing, and the production of coconut wood-processing manuals (Killmann and Fink, 1996; Bailleres *et al.*, 2010). Although coconut wood is not considered to be a mainstream wood source, these recent advances in wood technology are likely to enable a wider use. Timber laminating methods have the potential to be applied to coconut to produce a uniform, fit-for-purpose, structurally sound wood product that is resistant to humidity and reduces the need for nails, which may corrode over time.

12.6.6 Coir

Coconut husk fibre, or coir, was historically valued due to its water-resistant properties for rope making. It is likely that along with the use of coconut water for sea journeys, coir rope also played an important role in early voyages. Although it has largely been replaced by synthetic products, renewed interest in coir-based products has been sparked by people seeking clean, green alternatives to petroleum-based products.

A high lignin content and durability have ensured that coconut husk by-products are the first choice for many uses (Fig. 12.11). Modern applications of coir, alone or combined with other fibres, include cocopeat from processed coir (widely used in horticulture as a soil growth medium and soil conditioner), supercapacitors and conducting fibres, mattress pads, geotextile nets and biologs for environmental applications, and many more applications (Fig. 12.11).

12.7 Conclusion

In tropical coastal regions where coconut is grown, it is intertwined through life and culture. This chapter has covered a wide range of traditional coconut products generally produced at the household or community level. Although there is a growing global market for most coconut products, coconut production in many regions is in decline. Although palms are ageing and have declining annual fruit production, they are often not being replaced with young seedlings. This reflects unattractive prices for coconut as a bulk commodity, combined with growing disease and pest challenges. Smallholder farmers are resorting to more value-added

Fig. 12.11. Coconut husks waiting for processing into coir products.

products to preserve incomes. Many of these products have traditional origins. Without these traditional and new coconut product options, coconut production may not be sustainable for many farmers, as it is extremely difficult for smallholders to make a decent living from the crop as an unprocessed or bulk commodity.

In the context of increasing international interest in sustainable food, fibre, fuel and manufacturing options in the face of climate change, loss of biodiversity and environmental degradation, the coconut palm may well provide many of the solutions. However, it cannot be assumed that by the coconut industry doing well, smallholder farmers will automatically benefit. This has long been recognized, yet coconut farmers are still some of the poorest farmers worldwide. Overcoming this issue is going to require collaboration, careful engagement, and policies and structures to ensure that smallholder farmers benefit.

12.8 References

Agarwal, R.K. and Bosco, S.J.D. (2017) Extraction processes of virgin coconut oil. *MOJ Food Processing and Technology* 4: 00087.

Bailleres, H., Hopewell, G., House, S., Redman, A., Francis, L. *et al.* (2010) *Cocowood Processing Manual. From Coconut Wood to Quality Flooring*. Department of Employment, Economic Development and Innovation, Brisbane, Australia.

Bawalan, D.D. and Chapman, K.R. (2006) *Virgin Coconut Oil. Production Manual for Micro and Village Scale Processing.* FAO Regional Office for Asia and the Pacific, Bangkok. Food and Agriculture Organization of the United Nations, Rome.

Boateng, L. Ansong, R., Owusu, W. and Steiner-Asiedu, M. (2016) Coconut oil and palm oil's role in nutrition, health, and national development: a review. *Ghana Medical Journal* 50, 189–196.

Castilla-Caballero, D., Barraza-Burgos, J., Gunasekaran, S., Roa-Espinosa, A. and Colina-Márquez, J. (2020) Experimental data on the production and characterization of biochars derived from coconut-shell wastes obtained from the Colombian Pacific Coast at low temperature pyrolysis. *Data in Brief* 28: 104855.

Choeisunthon, R. and Tongchitpakdee, S. (2011) Effect of maturation on quality changes of pasteurized aromatic coconut water. In: *Proceedings of the 49th Kasetsart University Annual Conference,* Kasetsart University, Chatuchak, Bangkok, pp. 658–664.

Clement, C.R., Zizumbo-Villarreal, D., Brown, C.H., Ward, R.G., Alves-Pereira, A. *et al.* (2013) Cocoteros en las Américas. *Botanical Review* 79, 342–370.

Divya, P.M., Roopa, B.S., Manusha, C. and Balannara, P. (2022) A concise review on oil extraction methods, nutritional and therapeutic role of coconut products. *Journal of Food Science and Technology* 60, 441–452.

Dumancas, G.G., Viswanath, L.C.K., de Leon, A.R., Ramasahayam, S., Maples, R. *et al.* (2016) Health benefits of virgin coconut oil. In: Holt, B. (ed.) *Vegetable Oil: Properties, Uses, Benefits*. NOVA, Burleigh, Australia, pp. 161–194.

Foale, M. (2003) *Coconut Odyssey: The Bounteous Possibilities of the Tree of Life*. Australian Centre for International Agricultural Research, Canberra, Australia.

Ganeshram, R. (2017) Cracking Coconut's History. AramcoWorld, January/February 2017. Available at: www.aramcoworld.com/en-US/Articles/January-2017/Cracking-Coconut-s-History (accessed 1 September 2023).

Ghosh, D.K. (2015) Postharvest, product diversification and value addition in coconut. In: Sharangi, A.B. and Datta, S. (eds) *Value Addition of Horticultural Crops: Recent Trends and Future Directions*. Springer, New Delhi, pp. 125–165.

Gordon, A. and Jackson, J. (2017) Case study: application of appropriate technologies to improve the quality and safety of coconut water. In: Gordon, A. (ed.) *Food Safety and Quality Systems in Developing Countries*. Elsevier, pp. 185–216.

Harries, H. (2000) Coconut. In: Kiple, K. and Ornelas, K. (eds) *The Cambridge World History of Food*. Cambridge University Press, Cambridge, Massachusetts, pp. 388–397.

Hebbar, K.B., Ramesh, S.V., Ghosh, D.K., Beegum, P.P.S., Pandiselvam, R. *et al.* (2022) Coconut sugar – a potential storehouse of nutritive metabolites, novel bio-products and prospects. *Sugar Technology* 24, 841–856.

Hossain, M.A., Chowdhury, S.M., Rekhu, Y., Faraz, K.S. and Islam, M.U. (2012) Biodiesel from coconut oil: a renewable alternative fuel for diesel engine. *International Journal of Chemical and Molecular Engineering* 6, 524–528.

Jangchud, K., Puchakawimol, P. and Jangchd, A. (2007) Quality changes of burnt aromatic coconut during 28-day storage in different packages. *Food Science and Technology* 40, 1232–1239.

Keanu, W.K.W. (2020) Coconut lumber: an assessment of old and new approaches and methods of using coconut wood. PhD thesis, University of Hawaii, Manoa, Hawaii.

Keizer, M. and Le Thuy, N.T. (2006) *Building the Market Chain. Coconut in Vietnam. Midrib Baskets: The Art of Giving*. Bioversity International, Rome. Available at: https://hdl.handle.net/10568/105315 (accessed 1 September 2023).

Killmann, W. and Fink, D. (1996) *Coconut Palm Stem Processing: Technical Handbook*. Deutshe Gesellschaft fur Technische Zusammenarbeit (GTZ) GmbH, Frankfurt, Germany.

Madhavan, K., Arumuganathan, T. and Mathew, A. (2010) *Commercial Production of Coconut Chips*. CPCRI Technical Bulletin No. 62. Indian Council of Agricultural Research/Central Plantation Crops Research Institute, Kasaragod, Kerala.

Manikantan, M.R., Pandiselvam, R., Beegum, S. and Mathew, A.C. (2019) *A Guide to the Rotary Veneer Processing of Coconut Palms*. ACIAR Monograph Series. Australian Centre for International Agriculture, Canberra.

Mysore, V. and Arghya, A. (2022) Hair oils: indigenous knowledge revisited. *International Journal of Trichology* 14, 84–90.

Neogi, S., Sharma, V., Khan, N., Chaurasia, D., Ahmad, A. *et al.* (2022) Sustainable biochar: a facile strategy for soil and environmental restoration, energy generation, mitigation of global climate change and circular bioeconomy. *Chemosphere* 293: 133474.

Pandiselvam, R., Manikantan, M.R., Subhashree, N., Mathew, A.C., Balasubramanian, D. *et al.* (2019) Correlation and principal component analysis of physical properties of tender coconut (*Cocos nucifera* L.) in relation to the development of trimming machine. *Journal of Food Process Engineering* 42: e13217.

Pandiselvam, R., Hebbar, K.B., Manikantan, M.R., Prashanth, B.K., Beegum, S. *et al.* (2020) Microwave treatment of coconut inflorescence sap (Kalparasa®): a panacea to preserve quality attributes. *Sugar Technology* 22, 718–726.

Prades, A., Dorner, M., Diop, N. and Pain, J.P. (2002) Coconut water uses, composition and properties: a review. *Fruits* 67, 87–107.

Pravitha, M., Manikantan, M.R., Ajesh Kumar, V., Beegum, S. and Pandiselvam, R. (2021) Optimization of process parameters for the production of jaggery infused osmo-dehydrated coconut chips. *Food Science and Technology* 146: 111441.

Prithviraj, V., Pandiselvam, R., Babu, A.C., Kothakota, A. and Manikantan, M.R. *et al.* (2021) Emerging non-thermal processing techniques for preservation of tender coconut water. *Food Science and Technology* 149: 111850.

Sringam, S. (1997) *The Development and Testing of a Coconut Cheese Production Technology*. United Nations Report No. 97. Project US/RAS/90/132, United Nations Industrial Development Organization, Vienna.

Suprobo, F.P. and Santosa, A. (2017) Models of furniture design using coconut wood based on local culture for global and domestic markets. *Creative Industries Journal* 10, 89–99.

Zhao, F., Zou, G., Shan, Y., Ding, Z., Dai, M. *et al.* (2019) Coconut shell derived biochar to enhance water spinach (*Ipomoea aquatica* Forsk) growth and decrease nitrogen loss under tropical conditions. *Scientific Reports* 9: 20291.

Zope, S.A. and Varma, S. (2020) Effect of oil pulling on oral health. In: Chauhan, D.N., Singh, P.R., and Chauhan, N.S. (eds) *Natural Oral Care in Dental Therapy*. Scrivener Publishing, Beverley, Massachusetts, pp. 146–152.

13 Global Coconut Trade, Economy and the Value Chain

Sundaravelpandian Kalaipandian[1]*, Julianne Biddle[2] and Steve W. Adkins[1]
[1]*School of Agriculture and Food Sustainability, The University of Queensland, Gatton, QLD, Australia;* [2]*Australian Centre for International Agricultural Research, Canberra, ACT 2617, Australia*

Abstract

Coconut is an important crop grown in over 90 countries with the majority on smallholder lands. Demand for its products increases every year; however, the coconut industry is facing constraints due to fruit production shortfalls, price fluctuations, absence of accurate international market forecasts, poor public–private partnerships, and a lack of skilled labour and product processing facilities, all affecting the value chain from production and processing to consumption. With such instability, the industry may not meet the growing demand for coconut products unless improvements are made. The value chain is complex, with a seemingly endless abundance of coconut product options, but the global trade, economy and value chain of only a subset of products have been well studied, and in only a few countries. Studies on the full suite of coconut products are necessary to understand the market and to inform decision making to minimize industry instabilities and increase the income of coconut smallholders.

13.1 Introduction

Coconut (*Cocos nucifera* L.) is called the 'Tree of Life' in most parts of the world due to its versatility and endless uses. Coconut has been utilized by humankind for about 500, 000 years as a source of food, drink, oil, shelter and wood (Foale, 2003). The palm is grown on approximately 12 million ha around the world, with an annual production of around 65 billion fruit (ICC, 2022). Although over 90 countries cultivate coconut, more than 75% of the production comes from just three countries: India, Indonesia and the Philippines. Recently, the demand for coconut fruit and its products has increased substantially around the world as people have become aware of its health benefits, especially those of coconut water and virgin coconut oil (VCO; Rethinam, 2005). In addition, the demand for numerous other coconut products is also increasing exponentially (e.g. coconut milk, cream, flour and sugar), while the traditional products, such as desiccated coconut and copra, are still experiencing constant demand (Sri Lanka Export Development Board, 2017).

In the past, the coconut industry was dominated by the trading of copra, which is the raw endosperm material used to produce crude coconut oil. In the early commercial years, the international market was highly dependent on coconut oil for the manufacture of soap and candles (Prades *et al.*, 2016). In recent years, many new products have been introduced to the market, often requiring more complex processing.

* Email: s.kalaipandian@uq.edu.au

DOI: 10.1079/9781789249736.0013

These new products have dramatically altered coconut production and trading systems (Lin, 2018), but there is still plenty of room for the industry to evolve further.

Traditionally, the coconut palm has been strongly connected to the social, economic and cultural lives of around 80 million people in 92 countries (Naveena *et al.*, 2014). Although the demand for coconut products is constantly increasing, the price fluctuations of its fruit are a constant worry for growers. No reliable price modelling and forecasting approaches are available to minimize the burden on the coconut value chain. A recent study has suggested that other interacting factors, such as price, production, climatic variables, their interactions and various external factors, also need to be considered in the existing price modelling studies (Abeysekara and Waidyarathne, 2020). Most coconut farming is undertaken on plots of land no bigger than 0.5–4.0 ha (Prades *et al.*, 2016), and it is on such plots that 90% of the world's production takes place (OECD/FAO, 2017). Hence, the coconut supply chain plays an important role for smallholder farmers. Value-added coconut products are now increasingly exported around the world as healthy snacks and ingredients for cosmetics and other products. Such products are becoming increasingly popular and are being marketed in most supermarkets and specialty shops. For example, VCO and coconut water, sugar and flour are available in most shops around the world. Coconut is considered one of the top ten trending commodities in the world food market (Khaw *et al.*, 2018).

International coconut export trade is dominated by four countries, which contribute 80% of the trade: Indonesia, the Philippines, India and Sri Lanka (Jayasekhar and Thampan, 2016). These four countries also represent about 80% of the production area of coconut cultivation around the world. Although these countries have substantial industries, much of the current focus is on coconut as a bulk commodity. As higher-value coconut products become increasingly popular, it will be important for the coconut trade industry to concentrate on these products, as well as developing new products to enhance the prospects of creating a sustainable coconut sector, rather than one that is overly dependent on copra and crude oil (Jayasekhar and Thampan, 2016).

The *Coconut Statistical Yearbook for 2020* (ICC, 2022) reports on several important coconut trade data sets from around the world. In 2020, the total world area under coconut cultivation was over 12 million ha, with the Philippines continuing to dominate in terms of the area cultivated with more 3.6 million ha, followed by Indonesia and India, with over 3.3 and 2.1 million ha, respectively (Table 13.1; ICC, 2022). In addition to these countries and other South Asian countries, Tanzania, Papua New Guinea and Brazil are in the upper part of the producers list, with approximately 0.60, 0.22 and 0.18 million ha, respectively (Table 13.1; ICC, 2022). The previous top 10 producing countries, from 2015 to 2020, are also presented (Table 13.1; ICC, 2022). During this time, more than 65 billion fruit were produced from all coconut-producing countries, and India

Table 13.1. Area of land under coconut production in the top 10 countries and around the world from 2015 to 2020 (× 1000 ha). Data from ICC, 2022.

Country	2015	2016	2017	2018	2019	2020
Philippines	3,518	3,565	3,612	3,628	3,652	3,651
Indonesia	3,586	3,664	3,473	3,418	3,402	3,397
India	1,975	2,088	2,082	2,097	2,150	2,173
Tanzania	590	617	628	642	638	601
Sri Lanka	440	440	440	440	444	444
Papua New Guinea	221	221	221	221	221	221
Brazil	242	234	209	199	187	187
Vietnam	162	165	165	155	159	159
Thailand	192	186	179	121	124	125
Mexico	124	124	125	124	124	124
Total area in the world	12,200	12,447	12,284	12,223	12,276	12,258

dominated production, with about 20 million fruit, followed by the Philippines and Indonesia with 14.5 and 14.0 million fruit, respectively (ICC, 2022). These three countries alone contributed to 75% of world coconut production in 2020. Brazil, Papua New Guinea and Mexico were the major contributors other than the Asian countries, and the top 10 producers are listed in Table 13.2 (ICC, 2022).

The Philippines is the top producer of copra, with 1.8 million t, followed by Indonesia and India, with 1.5 million and 1.1 million t, respectively. In addition to these three major producers, the Pacific countries (e.g. Papua New Guinea and Vanuatu) and African countries (e.g. Ivory Coast and Mozambique) also contribute to copra production (Table 13.3; ICC, 2022). In 2020, the world consumption of desiccated coconut was more than 0.3 million t, and the export trends of various coconut products, such as fresh and dried coconut kernel, copra, coconut oil, copra meal, desiccated coconut, coir products, coconut charcoal, and coconut milk, cream and powder, also increased (Table 13.4; ICC, 2022). This growing trend in coconut product exports indicates a clear interest in coconut products, especially for value-adding products. Demand for coconut products has increased even in non-coconut-producing countries, which was clear from the recent statistical data published by the International Coconut Community (ICC, 2022). The global coconut market is estimated to be growing at approximately 11.3% annually, and the volume of products by about 79.9% (Alouw and Wulandari, 2020).

Broadly speaking, coconut products are marketed based on four core value-chain strands: food, oleochemicals, water and by-products. Food is predominantly based on oil from copra, and more recently cream, milk and VCO from the fresh kernel. Oleochemicals derived from oil are used for making soaps, personal care products and cosmetics. Water is a more recently emerging market, and by-products are derived from the husks, shells, timber and leaves of the coconut palm (McGregor and Sheehy, 2017). According to recent data from the International Coconut Community, coconut products are important commodities in the global market and contribute financially to the livelihoods of smallholders and the economy of the low- and lower–middle-income countries (ICC, 2022). Although coconut is a very important agricultural commodity, the trade and economics of its market are poorly studied. It is also essential to study the role of individual political systems and the impact of international institutions to help coconut-producing countries improve their industries and their smallholder farmers (Lin, 2021). This chapter focuses on the value chain and the economic status of the coconut industry from key coconut-producing countries. In addition, this chapter covers various constraints in smallholder coconut production, the processing industries, price fluctuations, and market opportunities for traditional and new products in the global market, and explores how these constraints can be overcome to make coconut production a profitable farming endeavour for smallholder farmers.

Table 13.2. Production of coconut fruit from the top 10 countries and from around the world from 2015 to 2020 (× 1000 fruit). Data from ICC, 2022.

Country	2015	2016	2017	2018	2019	2020
India	20,440	22,167	23,904	23,798	21,288	20,309
Philippines	14,735	13,825	14,049	14,726	14,765	14,491
Indonesia	14,603	14,521	14,272	14,201	14,140	13,994
Sri Lanka	3,056	3,011	2,450	2,623	3,086	2,792
Brazil	2,679	2,634	2,210	2,345	2,349	2,459
Vietnam	1,434	1,471	1,499	1,572	1,677	1,677
Papua New Guinea	1,483	1,483	1,483	1,483	1,483	1,483
Mexico	893	926	927	926	908	895
Thailand	723	708	666	686	645	591
Malaysia	506	505	517	495	537	561
Total production of fruit in the world	66,758	67,534	68,255	69,325	67,308	65,671

Table 13.3. Production of copra from the top 10 countries and from around the world from 2015 to 2020 (× 1000 Mt). Data from ICC, 2022.

Country	2015	2016	2017	2018	2019	2020
Philippines	1,934	1,719	1,913	2,033	2,372	1,869
Indonesia	1,316	1,280	1,170	1,640	1,560	1,550
India	1,072	1,244	1,306	1,327	1,141	1,161
Papua New Guinea	87	90	114	106	106	106
Malaysia	59	59	69	73	74	79
Thailand	7	23	34	70	98	43
Mexico	42	41	41	44	44	40
Vanuatu	19	34	30	30	30	30
Ivory Coast	29	29	29	30	30	30
Mozambique	20	20	20	20	20	20
Total production of copra in the world	4,930	4,757	4,873	5,511	5,642	5,056

Table 13.4. Exports of various coconut products from around the world from 2015 to 2020 (in Mt). Data from ICC, 2022.

Products	2015	2016	2017	2018	2019	2020
Fresh/dried coconut	841,711	902,341	1,081,760	650,712	681,486	631,874
Copra	227,511	207,344	180,471	209,486	318,796	215,457
Coconut oil	2,183,275	1,872,083	1,978,054	2,125,918	2,365,192	1,982,021
Copra meal	706,767	543,498	564,615	756,183	588,941	433,687
Desiccated coconut	376,505	391,880	398,652	401,060	427,834	458,296
Coir and coir products	229,054	238,553	267,282	254,069	292,254	358,252
Coconut shell charcoal and activated carbon	398,759	347,736	442,155	570,249	620,883	571,929
Coconut milk/cream	230,941	264,065	315,147	346,679	360,833	370,281
Coconut milk powder	14,256	14,267	13,824	13,091	10,201	9,526

13.2 Indonesia

Coconut is one of the important strategic commodities of the Indonesian agriculture sector as it has a high export potential. About 6.3 million farmers depend on coconut for their livelihood, and most of them (98%) are smallholders (Alouw and Wulandari, 2020). Although coconut exports increased to more than US$1.3 billion in 2018, Indonesian coconut export revenues were not able to reach the level of India or the Philippines. In Indonesia, the top two export commodities were coconut oil and copra for crude coconut oil, which contributed US$371.9 and US$354.7 million, respectively, in 2018. Coconut cultivation was carried out on around 3.4 million ha, with approximately 14.3 billion fruit year^{-1} produced in 2017 (Alouw and Wulandari, 2020). At least 11.7% of the smallholder cultivations are now thought to be senile, and this contributes directly to the low fruit production. In addition to an increasing number of senile palms, there is a lack of appropriate capacity in rural areas to develop value-added products, which compounds the decreasing income of smallholders.

Although Indonesia is one of the largest coconut producers and exporters in the world, little research has been undertaken to develop appropriate supply chains. Recently, Gunawan *et al.* (2021) reported that the low production and productivity are major issues in the Indonesian coconut value chain. Although Indonesian coconut products are in demand in international markets, no national companies have stimulated interest in developing new areas of coconut cultivation. Currently, VCO is selling at a premium price and those companies producing it could take a lead and increase the area of coconut cultivation in Indonesia for this product. The results from a simulation model developed by Gunawan *et al.* (2021) showed that the supply and demand ratio would fall below 1.0 by 2023, indicating that supply will no longer meet demand. This model based on the system dynamics approach suggests that the most effective policy to meet demand would be to double the coconut cultivation area and increase the yield up to 2 t of copra ha^{-1} (Gunawan *et al.*, 2022).

The majority of the coconut producers are smallholders, and processing is often operated on a small scale by their family or with a few labourers. These businesses are usually small to medium in size, and they process the coconut products before selling to wholesalers or retailers. Such small to medium enterprises (SMEs) are very important in coconut business systems. For example, about 99% of labourers work in SMEs and contribute 55% of the gross domestic product in Indonesia (Suselo, 2012). Although SMEs are the backbone of the economy, they do not actively participate in international business. For example, Indonesia is the largest producer of coconut sugar, driven by SME production, but the 120,000 t $year^{-1}$ produced is marketed by international business agents. Indonesia needs to address several problems including those of production, human resources, financial and marketing. Suselo (2012) pointed out that these SMEs need support for capacity building, operational training, workforce building, marketing and promotion to become better involved in international business.

In 2017, Indonesia dropped down to the world's second largest area under coconut cultivation, losing first position to the Philippines. This was mainly due to ongoing productivity loss from inferior and senile palms, and an increased pest and disease attack, all causing financial constraints for smallholder farmers. Such income losses could be enhanced through replanting programmes, producing high-value-added products, access to better processing infrastructure and export markets for the high-value products, and introducing intercrops within their coconut plantations. Such improvements in cultivation practice and value-adding will not only increase the income of smallholders but also enhance the national income, generate employment opportunities in rural areas, and ultimately meet the demand of consumers for food, oil and beverages. The Indonesian Ministry of Agriculture is implementing several programmes to increase coconut production by investing in research and development to produce superior varieties, expand cultivation areas and revitalize the coconut industry in urban areas by utilizing high-yielding varieties. In addition, the Ministry is investing in the implementation of good agricultural practices, improvement of partnerships between farmers and the broader industry, developing plant protection, intensification and diversification of cultivation, and in micropropagation through embryo and tissue culture for increased planting (Alouw and Wulandari, 2020).

13.3 The Philippines

Coconut is the major agricultural industry employing the most labourers in the Philippines, and its export market is very important to the country's economy (Moreno *et al.*, 2020). From 1961 to 2017, coconut products such as raw coconut kernel, coconut oil, desiccated coconut and coconut cake experienced a decline in production, which correlated directly with a reduction in fruit production (FAO, 2018). The situation has improved since then and the Philippines is currently ranked as the second largest producer of coconut in the world. At least one-third of the Filipino population depends on the coconut industry for their livelihood. The Philippines has 338 million palms occupying 3.5 million ha of arable land and producing 15.3 billion fruit annually, although 51 million palms are senile and need to be replanted immediately. This industry employs about 25 million people with exports mainly going to the USA, the Netherlands, Japan, Germany and China (Ap, 2018). In the period 2008–2013, crude coconut oil was the largest exported product, followed by coconut cake and desiccated coconut. During the 1990s and 2000s, the Philippines exported about 75% of their total coconut products, and only 25% were used for domestic consumption.

The coconut supply chain in the country is very complex as several products are produced from coconut. Overall, the coconut supply chains fulfil the customer's demand by operating from producer to intermediaries and then to the consumer. For example, the copra supply chain involves about six stakeholders: the producers, traders, processors, warehouse storage and distribution, wholesaler and retailer, and consumer (Pabuayon *et al.*, 2009). In the coconut supply chain, the price of one product influences the price of another. For example, coconut oil prices influence copra prices. However, the farmers lack technical and entrepreneurial skills, and sell their copra to community intermediaries without knowledge of the true value of their products on the world market. The annual growth rate for VCO, coconut sugar and coir showed an increase of 25%, 35% and 10%, respectively, over the period of 2017–2020, with the market forecast for the compound growth rate of organic coconut sugar being 41% (Bello *et al.*, 2020). In terms of the value chain for coconut sugar, production in the country is affected by palm senility, a lack of skilled tappers and a lack of mechanization for the collection of sap without affecting its quality. Furthermore, marketing and promotion are important once the local niche or an international market has been identified. For VCO, the current estimate for the industry is the highest among all coconut derivatives at US$40.5 million year^{-1} (Bello *et al.*, 2020). The coconut value chain for VCO includes producers and/or intermediate processors, consolidators, tertiary processors/exporters and wholesalers/retailers. Among these, the important stakeholders in the VCO value chain are the processors/exporters who determine the quality of the VCO that is marketed. Production of VCO is influenced by the lack of financial knowledge at the small-enterprise scale and a shortage of skilled labourers for processing of the coconut kernel (Bello *et al.*, 2020).

The demand for coconut coir is limited and sporadic. Production of coconut coir is highly laborious, with a shortage of skilled workers who also lack knowledge concerning demand, and all these factors impede coir production and marketing in the Philippines. Recently, the Philippines government has been helping smallholder farmers by developing lending programmes to help to sustain them during times of a fluctuating market. However, such lending programmes may not be successful if financing and production are not viewed in the broader context of value-chain financing (Bayudan-Dacuycuy *et al.*, 2022). It is important to interlink financial institutions with the collaborative efforts among farmers, and with public and private industries to better facilitate and strengthen the coconut sector in the Philippines.

In the Philippines, about 60% of coconut smallholder farmers are living at or below the poverty level. This means that they are less likely to be resilient to environmental and economic pressures. Their ability to build assets and to manage and mitigate risk is constrained by the following four barriers: (i) low productivity; (ii) lack of direct market access and low market prices; (iii) lack of access to appropriate financial services; and (iv) unpredictable weather extremes and pest and disease outbreaks (Herrera *et al.*, 2019). The Philippine Coconut Authority (PCA) recommends nine good agricultural practices (GAPs) for coconut farmers to increase their productivity. In this regard, the Grameen Foundation, in collaboration with the PCA, launched the FarmerLink programme to implement these GAPs, with the aid of the largest coconut buyer (Franklin Baker Company of the Philippines) and a farmer financial services provider (People's Bank of Caraga). The programme's primary goal was to increase the coconut smallholder's income by (i) increasing their productivity; (ii) facilitating farmers' access to financial services; (iii) providing market connectivity; (iv) providing early warnings to minimize losses due to weather, and pest and diseases; and (v) using new mobile technologies and digital solutions, in partnership with field officers and agents (Herrera *et al.*, 2019).

As this programme has only been implemented for a short period of time, it cannot be expected to have increased productivity and income yet. However, the implementation of GAPs and farmers' behavioural changes are an early outcome of this programme. There were two forms of communication adopted in the programme, first as text-message-based agricultural extension activities to the farmers, and second by text-message-based agricultural extension activities coupled with face-to-face interviews and training from field officers. This programme has already gained short-term positive outcomes by improving farmers' knowledge on the application of GAPs, by providing more direct access to markets and affordable inputs, and by providing better financial knowledge. The long-term implementation of GAPs is expected to produce more positive outcomes on a long-term basis by increasing productivity, income and resilience to any natural, abiotic and biotic stressor, and decreasing the vulnerability to those stressors (Herrera *et al.*, 2019). As the coconut value chain is very complex, it is important that the public and private sectors collaborate to find a solution to falling productivity. The FarmerLink programme has established a connection between the public sector and the private buyers, and financial providers, to build new technologies into the programme. The PCA has good capacity already, so is considering implementing these programmes for longer-term improvement of the quality of life of the coconut smallholder.

13.4 India

In terms of coconut production and productivity, India ranks third in the world with 23.8 billion fruit from an area of 2.09 million ha, with a productivity of 11,350 fruit ha^{-1} (CDB, 2018). Overall, 17% of the global coconut-growing areas and 31% of world production is in India (Jayasekhar and Jacob, 2021). Internationally, India provides little towards the export market compared with other major producing countries (e.g. the Philippines and Indonesia). This is mainly because of a large domestic consumption of the fruit and its products. Although India is at the forefront in terms of area planted to coconut and fruit production, the farmers face various challenges, mainly in production and marketing. In India, the coconut value chain consists of numerous stakeholders including farmers, harvest contractors, commission agents, exporters, processors, wholesalers and retailers (Kalidas *et al.*, 2020).

In a recent analysis of the value chain, undertaken at the stakeholder level, several problems were identified in production and marketing. At the production level, farmers are encountering severe yield losses due to pests and diseases, and this could be amended by developing new varieties with tolerance to these biotic and other abiotic stresses (Kalidas *et al.*, 2020). At the marketing level, various stakeholders are facing daily fluctuations in the price of fruit due to a lack of market information and proper forecasting (Kalidas *et al.*, 2020). This could be improved by developing innovative models to bridge the gap between demand and supply, and with accurate forecasting for the stakeholders (Kalidas *et al.*, 2020).

Indian coconut products were contributing only a minor portion to the export market before 1994. However, after 1994, trade liberalization and the upgrading of the status of the Coconut Development Board to that of an Export Promotion Council increased the export growth of most coconut products (Thasnimol *et al.*, 2020). Thasnimol *et al.* (2020) also showed that coconut oil and desiccated coconut did not have a comparative advantage in the world export market but that fresh and dried coconut kernel and copra did (Thasnimol *et al.*, 2020). Thus, they concluded that it is important to produce the products that possess the greatest comparative advantage in the global markets if export trade is to be successfully increased. The value chain for coconut oil consists of the farmer, copra maker, oil miller, wholesale seller and consumer. Typically, for the smallholder farm in India, the production and consumption shares are only 64%, while around 36% is shared by intermediaries in the value chain. A lack of interest in coconut oil in the domestic market has affected the coconut oil sector and has been replaced by imported oil palm (*Elaeis guineensis* Jacq.) oil.

Value-added coconut products and fair region-specific trade agreements could increase the coconut export trade and increase competitiveness within the domestic market (Jayasekhar *et al.*, 2019). Diversification of value-added products is likely to be necessary to stabilize the domestic and export markets. Currently, only 8% of the possible value-added products are in the national market, and it has been predicted that this needs to increase to 25% before a fair and stable price is achieved that results in an increase in income to farmers (Sairam and Jayasekhar, 2018). Products that have good potential in domestic markets include VCO, desiccated coconut and coconut sap. For example, the commercial production of VCO has shown a 1.12 cost:benefit ratio and a 21.5% domestic rate of return in India (Sairam and Jayasekhar, 2018).

13.5 Sri Lanka

In Sri Lanka, coconut production contributes greatly to the national economy and has done so for a long period of time. Production contributes about 11% to all agricultural produce with a total area under cultivation of approximately 443,538 ha, with smallholders working 371,244 ha of this land. Production is around 2.5–3.0 billion fruit year^{-1} (Dilrukshi *et al.*, 2022). The export of coconut products contributes about 4% to total export earnings (Coconut Development Authority, 2017). The coconut sector in Sri Lanka contributes to the employment of about 6.5 million people (Bandara and Kumari, 2020). At least one-third of the coconut products produced are processed within the country and exported as value-added products. The remainder are consumed at the domestic level. In 2020, domestic consumption was about 1826 million fruit annually (Dilrukshi *et al.*, 2022). In 2021, the Sri Lankan coconut industry gained a historic high foreign trade figure of US$834.12 million (Coconut Development Authority of Sri Lanka, 2021). However, price volatility remains a major issue in the Sri Lankan coconut industry.

Price volatility drives fluctuations in the price of a product and is an important factor for coconut industry stakeholders as it has an impact on both the national and export sectors. A sudden increase or decrease in the price of a product can occur at any time and is due to a disruption in the coconut supply chain. Changes in production, consumption and marketing are the major factors that affect the price in the market. Recently, Dilrukshi *et al.* (2022) investigated the price volatility of the Sri Lankan coconut sector using several estimation and forecasting models, and were able to minimize uncertainties in the market. This enabled the government, producers and consumers to make appropriate decisions and make the coconut sector profitable for everyone.

The market demand for coconut oil in Sri Lanka is strong, as it is the most widely used domestic cooking oil (Nipunika and de Silva, 2022). Although Sri Lanka was the fourth major coconut producer in the world in 2020, domestic coconut oil production declined, mainly due to drought and supply chain disturbances. From the mid-1970s onwards, coconut oil has been Sri Lanka's major processed export product (~28% of the annual output processed for export, involving 683 million fruit), followed by dry coconut kernel (~15%), copra and finally desiccated coconut (Pathiraja *et al.*, 2019).

Climate variability has been found to substantially influence the economic value of coconut products in Sri Lanka. About 60% of the variation seen in coconut production is ascribed to climatic variability, evident from the analysis of climatic data and coconut production figures in the period 1971–2001 (Fernando *et al.*, 2007). The economic loss associated with climatic variability during this period was estimated to be between US$32 million and US$73 million (Fernando *et al.*, 2007). When analysing the market, Nipunika and de Silva (2022) described a need for Sri Lanka to develop policies to acquire both domestic and international market competitiveness in the coconut oil value chain. This study recommended that policies be developed to facilitate public–private collaboration to attain a better position for Sri Lanka in the global value chain for coconut. They suggested that attention needs to be paid particularly to safety and quality standards in the domestic value chain, which need to be improved to meet international standards. It has been suggested that if international markets are to be reached, this would require an improved industrial framework for producers and exporters (Nipunika and de Silva, 2022).

In Sri Lanka, strong demand exists for value-added coconut products, in both local and international markets; however, the coconut sector struggles to meet the demand to produce such products mainly due to price fluctuations, production costs, scarcity of labour, issues in the supply chain, and poor access to markets and market intelligence (Jayewardena, 2018). Value-chain management is also key to business activity and performance (Jayawardhana and Warnakulasooriya, 2020). In most low- and lower–middle-income countries, traditional coconut value chains involve producers, wholesalers and exporters. The wholesalers are the important players who buy the raw materials from the coconut producers and gain extra value by taking them to the next level in the value chain. Any disturbances that occur in one part of the value chain can have knock-on effects. For instance, the use of poor agricultural practices and the adverse effects of urbanization have been demonstrated to affect production, which in turn influences the supply of raw materials to the wholesalers (Jayawardhana and Warnakulasooriya, 2020).

In Sri Lanka, coconut export values showed an increasing trend in the period 2009–2018, and the percentage share of coconut exports out of total exports increased from 3.5% to 5.5% (Prasannath, 2020). The export of coconut fibre and fibre-related products has also recently increased, but the processing industries are constrained by a shortage of raw materials, and this is likely to continue (Bandara and Kumari, 2020). Another major challenge is the fluctuation of coconut fruit prices, which at times results in grower, manufacturer and consumer demands not being met (Bandara and Kumari, 2020). Recently, the Sri Lankan coconut industry established several programmes to help increase yield, combat the effects of climate change, improve soil fertility, replace senile palms and better manage pests and diseases.

Rainfall is closely linked to coconut farming as most plantations are rainfed, so rainfall patterns can have a major impact on production (Bandara and Kumari, 2020). To truly understand all the factors resulting in market fluctuations, rainfall distribution patterns should also be included in future studies. Overall, the industry needs production stability and an accurate forecasting system for the coconut markets, at both the domestic and international levels.

13.6 The Pacific Region

For Pacific Island countries (PICs), coconut flesh is generally extracted from mature fruit and dried to produce copra locally; the oil is then extracted locally or in Asian mills via the Pacific commodity copra traders. The refined oil is used in factories as a principal ingredient in culinary oils, soaps, shampoos, moisturizers and lotions (PARDI, 2011). About 5% of the world's copra production is from PICs, and the export of coconut oil contributes 3–6% to world export trade annually (McGregor and Sheehy, 2017). According to data collected in 2017, Papua New Guinea is one of the largest exporters of copra, contributing approximately 50% of global copra export trade (McGregor and Sheehy, 2017). Thus, copra export is dominated by PICs, especially by Papua New Guinea and Vanuatu, but their contribution is declining due to the growing number of senile palms, the lack of a replanting programme and changes in land use.

Coconut products such as copra and coconut oil were historically exported to large-scale processors in Europe. These processors remain the leading buyers of PIC coconut products, but not all PICs continue to utilize this major export opportunity. Only Fiji, Samoa and the Solomon Islands have been early participants of a VCO export market. Women in Business from Samoa and the Wainiyaku Estate on Taveuni in Fiji were among the first to secure organic certification, which provided greater recognition for their VCO in the European market (McGregor and Sheehy, 2017). Kokonut Pacific Solomon Islands Ltd obtains VCO from small, village-based communities using the direct VCO microexpeller process. Such practices have now been duplicated by several other PICs for VCO production, including the Marshall Islands, Papua New Guinea and Vanuatu. However, it is a major challenge for these small producers to meet the international standards required, and to cope with price fluctuations and to cover transport costs. Inconsistent inputs, access to packaging, sustainable marketing and marketing linkages also remain challenges. Hence, many small VCO enterprises have struggled to remain viable.

If PIC producers want to remain in the competitive export market, they must produce a quality product, generally with both organic certification and fair-trade labelling. Coconut water consumption is increasing globally, but as most traditional coconut palms are the tall varieties in PICs, they tend to wait until the coconut matures and has fallen to the ground before using them, so they are unsuitable for drinking water. To collect immature fruit for drinking water requires climbing equipment, except for dwarf varieties, but this type is less common and often only exists as small groups of palms planted for household use (McGregor and Sheehy, 2017). Consequently, commercial coconut water production is practically non-existent, and represents a lost opportunity for the region.

Coconut milk and cream are products that have also recently experienced a rapidly growing demand globally. Gosling (2017), the founder of COYO Australia, a company with an international footprint that produces organic, dairy-free coconut products, has recently expressed a willingness to collaborate with PICs in the production of these coconut products. The growing tourism industry in the PICs such as Fiji, Vanuatu and

Samoa may also offer particularly good opportunities for these types of products.

To maintain their position in the global coconut market, PICs need to focus on the appropriate market products by researching trade and consumer preferences. This will provide an understanding of the potential market opportunities for the various coconut products (e.g. VCO, coconut water, coconut milk and by-products made from the shell, husk and cake; PARDI, 2011). PICs also need to assess value-adding opportunities and quality processing to improve profits at all levels within the supply chain. Lastly, research is necessary to understand the capability and constraints of smallholders to participate in the supply chain.

Lin (2021) identified that in Fiji the existing coconut value chain is underdeveloped and limited to low-value products such as copra and crude oil, which give little return to the coconut producers. In addition, the land tenure system in Fiji and the lack of diversified coconut products restricts them from global market access. The major stakeholder groups are fragmented, and the international funding agencies are not directly communicating with the coconut producers. Hence, smallholders are not aware of the full range of market opportunities and consequently show less interest in continuing coconut cultivation and are moving towards cultivating other crops. The PICs are naturally enriched with coconut palms and smallholders have a wealth of knowledge in coconut production, but this is not necessarily used to its full extent. It is important to understand the value chain and to work together so that the PIC coconut industry can widen its engagement with the global coconut market. A large component of what is needed is education of smallholder farmers on the potential opportunities and benefits of coconut production.

13.7 General Comments

Muralidharan *et al.* (2019) conducted a study that identified the constraints existing in the coconut value chain, at several steps, including at production, postharvest, processing and marketing of products. Some suggestions they proposed to improve the coconut value chain included: (i) maintenance of a coordinated approach among important suppliers: (ii) reducing transaction costs; (iii) better management of human resources; (iv) upgrading processing to improve operational productivity; and (v) upgrading of the types of products produced. Additionally, they suggested that enhancement of the coconut value chain could be achieved by adopting improved production technologies on farm, with better coordinated procurement of fruit and more community-level processing. It has been demonstrated that organizing farmer meetings to train farmers to make appropriate decisions, adopt improved practices and technologies, and collectively obtain farm equipment to reduce costs can together increase returns by over 100% to the participating farmers (Muralidharan *et al.*, 2019).

13.8 Conclusion

The coconut trade industries are influenced by a multitude of factors, which are channelled through complex value chains involving steps from production through processing to consumption. As well as the fresh consumption of coconut and its products at a domestic level, coconut products are processed at various points in the chain to prepare them for export. The export value chain is often long and convoluted, with many players involved. Hence, improvements in the analysis of these systems would no doubt make it easier to implement improvements for the industry.

Although demand for coconut products has increased dramatically in recent years, production has not kept up with this demand. The current constraints on production include a growing proportion of senile palms, increased climate variability, increased biotic and abiotic stresses, and a lack of incentives for producers. Some of these problems could be overcome by replanting land with superior varieties and improved agronomic practices, but others are more difficult to address. Coconut markets face great challenges throughout the supply chain including shortages of raw materials, product price fluctuations, poor infrastructure, a lack of skilled labours for processing, and a lack of public–private partnerships to better understand and improve the market.

Although the demand is very high for most coconut products, little is known about the value chain of these products. Only for some traditional coconut products have the markets been studied. The markets for many new products and opportunities are yet to be documented. More research to improve production and gain a better understanding of the coconut value chain and value for smallholder farmers is needed. Alongside research, programmes are needed to help smallholders establish businesses, create favourable business environments, improve policies and infrastructure, provide access for smallholders to export opportunities, enable public–private partnerships, and provide training in the production of new products and quality assurance. Wider sharing of information on marketing opportunities across the world for wholesalers and retailers would probably also improve the income of coconut farmers and the economy of the low- and lower–middle-income countries.

Coconut is largely grown by smallholder farmers in low- and low–middle-income countries. These farmers often face severe financial problems and can be some of the poorest farmers in the world. For the industry to flourish, these growers need to be the focal point of industry improvement efforts. If smallholders cannot gain a reasonable income and standard of living growing coconut, then the industry will face long-term challenges.

13.9 References

Abeysekara, M.G.D. and Waidyarathne, K. (2020) The coconut industry: a review of price forecasting modelling in major coconut producing countries. *CORD* 36, 6–15.

Alouw, J.C. and Wulandari, S. (2020) Present status and outlook of coconut development in Indonesia. *IOP Conference Series: Earth and Environmental Science* 418: 012035.

Ap, T. (2018) How Philippines is battling to cash in on coconut craze. Cable News Network International, 30 November. Available at: https://edition.cnn.com/2015/11/30/asia/philippines-coconuts/index.html (accessed 18 October 2023).

Bandara, J.M.A.K. and Kumari, D.A.T. (2020) Factors affecting coconut price determinants in Sri Lanka: an integrative review. *Applied Economics and Business* 4, 61–71.

Bayudan-Dacuycuy, C., Magno-Ballesteros, M., Baje, L.K.C. and Ancheta, J.A. (2022) Sustainable value chain financing for smallholder agricultural production in the Philippines. *Journal of Philippine Development* 46, 19–40.

Bello, R.T., Pantoja, B.R., Tan, M.F.O., Banalo, R.A., Alvarez, J.V. and Rañeses, F.P. (2020) *A Study on Skills for Trade and Economic Diversification (STED) in the Non-traditional Coconut Export Sectors of the Philippines*. STRENGTHEN Publication Series, Working Paper No. 21. International Labour Organization, Geneva. Available at: www.ilo.org/manila/projects/strengthen/WCMS_742455/lang--en/index.htm (accessed 4 September 2023).

Coconut Development Authority (2017). *Annual Report 2017*. No. 2017. Coconut Development Authority, Colombo.

Coconut Development Authority of Sri Lanka (2021) *Coconut Statistics*. Coconut Development Authority, Colombo.

Coconut Development Board (CBD) (2018) *All India Final Estimates of Area and Production of Coconut–2017–18*). CDB, Kochi, India. Available at: https://coconutboard.gov.in/Statistics.aspx (accessed 4 September 2023).

Dilrukshi, K.B.M.T., Jayalath, K.V.V.N., Dissanayake, S.N. and Perera, S.M.S.D. (2022) Retail price volatility: the case of coconut in Sri Lanka. *Sri Lankan Journal of Agricultural Economics* 23, 22–34.

FAO (2018) FAOSTAT. Food and Agricultural Organisation of the United Nations, Rome. Available at: www.fao.org/faostat/ (accessed 4 September 2023).

Fernando, M.T.N., Zubair, L.M., Peiris, T.S.G., Ranasinghe, C.S. and Ratnasiri, J. (2007) *Economic Value of Climate Variability Impacts on Coconut Production in Sri Lanka*. The AIACC Project Office, International START Secretariat, Washington, DC.

Foale, M. (2003) *Coconut Odyssey: The Bounteous Possibilities of the Tree of Life*. Australian Centre for International Agricultural Research, Canberra.

Gosling, H. (2017) Thinking outside the square. Keynote address to *Pacific Coconut Sector Value Chain Workshop*, 11–13 July 2017, Nadi, Fiji.

Gunawan, I., Trihastuti, D. and Mulyana, I.J. (2021) Sustainability issues of the coconut supply chain in Indonesia. In: *2021 IEEE International Conference on Industrial Engineering and Engineering Management (IEEM)*, 13–16 December 2021, Singapore, pp. 158–162.

Gunawan, I., Trihastuti, D., Mulyana, J. and Limbara, C. (2022) Securing coconut availability in Indonesia. *International Journal on Food System Dynamics* 13, 294–307.

Herrera, A., Brul, B., Cadavid, R. and Chavarría, E. (2019) *Building Resilience of Coconut Smallholder Farmers in the Philippines: Final Evaluation Report of the FarmerLink Program*. Grameen Foundation, Washington, DC.

ICC (International Coconut Community) (2022) *Coconut Statistical Yearbook 2020*. International Coconut Community. Compiled by Alit Pirmansah, Market and Statistics Officer ICC, Jakarta, Indonesia.

Jayasekhar, S. and Jacob, N. (2021) *Coconut Sector in India at the Crossroads: A Brief Narrative*. Coconut Development Board, Kochi, India.

Jayasekhar, S. and Thamban, C. (2016) Coconut sector: global scenario and Indian perspective. *Indian Coconut Journal* 59, 17–21.

Jayasekhar, S., Chandran, K.P., Thamban, C. and Muralidharan, K. (2019) Coconut sector in India experiencing a new regime of trade and policy environment: a critical analysis. *Journal of Plantation Crops* 47, 48–54.

Jayawardhana, M.B.S.M.T. and Warnakulasooriya, B.N F. (2020) Impact of problems associated with supply chain management practices of wholesalers on their business performance in the coconut industry in Sri Lanka with special reference to Kurunegala District. *Vidyodaya Journal of Management* 6, 59–87.

Jayewardena, J. (2018) Sri Lankan coconut industry: Performance and challenges for the future. *Sunday Observer*, 2 December. Available at: www.sundayobserver.lk/2018/12/02/news-features/sri-lankan-coconut-industry-performance-and-challenges-future (accessed 4 September 2023).

Kalidas, K., Mahendran, K. and Akila, K. (2020) Constraints in coconut value chain – a framework for analysis using response priority index. *Current Journal of Applied Science and Technology* 39, 76–82.

Khaw, K.T., Sharp, S.J., Finikarides, L., Afzal, I., Lentjes, M. *et al.* (2018) Randomised trial of coconut oil, olive oil or butter on blood lipids and other cardiovascular risk factors in healthy men and women. *British Medical Journal Open* 8: e020167.

Lin, J. (2018) *The Role of Institutions in International Coconut Trade: A Gravity Model Approach*. AgEcon Search, Research in Agricultural and Applied Economics, University of Minnesota, Minnesota. Available at: http://ageconsearch.umn.edu/record/277012/files/568.pdf (accessed 4 September 2023).

Lin, J. (2021) Fiji's participation in the global coconut value chain: opportunities and challenges. *Journal of Agribusiness in Developing and Emerging Economies* 11, 345–365.

McGregor, A. and Sheehy, M. (2017) *An Overview of the Market For Pacific Island Coconut Products and the Ability of Industries to Respond.* Secretariat of the Pacific Community, Suva, Fiji. Available at: https://pacificfarmers.com/wp-content/uploads/2018/02/CIDP_Report_Final_e-copy_resized.pdf (accessed 4 September 2023).

Moreno, M.L., Kuwornu, J.K.M. and Szabo, S. (2020) Overview and constraints of the coconut supply chain in the Philippines. *International Journal of Fruit Science* 20 (Suppl. 2), S524–S541.

Muralidharan, K., Subramanian, P., Mathew, A.C., Thamban, C., Jayasekhar, S. *et al.* (2019) Upgrading a coconut value chain: empirical evidence from North Kerala. *International Journal of Innovative Horticulture* 8, 72–80.

Naveena, K., Santosha, R., Garima, S. and Yogish, K.J. (2014) Forecasting of coconut production in India: a suitable time series model. *International Journal of Agricultural Engineering* 7, 190–193.

Nipunika, M.L.D. and de Silva, D.A.M. (2022) An institutional environment and its involvement in coconut oil value chain. *Ag Insights* 2022, 400–403.

OECD/FAO (Organisation for Economic Co-operation and Development/Food and Agriculture Organization of the United Nations) (2017) *OECD-FAO Agricultural Outlook 2017–2026*. OECD, Paris. Available at: www.oecd-ilibrary.org/agriculture-and-food/oecd-fao-agricultural-outlook-2017-2026_agr_outlook-2017-en (accessed 4 September 2023).

Pabuayon, I., Cabahug, R., Castillo, S.V. and Mendoza, M. (2009) Key actors, prices, and value shares in the Philippine coconut market chains: implications for poverty reduction. *Journal of the International Society for Southeast Asian Agricultural Sciences* 15, 52–62.

PARDI (Pacific Agribusiness Research & Development Initiative) (2011) Coconut value chain review. available at: www.adelaide.edu.au/global-food/ua/media/239/pardi-coconut-chain-review-nov-2011.pdf (accessed 4 September 2023).

Pathiraja, E., Griffith, G., Farquharson, R. and Faggian, R. (2019) The cost of climate change to agricultural industries: coconuts in Sri Lanka. *International Journal on Food System Dynamics* 10, 428–457.

Prades, A., Salum, U.N. and Pioch, D. (2016) New era for the coconut sector. What prospects for research? *Oilseeds and Fats, Crops and Lipids* 23: D607.

Prasannath, V. (2020) Export performance of coconut sector of Sri Lanka. *South Asian Journal of Social Studies and Economics* 6, 35–43.

Rethinam, P. (2005) Asian and Pacific coconut community activities, achievements and future outlook. In: Adkins, S., Foale, M. and Samosir, Y. (eds) *Coconut Revival: New Possibilities for the 'Tree of Life'. Proceedings of the International Coconut Forum*. Australian Centre for International Agricultural Research, Canberra, pp. 15–21.

Sairam, C.V. and Jayasekhar, S. (2018) World coconut economy: sectoral issues, markets and trade. In: Krishnakumar, V., Thampan, P. and Nair, M. (eds) *The Coconut Palm (*Cocos nucifera *L.) Research and Development Perspectives*. Springer, Singapore, pp. 801–820.

Sri Lanka Export Development Board (2017) *Growth of Demand for Coconut in the Global Market*. Sri Lanka Export Development Board, Colombo. Available at: www.srilankabusiness.com/blog/growth-of-global-demand-for-coconut.html (accessed 4 September 2023).

Suselo, S.L. (2012) Peranan Bank Indonesia dalam Pengembangan Usaha Mikro Kecil dan Menengah [The role of Bank Indonesia in the development of micro, small and medium enterprises]. In: *11th Sustainable Competitive Advantage 2021 International Conference*, Jenderal Soedirman University, Jawa Tengah, Indonesia.

Thasnimol, F., Prema, A. and Jayasekhar, S. (2020) Impact of trade liberalisation policies on the coconut export from India. *Economic Affairs* 65, 603–609.

14 The Future of Coconut as an Economic Crop WorldWide

Vincent B. Johnson[1,2*], Edna A. Aguilar[3], Jelfina Alouw[4,5], Julianne Maree Biddle[6,7], Julieane Camile Lacsina[8] and Erlene C. Manohar[9]

[1]GreenQuills International (GQI), Prades-le-Lez, France; [2]Alliance of Bioversity International and CIAT, Montpellier, France; [3]Institute of Crop Science, College of Agriculture and Food Science, University of the Philippines, Los Baños, Philippines; [4]International Coconut Community (ICC), Jakarta, Indonesia; [5]Horticulture and Estate Research Center, National Research and Innovation Agency, Cibinong Science Center, Cibinong Kabupaten Bogor, Indonesia; [6]Australian Centre for International Agricultural Research (ACIAR), Canberra ACT, Australia; [7]School of Agriculture and Food Sustainability, The University of Queensland, Gatton, QLD, Australia; [8]Independent Consultancy, Tarlac City, Philippines; [9]International Coconut Genetic Resources (COGENT), ICC, Quezon Province, Philippines

Abstract

Coconut offers an increasingly relevant, climate-resilient component of any agricultural production system, but unfortunately coconut smallholder farmers remain some of the poorest farmers in the world. The global industry could exceed US$30 billion by 2026. As well as dietary uses, coconut provides raw materials for the automotive, beverage, biofuel/biomass, confectionery, cosmetics, housing, oleochemical and pharmaceutical industries. For the industry to thrive, producers must better harness economically meaningful opportunities. Coconut supply chains must be stabilized. Market and policy incentives must support long-term commitments and sustainable initiatives. Expanding emerging niche markets will help coconut-producing countries compete globally. Addressing supply requirements and product quality through modernization is essential for industry prosperity. Effective crop management, including intercropping and harnessing improved germplasm, will ensure higher productivity. Many coconut products will only be profitable where copra production, processing and quality challenges are addressed. Optimizing the combination of genetic, agronomic and commercial influences will be required to achieve industry-wide prosperity.

14.1 Introduction

The economic outlook for coconut and its value-added products such as water, coconut sap sugar and virgin coconut oil (VCO) is potentially extremely positive, as global demand continues to increase (see Chapters 12 and 13, this volume). The crop offers an increasingly relevant, climate-resilient component of agricultural production systems worldwide. All parts of the palm are useful (see Chapter 12, this volume), and it is an attractive component of mixed farming systems (see Chapter 3,

* Email: vinbjohnson@yahoo.com or v.johnson@cgiar.org

DOI: 10.1079/9781789249736.0014

this volume). As a functional food, coconut is now recognized for its many health benefits (see Chapter 11, this volume), and global demand for a range of coconut products continues to grow. Indeed, according to Sushant (2020), the global coconut industry could easily exceed US$30 billion by 2026 (tripling the 2018 market values), although market expansion forecasts should be treated with some reservation, as there is considerable variation in predictions of the global market size. As well as its dietary and sustainable production roles, coconut also provides raw materials for the automotive, beverage, biofuel/biomass, confectionery, cosmetics, housing, oleochemical and pharmaceutical industries (see Chapter 12).

Regardless of the utility and widespread use of coconut, the industry must overcome various barriers to profitability, including unproductive, senile palms and poor-quality planting material; virulent pests and pathogens (see Chapter 5); increasingly challenging climate conditions and extreme weather events (see Chapter 4); and a lack of accessible genetic diversity (see Chapter 9), agronomic knowledge (see Chapter 3) and economically meaningful opportunities for harvested coconut. Coconut-product supply chains must be stabilized, as fragility can cause price and relationship fluctuations. Market and policy incentives would help support long-term commitment and sustainable initiatives. The socio-economic survey by Turner *et al.* (2022) outlined an uncertain economic horizon for coconut production. Unless value-chain investors can better understand and meet producers' needs, coconut oil's exposure to volatile or falling global commodity prices decreases its power to supplement incomes, discouraging investment in replanting and production expansion. Labour-intensive coconut production also faces stiff 'cash-crop' competition (e.g. from cacao (*Theobroma cacao* L.), durian (*Durio zibethinus* L.), kava (*Piper methysticum* G. Forst), and oil palm (*Elaeis guineensis* Jacq.)) which can be less labour intensive and more profitable. Aquaculture and livestock production offer yet other more attractive alternatives. Additional barriers include the fragmented nature of production, poor transport infrastructure and the geographical remoteness of many coconut-growing regions.

In addition to value-chain challenges, breeders and farmers must be able to access the genetic diversity of coconut to ensure higher productivity, offering attractive, marketable quality. The International Coconut Genetic Resources Network (COGENT) should play a more active role in accessing and managing the 400-plus genotypes held in 24 collections (see Chapter 9, this volume) and harnessing *in situ* diversity elsewhere. For example, some Polynesian compact dwarf varieties offer several advantages over other varieties but are very difficult to obtain and are only found in limited locations. Improved access to pathogen/disease-free, elite planting material is needed, and this can be achieved with the effective use of recently developed tools such as coconut tissue micropropagation (see Chapter 10, this volume), in partnership with strict biosecurity protocols to enable international germplasm dissemination (Bourdeix and Sourisseau, 2021).

More effective agronomic and eco-friendly crop management (see Chapter 3) could ensure sustainable and high productivity, which could enhance the economic competitiveness of coconut. In augmenting coconut's largely organic status, there is great scope to introduce agroecologically intensive and regenerative approaches, including mixed farming. There are likely to be substantial economic gains made from these systems, although intercropping needs to be carefully considered in drought-prone regions to manage risk.

This chapter addresses industry status and future trends. It also covers possibilities for coconut-based production systems in terms of: (i) their ecological and economic valuation; and (ii) their innovation for agroecological intensification, including improved resilience. This chapter articulates the supply, demand and market prospects for new and old coconut products in terms of volume, price and market trends, and outlines key export and import variations. Finally, the chapter briefly re-examines emerging coconut products.

14.2 A Bright Future for Sustainable, Coconut-based Production Systems

Climate-resilient, diversified, market-driven yet sustainable farming systems should be the norm for coconut farms of the future. Ideally, farmers should be applying agroecological and regenerative practices in their diversified farming system that consider both indigenous and modern technologies.

Coconut production in Indonesia, the Philippines and India is currently dominated by smallholder farmers, who often rely on coconut as their main source of income, processed mostly into copra and crude coconut oil. While intercropping under coconut is an age-old practice worldwide, more than 90% of farmers generally follow traditional production practices of unsystematic and undermanaged farming systems. This results in low-volume, low-quality production, which is individually marketed or not marketed at all. Generally, low productivity results from growers treating coconut as a supplementary, lower-priority crop (Hawkins, 2019).

Any sustainable coconut-based farming system should integrate managing natural resources, agricultural inputs and human resources in more productive and profitable ways (Alouw and Wulandari, 2020). Producers, small and large, must shift to approaches that consider total coconut farm productivity, farmers' welfare and income, integrating crops and/or animals, and value-adding, while minimizing environmental impacts and increasing resilience through sustainable and regenerative practices.

Farmers need to be trained and empowered to make informed decisions, for example via season-long farmer field schools. Using experiential learning, the curricula farmer field schools can teach farmers adaptive management to grow healthy, productive and profitable coconut varieties, intercrops and livestock (directly linked to their processed products). The curricula can also include managing risks and resource limitations and applying agroecosystem analysis principles to regular farm monitoring, as well as integrating climate forecasting in their farm management and decision making (FAO, 2022).

Farmers should be encouraged to utilize innovative decision-aid tools such as mobile phones and web-based applications that can identify pests for real-time monitoring and reporting to crop-protection experts for early warning and a swift response. Khaeruddin (2022) presented a successful example of using mobile-phone application technology for smart irrigation systems in sustainable Malaysian coconut nurseries. Use of indigenous and modern techniques following agroecological principles and regenerative farm practices can increase farm productivity and income, while maintaining soil health (Gliessman, 2011; Wezel, 2014). Such practices may include intercropping, intercrop rotation, livestock integration, reducing agrochemical use, use of biofertilizers and natural pesticides, and harnessing nitrogen-fixing or allelopathic plant relationships.

Coconut researchers have demonstrated the benefits of recycling nutrients from coconut by-products (e.g. the husk and water) and farm wastes (e.g. coconut fronds, weeds, intercrop biomass, fodder and manure from animals using various composting, vermiculture and biochar technologies) or by using farm waste as mulching materials or for incorporation into planting pits (Aguilar *et al.*, 2022). This high-organic-matter material can support highly productive intercrops and a dynamic microflora and fauna, and is a key component of climate-resilient, coconut-based farming systems. A series of consolidated agroecology principles has been developed from several high-level panel discussions initiated by the Food and Agriculture Organization of the United Nations (Wezel *et al.*, 2020; Fig. 14.1).

Future coconut-based farming systems should address multiple and interacting concerns, both from the supply and the demand (market and consumption) sides for the sustainable growth of the various coconut value chains (see Chapter 13, this volume). On the supply side, there is a need to increase the sustained volume and quality of coconut fruit entering the market. However, this is challenging, given the socio-economic circumstances of a typical smallholder farmer whose income from coconut is low and fluctuating. Coconut farmers and their livelihoods are repeatedly, and now more frequently, threatened by climate-related hazards and risks including the increased frequency of typhoons, changes in rainfall patterns, flooding, severe and prolonged droughts, and changing dynamics of pests and disease attack (see Chapter 5, this volume). On the market side, there is a demand for an increased volume and quality of fruit that comply with requirements for traceability, sustainability of farm practices, workers' safety and welfare, and stricter quality standards, such as elimination of the cancer-causing polycyclic aromatic hydrocarbons and toxic aflatoxins found on some copra-based products. The concerns of low coconut production must be addressed with farmers'

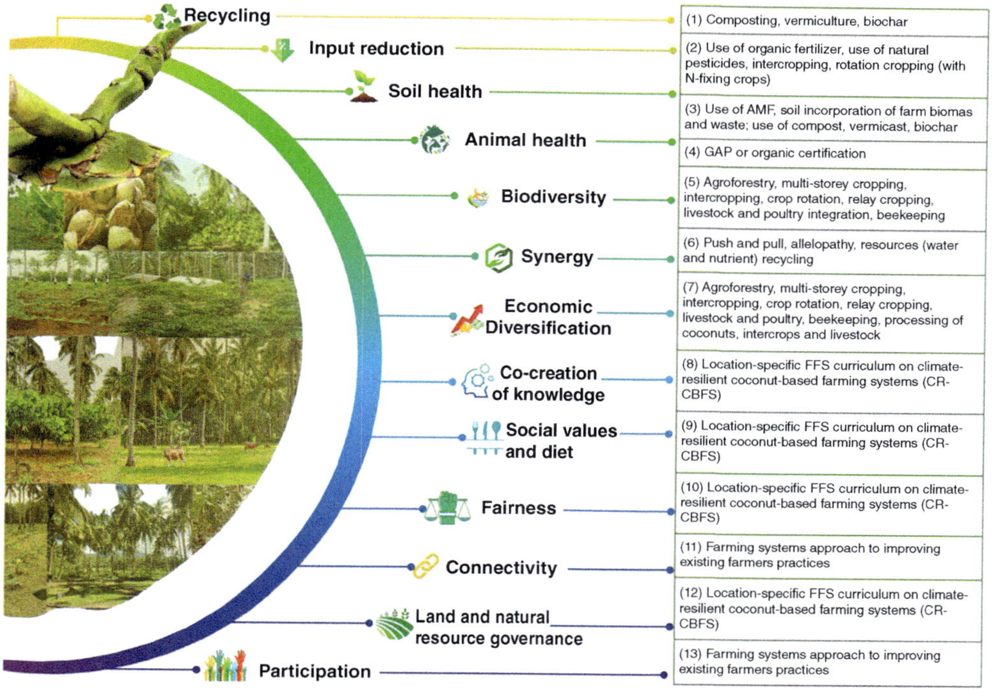

Fig. 14.1. Application of agroecology principles to coconut-based farming systems. AMF, arbuscular mycorrhizal fungi; FFS, farmer field schools; GAP, good agricultural practices; N, nitrogen. Adapted from Wezel *et al.*, 2020.

participation as the ultimate decision makers on how to operate their coconut farms. Expanded agroecology principles (Wezel *et al.*, 2020), which are compatible with the farm household-centred farming-systems approach, will be a useful aid to the development of future farming systems. Smallholder coconut farmers of the future should be empowered and prosperous, as they become more active participants in higher levels of coconut and intercrop value chains.

14.3 Coconut in Mixed Farming Systems

Coconut has a wide environmental adaptability and where it grows well in upland conditions, most other upland crops grow too. Coconut is a single-stemmed palm, planted at 7–10 m apart. Its unique canopy architecture allows 40–70% light transmission to the understorey, depending on planting density, palm age and variety. Utilizing the growing area under coconut could substantially increase agricultural production without opening new lands. Intercropping 40% of the coconut area could add at least 2.4 million ha of land to the existing global agricultural production area, without increasing the total net area of planting.

Diversification in coconut farms is an often-cited strategy for increasing sustainable food production and for boosting livelihoods for a growing world population estimated to reach 9.1 billion by 2050. Well-managed coconut intercrops and/or associated livestock can substantially contribute to increasing land-use efficiency and improving total farm productivity and income via additional intercrop yields and livestock production, as well as increased coconut yields (Aguilar *et al.*, 2022).

Field research studies on coconut intercropping at the Coconut Research Institute (Baddegama, Sri Lanka) have demonstrated the following increases in coconut yield only 1 year after intercropping started: cocoa 22%; coffee (*Coffea* sp.) 34%; pepper (*Piper* sp.) 5%; cloves (*Syzygium aromaticum* (L.) Merr. & L.M. Perry) 17%; cinnamon (*Cinnamomum* sp.) 26% (Liyanage *et al.*, 1984). In Kerala, India, fruit yield increased by 18% in a coconut and cocoa intercropping system (Phukon *et al.*, 2021). In Zamboanga, the Philippines, coconut fruit yield doubled after 2 years of banana (*Musa* sp.) intercropping, where the coconut management practice was unchanged, but the banana intercrop was well managed. The computed per hectare land-equivalent ratio of coconut farmed with banana equated to 2.13 compared with a coconut only crop (Espaldon *et al.*, 2017). A coconut with citrus intercropping study in Ghana gave a land-equivalent ratio of 1.07 (100 citrus trees intercropped in 128 coconut palms spaced 9.5 m apart). The growth and fruit weight of both coconut and citrus were not affected in the intercropping system (Andoh-Mensah and Ofosu-Budu, 2012).

Agroforestry can boost land productivity and provide numerous economic and environmental benefits, particularly on sloping land (Murthy *et al.*, 2013). In their value-chain analysis of coconut-based farming systems in Mindanao, the Philippines, Mendoza *et al.* (2018) reported that even at the farm gate, the added income from intercropped fruit trees could significantly increase the net income over copra alone. The following is an enumeration of intercrops and their corresponding increase in revenue: banana (12.6%), mango (*Mangifera indica* L., 9.4%), durian (7.8%), mangoseen (*Garcinia mangostana* L., 5.3%), cocoa (2.8%) and coffee (2.4%). They also noted a progressive increase in income as the farmers' products move through the value chain.

Biodiversity is essential for maintaining interacting organisms' functional balance in any agroecosystem. Coconut provides regulatory ecosystem services including nutrient cycling, community diversity and soil conservation, and these have complementary relationships with other ecosystem services such as carbon sequestration, shelter, and food and fuel provisioning (Costales, 2019). Carbon sequestration is a function of plant biomass accumulation. Lasco *et al.* (2000) showed that for above-ground biomass, coconut sequesters an estimated 4.78 t of carbon ha^{-1}. Coconut with coffee intercropping in Makiling, Laguna, the Philippines, was estimated to store considerably more – possibly as much as ten times more carbon per hectare (Zamora, 1999, cited by Lasco *et al.*, 2000). A study on coconut-based cropping systems with different vegetable crops and integrated nutrient management (using combinations of inorganic fertilizer and/or compost/*Azotobacter* spp.) in Karnataka, India, demonstrated significant increases in above- and below-ground carbon stocks (Kumar and Maheswarappa, 2019).

Purbopuspito (2022) compared the earning potential from a coconut mixed-farming system with a coconut monoculture in Indonesia (Kawau *et al.*, 2015). This study corroborates the higher earning potential of a coconut-based mixed-farming system when compared with a monoculture (Fig. 14.2). Although production efficiencies are still needed to achieve the break-even point in this example, the study demonstrates clear economic gains.

Considerations for selecting intercrops include their biophysical suitability, age, coconut canopy density, coconut planting system and climate-related risks. It is important to prioritize native intercrops as these can be easier to maintain and returns are likely to be higher (AMRA Farms, 2020). Market availability and targeting farmers' participation in higher levels of the value chain (through cooperative production and marketing, processing and value adding) are also important considerations for any coconut-based farming system of the future. Although generally successful, intercropping performs less well in certain more drought-prone situations, where some crops are not viable without additional irrigation. This can be especially true where the intercrops compete with the palms for water and nutrients (L. Perera, 2022, personal communication).

Light transmission under coconuts varies according to environment, crop management practices, the age of palms and the planting system used. Recommendations for intercrops differ for each growth stage. The best period for intercropping is during the first few years after coconut planting, before the canopy closes in. At this time, any suitable farmer-preferred or commercially suitable crop can be planted, including

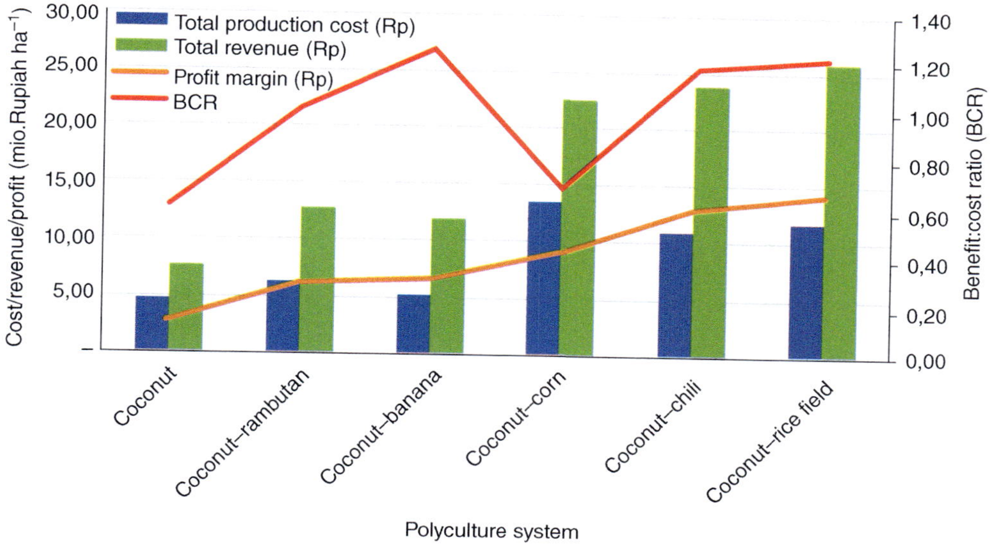

Fig. 14.2. Comparing coconut monoculture and polyculture. From International Coconut Community (ICC) Ministerial Session Documents, 2020, by kind permission of the ICC.

cereals, legumes, root crops, vegetables and biennials such as banana, papaya (*Carica papaya* L.), and pineapple (*Ananas comosus* L. Merr.). When the canopies of adjacent palms overlap (starting at 4–7 years and up to about 20 years from planting), shading can severely constrain many crops. From this point (or earlier), shade-tolerant crops such as black pepper, ginger (*Zingiber officinale* Roscoe), taro (*Colocasia esculenta* (L.) Schott.) and vanilla (*Vanilla* Plumier ex. Mill., 1754) can be intercropped. Perennials that require nurse crops such as cocao, coffee, durian and mangosteen can also be intercropped. In some cases, livestock integration is suitable with intercrops such as nitrogen-fixing leguminous fodder and grasses such as hybrid Napier grass (*Cenchrus purpureus* (Schumach.) Morrone) or Guinea grass (*Megathyrsus maximus* (Jacq.) B.K. Simon & S.W.L. Jacobs, 2003) along with legumes and has proven profitable (JICA, 2017). Nitrogen-fixing leguminous trees such as gliricidia (*Gliricidia sepium* (Jacq.) Steud.), can provide a sustainable organic nitrogen source and extra biomass. As the palms increase in height with age, light transmission through the canopy increases and after year 20, both shade- and sun-loving crops can again be intercropped.

14.4 Key Drivers of Market Trends and Future Prospects for Coconut Products

This section presents some key drivers and demand-changers that influence market trends for traditional and non-traditional coconut products. These include consumer awareness of, and requirements for, nutritional and environmental benefits, changing diets and living standards, demographic trends, increased product diversity and the increasing desire to address certain global challenges (unhealthy foods, non-degradable/non-renewable products). The key factors that can negatively influence demand for some products are also explored. Future market trends and prospects are also thoroughly discussed, providing background on the economic outlook for coconut. The section also considers both existing coconut products and emerging products with a promising future.

Meaningful coconut industry forecasting must be based on market trend analyses. Trends are shaped by market key drivers or demand factors. Accurate forecasting is also informed by thorough market research and considers key assumptions. Such forecasts also examine production issues, logistical support for product development, and processing and market requirements. Certain traditional and non-traditonal coconut products have persisted in global markets over decades, due to their valued uses. Additionally, diversified product lines and new products have had increased global market demand due to evolving trends in consumption. In this way, with these two areas of focus, coconut can be perceived as an economic crop of the future, although for this to happen will require support and innovation related to production, processing and marketing.

For the major producing countries such as India, Indonesia and the Philippines, coconut can help shape the future of their national economies, and could offer an antidote to poverty that will boost wellbeing for those millions of people directly and indirectly dependent on coconut production.

14.4.1 Coconut-Product Market Drivers and Constraints

According to Shankar (2019), the global coconut products market was valued at US$11.5 billion in 2018 and is estimated to increase to US$31.1 billion by 2026, registering a compound annual growth rate of 13.6% from 2019 to 2026. Associated with this high compound annual growth rate, coconut products are linked to their increasing use in the food, snack, beverage, cosmetic and personal-care industries, where health-conscious consumers have shifted preferences towards natural alternatives, for example away from caffeinated and sugar-based processed drinks to coconut water as it is a natural and nutritious energy drink.

Some coconut oil market drivers include a demand for its emollient properties, rich flavour and fatty acid profile, especially for the cosmetics and food and beverages industries. Cosmetic market drivers for coconut milk and cream are linked to their antioxidant, anti-ageing and anti-inflammatory properties beneficial for hair and skin. The coconut milk market is largely driven by its consumption as a healthy dairy alternative. Increasing demand for coconut-based snacks is driven by the desire for healthy alternatives to conventional snack products. The coir coconut fibre market drivers come from horticulture and textile demands. Another set of market drivers is provided by around a dozen key coconut-product market players (e.g. Vitacoco and the Sambu Group) who have adopted product launch and business expansion strategies to expand their market share, increase profitability and remain competitive (Shankar, 2019).

Coconut currently has a cultural place in many countries, but sustaining any meaningful part of farmers' livelihoods may require coconut products and business models to be innovative and capture a higher economic value for smaller-scale producers. Turner *et al.* (2022) described how reconfiguring the coconut industry could increase production and achieve impact, demonstrating how jobs can be created, export and tourism promoted, and untapped value created. The approach combines 'western' consumer insights, product and business model innovation expertise, and global capital and market access with donor-country owned and operated start-ups. This approach could be scaled to escape commoditized stagnation (Turner *et al.*, 2022).

The market trends (mostly export) of the main coconut products are outlined below (Sangamithra *et al.*, 2013; Aguilar *et al.*, 2022).

Coconut oil

Coconut oil ranks as the second highest export product (after coconut water) in terms of value, with a 2% share of the global vegetable oil market. The coconut oil market is driven by its increased use as a base product in many high-value products, especially cosmetics. It also offers health, energy and other benefits. Coconut oil export volumes from 2016 to 2021 ranged from 1.8 to 2.4 million t year $^{-1}$ (Table 14.1). In 2020, the market was worth US$4.1 billion year^{-1} and has been predicted to increase to US$5.5 billion by the end of 2026, growing at a compound annual growth rate of 4.3% (Global Oil Market Research Report, 2020). Coconut oil exports from the Philippines for the same 6-year period peaked at 1.2 billion t in 2019, with Indonesia ranking second (Aguilar *et al.*, 2022). The growing demand for coconut oil as a base ingredient in hair and skin-care products is expected to further stimulate the share of coconut oil in the vegetable oil market.

Table 14.1. Crude coconut oil exports (× 1000 Mt), 2016–2021. From International Coconut Community Session Documents, 2022, reproduced by kind permission of the ICC.

Country	2016	2017	2018	2019	2020	2021
Philippines	693	943	951	1239	840	881
Indonesia	602	510	678	611	577	611
Malaysia	152	116	122	223	203	187
Papua New Guinea	17	16	22	21	33	36
India	7	33	7	8	7	12
Vietnam	18	10	14	8	7	10
Other countries	334	380	492	260	163	164
Total	**1823**	**2008**	**2286**	**2370**	**1830**	**1901**

Results from studies on the health benefits of coconut oil support its growing reputation as a leading coconut-export health product (see Chapter 11, this volume). Studies indicate that cardiovascular health benefits from dietary coconut oil are likely to include reduced heart-attack risk due to its medium-chain fatty-acid profile. These medium-chain fatty acids are metabolized differently from long-chain fatty acids, and evidence suggests that coconut oil does not pose the same health risks as other saturated fats (Eyres *et al.*, 2016). Studies have also shown that coconut oil can boost the immune system, regulate blood sugar levels and increase 'good' cholesterol levels (Fernando *et al.*, 2015; Haron *et al.*, 2018; Pruseth *et al.*, 2020; Takeishi *et al.*, 2021).

Although there are demonstrated health benefits, there are unwarranted concerns about health risks related to coconut oil saturated fats, perpetuated by organizations such as the American Heart Association. Health benefit considerations regarding coconut oil have become significant drivers for coconut oil to evolve as a premium, sustainable vegetable oil within the global market. Organizations like the Fairtrade Sustainability Alliance, the Rainforest Alliance and the International Coconut Community (ICC) are working to establish transparent standards (Gallagher, 2021; ICC, 2022). These issues should be strategically addressed via further clinical trials (see Chapter 11, this volume).

Coconut oil's 6-year stable market performance from 2016 to 2021 should continue to improve. Continuing coconut oil promotion and its potential for cosmetics and health products is likely to further drive market demand, along with emerging coconut oil-based products.

Copra

Copra is the foundation of coconut oil processing, so acceptable copra quality is critical. The Philippines is ranked as both the largest copra producer and consumer globally, generating about 45% of global production and consumption. Indonesia and India remain in second and third position, accounting for about 27 and 12%, respectively, in terms of both production and consumption (ICC, 2022). The coconut industry is export oriented, where historically most coconut-farming communities depended on low-value copra processing. The Philippines does not export copra because its many oil mills crush all their copra into coconut oil for export. In 2020, the top copra exporters were Indonesia (US$70.4 million), Thailand (US$30 million), Papua New Guinea (US$29.9 million), Vanuatu (US$8.7 million) and the Solomon Islands (US$4.41 million). In 2020, the top copra importers were the Philippines (US$25.9 million), China (US$19.5 million), Bangladesh (US$18.6 million), Pakistan (US$14.4 million) and India (US$11.1 million; ICC, 2022).

The copra-market and export-value trends are highly volatile (Fig. 14.3). In the Philippines, world-class oil-mill processing technology is available; however, poor on-farm drying can reduce the copra quality supplied to these mills. This in turn reduces coconut oil production volumes and milling efficiency. Traditional drying methods can generate high polycyclic aromatic hydrocarbon levels and aflatoxin contamination. This reduces prices and constrains exports.

Poor-quality copra and its multi-layered trading system constrain the supply-chain due to additional logistics and quality losses. New white copra production technology (the split kernal is

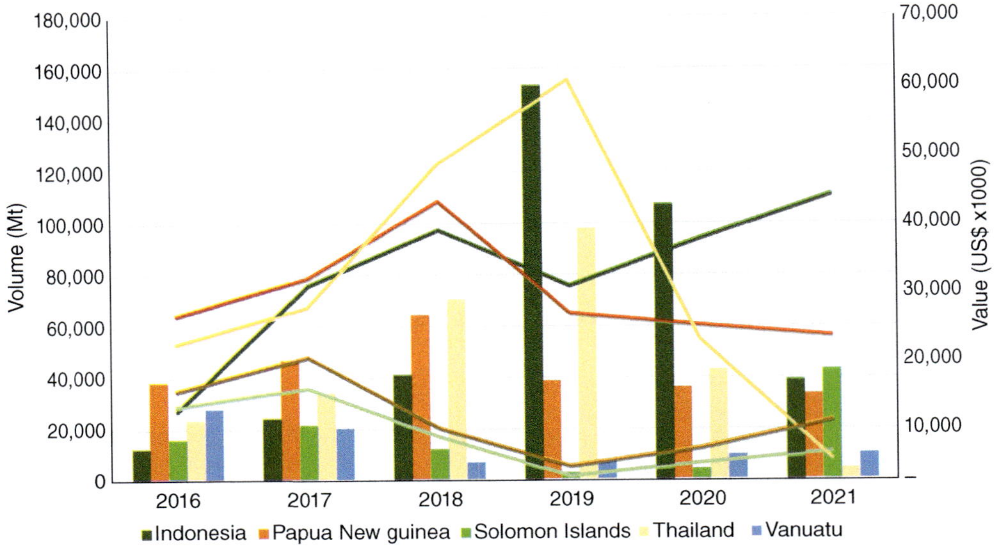

Fig. 14.3. Volume and value of exports of copra, 2015–2020. From International Coconut Community (ICC) Ministerial Session Documents, 2020, reproduced by kind permission of the ICC.

dried using an oven) and shifting to a direct copra-marketing system will help maximize copra-processing benefits (Pestaño and Wifredo, 2016). Compliance with the stricter copra export market regulations will promote the preparation of white copra. Farm-level transformation coupled with direct marketing transactions within organized farmer groups should flatten or eliminate the current multi-layered marketing system.

Coir and coir products

As a coconut by-product, coir products can be used to make floor coverings, domestic and automobile upholstery, brushes, twines, geotextiles and insulation and as a packaging material. As a 100% natural and in most instances organic material, it offers an environmentally friendly source of biodegradable products. The key drivers for the coir market are linked to horticulture as it is most often used for gardening, as a soil conditioner and in hydroponics, although increased demand for geotextiles (e.g. for civil engineering) is anticipated. Continuing innovations for coir are likely to create a global demand for its use as an eco-friendly substitute for petroleum-based or less renewable products. This includes coir ply as a substitute for wood, rubberized coir mattresses for orthopaedic beds and automobile seat cushions, garden articles from coir fibre to replace plastic items, and coir pith for organic manure or plant-growth media for nurseries and reforestation programmes. The signs are good, but many of these emerging products still require market development (Loshali, 2022).

The global coir export trend is reflected by the two main exporting countries, Sri Lanka and India (Fig. 14.4). There has been no significant rise in market volume produced, although export values have grown by 9.4% during the last 5 years. Husk volume production in the Philippines, Sri Lanka and India has long dominated the coir export market, with China importing about 75% of the total global coir and cocopeat exports. Prices have and continue to remain volatile, with imports constrained by recent declines in the Chinese economy. Any global coir market growth is likely to depend on China's economic recovery or other countries becoming more substantial users.

In South Korea, cocopeat is already widely used in rice (*Oryza sativa* L.) farming, while in the Middle East, cocopeat could substitute for peat moss as a soil conditioner for imported soil for landscaping and horticultural applications. There is also a projected increase in the demand from the USA (Costales, 2019).

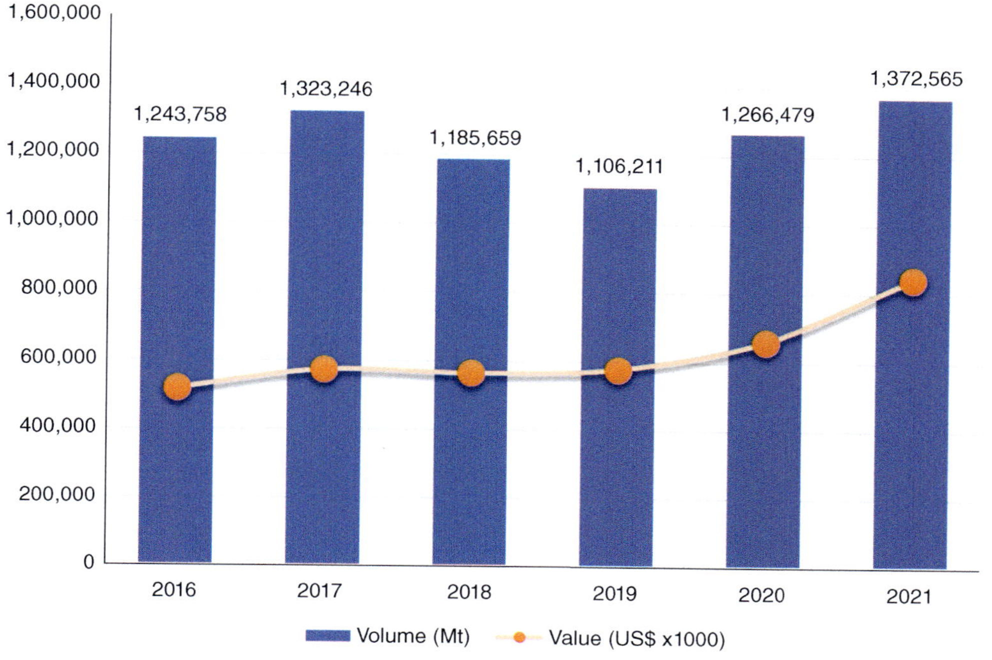

Fig. 14.4. Export of coir products from India and Sri Lanka, 2016–2021. From the Coir Board of India and Coconut Development Authority, Sri Lanka.

Significant growth in the global coir market in the period 2020–2030 is anticipated. A key contribution to market growth could be the projected increased demand from Europe and North America as these consumers are increasingly keen on hydroponics, which uses coir pith.

Desiccated coconut

Desiccated coconut is mostly used as a culinary ingredient. This is driven by consumer-demand for coconut-based products such as breakfast cereals and many Asian dishes, which use desiccated coconut as a key ingredient. As a healthy food product, desiccated coconut market demand steadily increased between 2012 and 2021 with an indicative compound annual growth rate of 5.3% (Fig. 14.5). The Philippines (37%) and Indonesia (31%) have dominated the global production of desiccated coconut and export markets along with other key players such as Singapore (8%), Sri Lanka (7%), Vietnam and Malaysia (each 8%). The Philippines remains the leading desiccated coconut supplier and commands a high price for its premium-quality desiccated coconut. Sri Lankan high-quality desiccated coconut has also garnered premium prices in recent years because most European buyers consider it to be of a high quality, giving the country an edge in the global desiccated coconut market in recent years (Aguilar *et al.*, 2022)

The major desiccated coconut-importing countries are the USA, Vietnam, the Netherlands, Benin and Singapore (Fig. 14.6). These imports are highly dependent on the confectionery and bakery industries, with positive forecasts. Desiccated coconut imports into the European Union are predicted to expand substantially in the longer term (Aguilar *et al.*, 2022). As the largest exporter of desiccated coconut to Europe, the Philippines is likely to benefit the most from the projected growth in European market.

Coconut water

Health consciousness is a major driver for the increasing market demand for coconut water. Currently, it is marketed as a healthy product in several ways. Because of its high mineral content

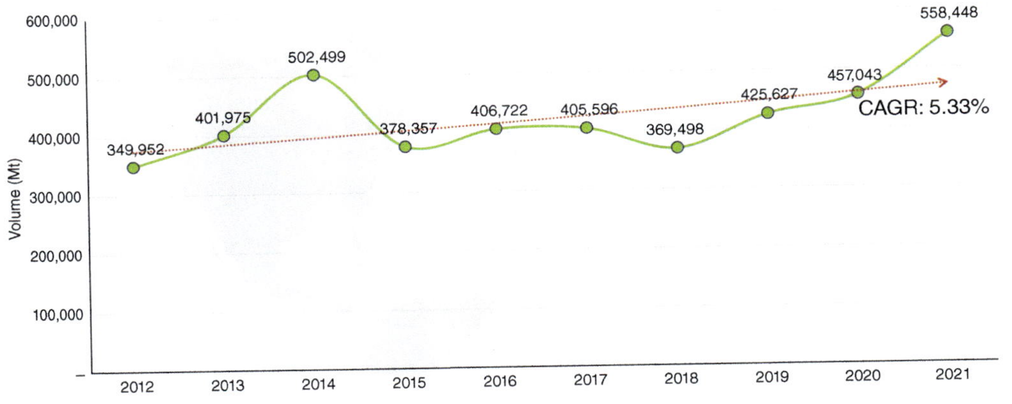

Fig. 14.5. Global exports of desiccated coconut (Mt), 2012–2021. CAGR, compound annual growth rate From International Coconut Community (ICC) Session Documents 2022, reproduced by kind permission of the ICC.

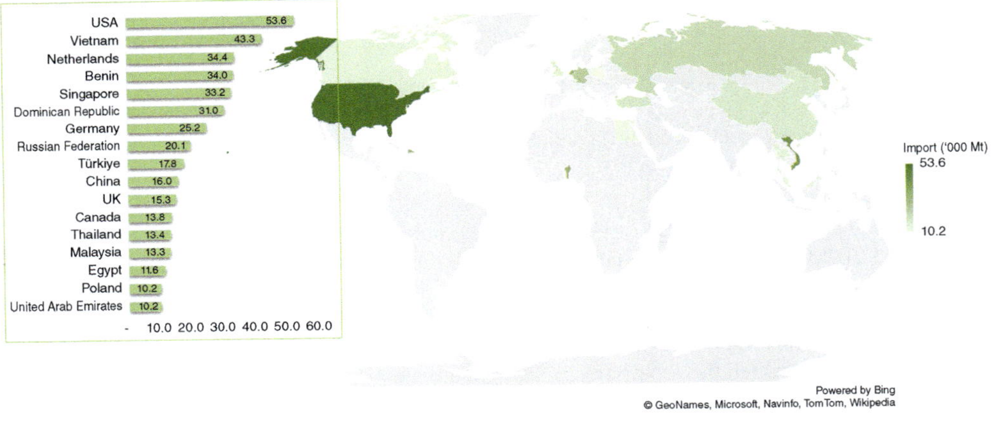

Fig. 14.6. Top importers of desiccated coconut (Mt), 2021. From International Coconut Community (ICC) Session Documents 2022, reproduced by kind permission of ICC.

and carbohydrates, coconut water is classified as an isotonic drink and a sports drink. The global coconut water market is projected to grow at a rate of more than 5% year^{-1}. It has a competitive advantage as it is considered a healthy natural drink compared with other manufactured products. As the fastest growing new beverage product in the USA, coconut water has also gained widespread international popularity (Aguilar *et al.*, 2022). It is predicted that the increasing demand for this drink is such that the coconut water market demand will escalate considerably from its current annual consumption of around 600 million l year^{-1}.

Thailand, Brazil, Sri Lanka, the Philippines and Indonesia were the top coconut water exporters in 2021 (ICC, 2022). The Philippines is likely to benefit most from the fast-rising demand, as it is second only to VCO as a major contributor to the Philippines' export earnings from non-traditional coconut products. The Philippines coconut water export volume has increased markedly, with the country now exporting around almost 40,000 l annually, worth an estimated US$41 million over 11 years (Aguilar *et al.*, 2022). The Philippines' export destinations for coconut water show the USA and Germany as the largest importers in 2020–2021 (Fig. 14.7).

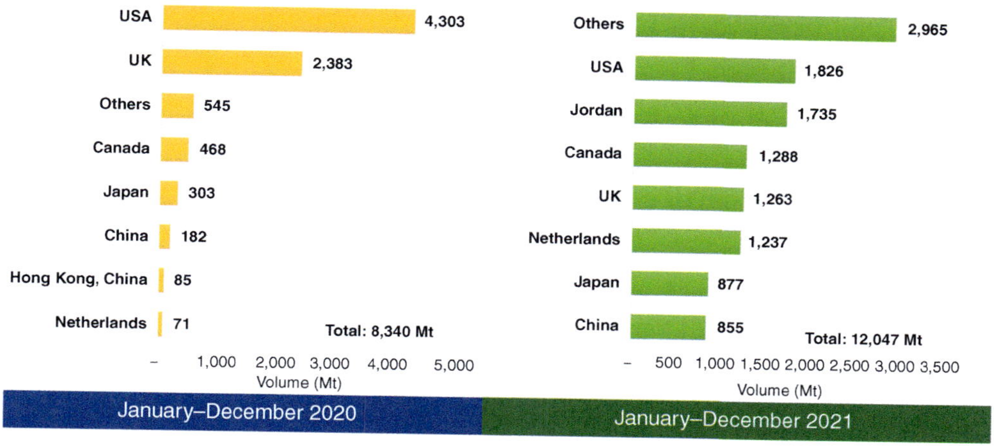

Fig. 14.7. Export of coconut water from the Philippines. From International Coconut Community (ICC) Session Documents 2022, reproduced by kind permission of the ICC.

Western Europe, Australia and New Zealand are large markets for packaged coconut water. The European coconut water market accounts for around 10% of total world sales volumes, representing 60–70 million l consumed annually. Annual European consumption is predicted to increase by 15–20% year^{-1} in the short term and then to around 10% year^{-1}. European coconut water market growth is shaped by aspirational alternative lifestyles expressed as demand for functional and plant-based drinks (CBI, 2020). It is worth noting that the actual coconut water import volumes into Europe are less than consumption volumes, as imported concentrated coconut water is often further repacked with water or used as a beverage ingredient. The UK is the largest European coconut water consumer, with the Netherlands serving as a trade hub and Austria having the greatest number of beverage processors. French coconut water consumption has increased at least five-fold over the last 5 years. Currently, the French consume around 12–14 million l year^{-1} and are expected to be of similar consumption levels to the UK soon. Germany was the first country to import coconut water, but volumes did not initially expand as they did in the UK. At present, the consumption rate increase is now comparable with that of Europe's top three coconut water-consuming countries, with German consumption in the range of 6–8 million l year^{-1} (CBI, 2020). Italy annually consumes 3–4 million l year^{-1}. In the Netherlands, coconut water demand is expanding rapidly with many new brands emerging. Coconut water is also traded from the Netherlands by specialized juice ingredient suppliers who have processing facilities in Brazil, Vietnam, Mexico and Poland (CBI, 2020). The Australian packed coconut water market is forecast to grow rapidly. However, competition with cheaper substitutes and the high packaging costs may constrain market growth (Report Linker, 2020).

Coconut sap sugar

The popularity of coconut sap sugar has recently escalated due to claimed health benefits attributed to its low glycaemic index. Research indicates that an increase in recorded diabetes rates could be a key driver when more evidence is provided, with the increasing coconut sugar market demand coupled with better promotion of its health benefits (WHO, 2022). It is anticipated that the market demand for coconut sugar will gradually increase if global trends match the Indonesian market performance from 2017 to 2021 in volume and value (Fig. 14.8).

Growing awareness of the adverse effects of refined sugar is likely to confer a competitive advantage to coconut sugar. As well as having a lower claimed glycaemic index than cane sugar and honey, coconut sugar is largely organic. The World Health Organization (WHO, 2022) estimated

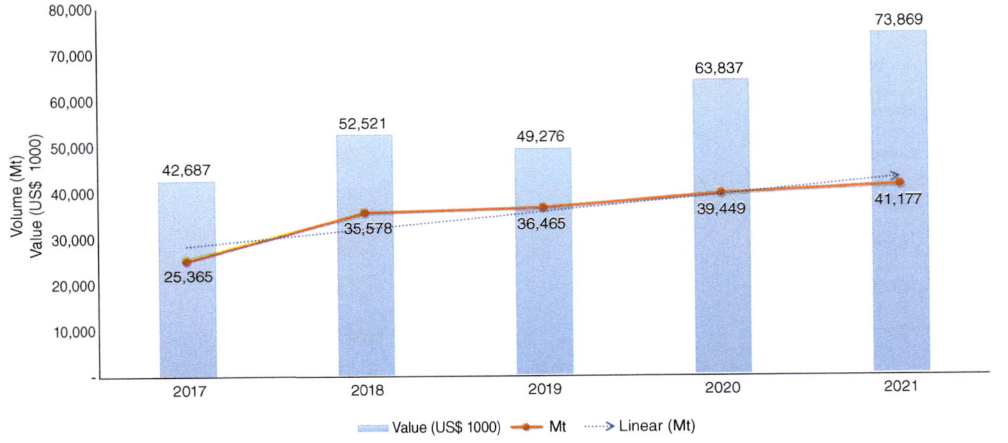

Fig. 14.8. Coconut sugar exports from Indonesia. From International Coconut Community (ICC) Session Documents 2022, reproduced by kind permission of the ICC.

that 1.6 million deaths in 2016 were caused directly by diabetes, while 1.9 billion adults were overweight and around 650 million of these were obese. The rising incidence of diabetes and overweight individuals may result in increased interest in the consumption of coconut sugar. Coconut sugar does not spike insulin levels like many other regular sweeteners, and it contains fibre which is known to reduce glucose absorption. It is predicted that increasing cases of cardiovascular disease and obesity will heighten the demand for coconut sugar as a healthier sweetener (Vijayakumar *et al.*, 2016; Asghar *et al.*, 2020; Arshad, *et al.*, 2022; WHO, 2022).

The use of coconut sugar in snacks, chocolate, breakfast cereals, bakery products, confectionery, food seasoning, juices, coffee and tea are predicted to have a 5–6% compound annual growth rate. Specifically, in European markets, consumers tend to aim to avoid non-organic products and coconut cultivation is usually organic. Increasing numbers of food and beverage companies have shifted to using natural sweeteners, where organic coconut sugar is often prioritized. Sale of coconut sugar can be lucrative as it often commands a higher price than cane sugar, particularly if it is produced organically (Samriddhi and Roshan, 2022). Even the chocolate industry is banking on healthier ingredients, which increasingly include coconut sugar. In Sweden, in 2017, there was a sales turnover of more than US$65 million in organic chocolates, sugar, jam and confectionery. Non-food utilization of coconut sugar includes skincare products such as body scrubs, shaving gels, and facial and body creams, as well as hair care products, and this is also contributing to coconut sugar demand growth.

A snapshot of coconut sugar exports from the Philippines for 2020 shows the main importers to be Sri Lanka, Europe, Taiwan and Australia (Fig. 14.9). It has been documented that the coconut sugar market will have a compound annual growth rate of 5.86% in the period 2019–2024 and should reach US$397 million by the end of 2024 (Market Research Future, 2021).

VCO

The value of VCO exports from the main exporting countries (e.g. Indonesia, the Philippines and Sri Lanka) in the period 2019–2021 continued to rise, even though production was constrained due to the coronavirus 19 (COVID-19) pandemic (Figs 14.10 and 14.11). The global VCO market is projected to grow from US$2.24 billion in 2021 to US$3.69 billion (65% increase) by 2028 at a compound annual growth rate of 7.35% (Fortune Business Insights, 2022).

The purity of VCO, often with organic status, is a key market driver, along with its health-giving properties. These health benefits have been demonstrated by numerous studies (Vijayakumar *et al.*, 2016; Haron *et al.*, 2018; Pruseth *et al.*,

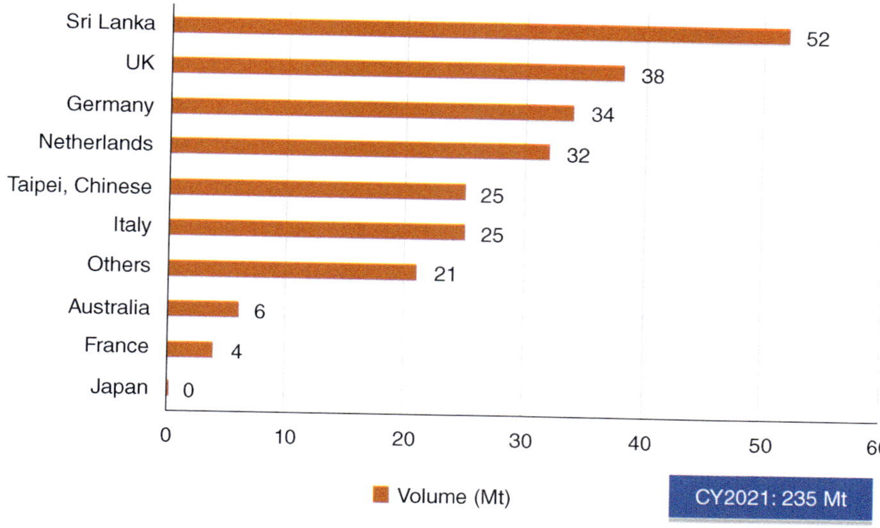

Fig. 14.9. Export of coconut sugar from the Philippines (Mt), 2020. From International Coconut Community (ICC) Session Documents, 2022, reproduced by kind permission of the ICC.

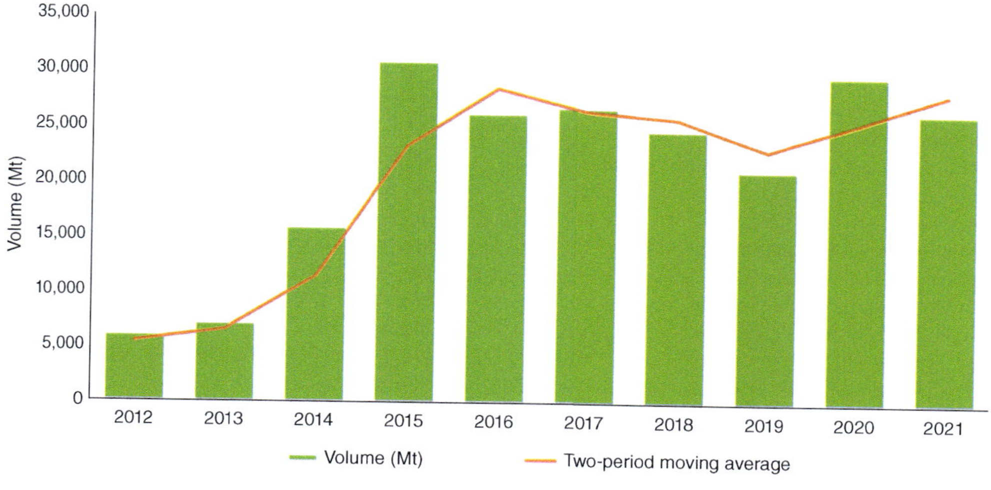

Fig. 14.10. Export of virgin coconut oil from the Philippines. From International Coconut Community (ICC) Session Documents, 2022, reproduced by kind permission of the ICC.

2020; Angeles-Agdeppa, *et al.*, 2021). The oil has been shown to help treat diabetes (Takeishi *et al.*, 2021) and reduce 'bad' cholesterol levels (Dehghan *et al.*, 2017). It contains more than 50% medium-chain triglycerides, including lauric acid and capric acid, which act as immune-system stimulants with potent antiviral, antibacterial and antiprotozoal properties that can help in the treatment of genital herpes, influenza, HIV/AIDS, hepatitis C and various forms of gum disease (Astrup *et al.*, 2020; Hewlings, 2020; Takeishi *et al.*, 2021). A recent dietary study conducted by the Department of Science and Technology in the Philippines revealed that VCO decreased COVID-19 viral load by 60–90%. Research has also revealed VCO anticancer effects (Praagman *et al.*, 2016; Pruseth *et al.*, 2020); however, these findings need to be corroborated by further research.

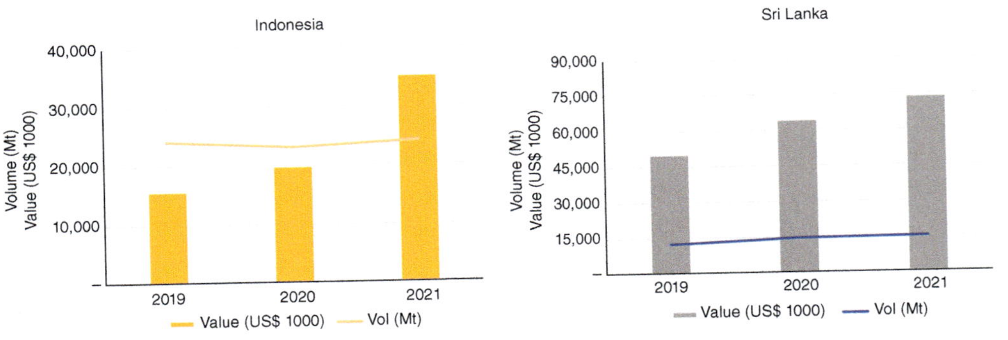

Fig. 14.11. Export of virgin coconut oil from Indonesia and Sri Lanka. From International Coconut Community (ICC) Session Documents, 2022, reproduced by kind permission of the ICC.

Other key drivers for VCO sales relate to beauty, cosmetics and personal care. Accordingly, 40% of VCO consumers are already using organic VCO-based beauty products, while 8% more are interested in shifting to these products (Aguilar *et al.*, 2022).

The VCO market is divided by global region: (i) Asia-Pacific; (ii) European; (iii) the Americas; and (iv) the Middle East and Africa (Costales, 2019). Market Research Future (2019) showed the Asia-Pacific region to have the highest regional share in the global VCO market and to be the largest regional VCO producer and exporter. Countries such as Japan, China, Singapore, South Korea and Australia are among the top importers of VCO, while the three major exporters (the Philippines, Indonesia and Sri Lanka) are also located in the Asia-Pacific region (Costales, 2019). The regional VCO market demand for culinary purposes is predicted to expand, especially in China, India and Japan. The Asia-Pacific region is also home to several VCO manufacturing companies (Market Research Future, 2019)

Europe's VCO market performance is driven by increasing consumer awareness of healthy lifestyles and diets, stimulating demand for organic beauty and health supplements. Former VCO niche markets are expanding into outlets including large supermarkets (Aguilar *et al.*, 2022). The rapid market growth in Europe is greatly influenced by the UK's growing VCO demand and consumption. Europe imports VCO from the Philippines, Sri Lanka, Indonesia, India, Thailand, Malaysia and Papua New Guinea (Costales, 2019).

In the Middle East and Africa, VCO continues to gain popularity, particularly in Dubai and the United Arab Emirates. Market research indicates that the market for coconut products in the African region will grow considerably, motivated by the entry of coconut processors into Africa, specifically the Rocac Group in Nairobi, Kenya, which has started to export VCO and coconut water to the United Arab Emirates and other countries (Costales, 2019).

In the Americas, greater awareness of VCO nutritional benefits has triggered interest, particularly as a therapeutic treatment for obesity, a prevalent problem in the region (Costales, 2019; Adeyemi *et al.*, 2020; de Vasconcelos *et al.*, 2022). For the Philippines' VCO exports, the top importers are the USA, Germany, Canada, the UK, Brazil, the Netherlands and China. The North American market remains the biggest market for Philippines-produced VCO, mainly provided by the desiccated coconut producers that have branches in the USA (Aguilar *et al.*, 2022). In 2015, the USA acquired more than 50% of the Philippines' and 41% of Sri Lanka's total VCO exports (Costales, 2019). The USA market began to slow in 2016 due to the American Heart Association's advice on the negative effects of coconut oil on health. Given VCO's demonstrated benefits for health, beauty and industrial product consumers, market confidence has been restored, following the example of some European users, including Get Fruity Bar (UK), Bio Paranuss-Guarana Riegel (Germany) and Lifefood Coconut Bar (Germany/Europe) (Costales, 2019).

Oleochemicals

There is growing demand for this coconut oil-based group of products (mainly as a sustainable source of fatty acids, fatty alcohols and glycerin; OFI, 2021) in the pharmaceutical and personal-care industries (Joseph *et al.*, 2022). Any increase in demand for cosmetics and personal-care products is likely to trigger a corresponding increase in demand for oleochemicals (naturally occurring fats and oils of vegetable or animal origin). This market is also driven by high demand from 'green' consumers who are aware of the environmental and economic benefits of using oleochemicals. This has driven an industry shift to oleochemical applications, including in the food and beverage sectors. Oleochemicals are tagged as natural, green, organic, safe, renewable and biodegradable. The increase in demand for biodegradable products is probably the main driver of the increase in demand for oleochemicals.

It is predicted that the total global oleochemical market size could reach US$31.4 billion by 2027, registering a compound annual growth rate of 5.8% (Global Oil Market Research Report, 2022). Oleochemical manufacturers are replacing many petroleum-based products with eco-friendly and renewable raw materials, including from coconut. Personal-care, soaps and detergent segments account for more than 51% of the global oleochemical market. Between 2018 and 2021, the total market grew by almost 9% (Fortune Business Insights, 2022), attributed to increasing product demand from industrial and institutional cleaning and textile sectors.

Currently, the highest share in the global oleochemical market is by far the Asia-Pacific region because of the abundant raw material availability and the region's expanding global-scale oleochemical production plants. Palm oil, palm kernel oil and crude coconut oil provide the feedstock for oleochemical production. Malaysia and Indonesia are the major producing and exporting countries of palm oil and palm kernel oil, and the Philippines is the largest exporter of crude coconut oil. Palm kernel oil and coconut oil have a wide mix of fatty acids (Beroe, 2018), which is the reason why these oils are the oleochemical industry's preferred feedstock. Palm kernel oil is usually used because of its lower bulk price compared with crude coconut oil. With the recent European ban on Indonesian and Malaysian palm and palm kernel oil imports for use as biofuel due to environmental issues caused by growing oil palm (e.g. deforestation and biodiversity loss), the use of Filipino coconut oil is likely to be favoured by the European market. In addition, the import ban issued by the USA for Malaysian palm oil is likely to create a competitive advantage for coconut methyl ester from the Philippines to be used as a biodiesel, as this country is the main exporter, as well as from other coconut oil-exporting countries with sustainable credentials.

Activated carbon from coconut shell charcoal

Activated carbon is an effective adsorbent widely used in industry due to its high surface area, well-developed reproducible microporous structure, and high degree of surface reactivity and adsorption capacity (Louarrat *et al.*, 2019). Charcoal from coconut shells has considerable export market potential. Activated carbon exports are generally driven by the increasing demand for activated carbon in treating water, sewage, human and industrial waste. Coconut activated carbon can remove particulates, chlorine and dissolved organic impurities in water and from air. Growing environmental concerns, stringent regulations and standards, as well as government initiatives to safeguard the environment are likely to boost the activated carbon market over time. Coronavirus and other virus protection matters are likely to lead to a rising demand for the product in air purification applications for domestic as well as commercial use.

The global activated carbon market is expected to strengthen during the next decade. The projected trend is based on positive global market growth over the last decade, where an 8.2% compound annual growth rate has been recorded. This can be attributed to existing environmental laws encouraging the use of activated carbon for industries that provide water filtration, air purification, mining services and healthcare, among other uses (Aguilar *et al.*, 2022).

In 2019, global imports of activated carbon exceeded US$1.92 billion (Aguilar *et al.*, 2022).

The USA and Japan were the largest activated carbon-importing countries globally, with an 11.9% (US$229 million) and 8.6% (US$165 million) market share, respectively. Other major importers of activated carbon in the same year were Germany (8.45% share or US$162 million), China (8.21% or US$157 million), South Korea (6.49% or US$124 million), and Belgium (4.86% or US$93 million). These six countries imported 48.5% of the activated carbon produced in 2019. The growing demand of these importing countries (e.g. Japan, the USA, Germany, China and South Korea) has contributed to the global market growth of activated carbon.

The Philippines, Indonesia, India and Sri Lanka are leading producers and suppliers of coconut-shell charcoal and activated carbon. The USA, Japan, Germany and South Korea were the top importers of activated carbon in 2021 (ICC, 2022). In 2019, China was the Philippines' largest coconut-shell charcoal buyer, accounting for 57% of the total coconut-shell charcoal volume, followed by Japan with 48% of the market share. China registered extremely strong growth, with a 149% compound annual growth rate, compared with Japan's 6.1% compound annual growth rate between 2009 and 2019 (Aguilar *et al.*, 2022).

Given their fast-growing development of infrastructure, emerging economies such as China and Brazil are likely to strengthen their demand for wastewater management, and the growing demand for processed foods and beverages is likely to help drive the activated carbon market. Other factors that may further boost demand for coconut activated carbon include pharmaceutical and medical applications in purifying raw materials, and use in kidney machines, nursing supplies and respirators. Use within the automotive industry is anticipated to stimulate an abrupt rise in market growth to at least 2028. However, products such as granulated rubber and coke breeze may continue to compete against coconut-shell activated carbon. There is increasing activated carbon demand from gold-mining companies to recover gold from cyanide solution (Gratuito *et al.*, 2008; Khosravi, *et al.*, 2017; Louarrat, *et al.*, 2019). Coconut-shell activated carbon is also used in remediating heavy-metal waste in soil around mining areas (Septiana *et al.*, 2019).

The Philippines' market growth for coconut-shell activated carbon is favoured by the largest activated carbon-importing countries (Aguilar *et al.*, 2022). In the last decade, the Philippines' leading activated carbon export markets were Japan, the USA, Germany, China and South Korea. Based on Philippines Coconut Authority data, these importing countries of coconut-shell activated carbon from the Philippines exhibited high annual growth rates in terms of export volume from 2009 to 2019. In 2019, Japan was the top buyer of Philippines-manufactured coconut-shell-based activated carbon, with an 18.0% market share, followed by the USA with a 12.3% market share. Germany trailed slightly behind at 11.9%, followed by China (7.9%) and South Korea (6.8%). From 2012 to 2015, the Philippines was the top-ranked exporter but was surpassed by India in 2019. In 2021, the Philippines' activated carbon exports were down by 4.1% from 2020, and Sri Lanka showed a substantial increase of 16.4% from 2020 to 2021 (Fig. 14.12). It is noted, however that USA imports have remained steady between 2016 and 2021 (Figure 14.13).

Coconut milk

The process of making coconut milk is explored in detail in Chapter 12 (this volume), along with food and other uses. Briefly, coconut milk, like other coconut products, is experiencing increasing demand. For coconut milk, this is partly being driven by the increased demand for vegan food and drink products, especially in Europe. Coconut products are an excellent substitute for animal-sourced products such as dairy and animal fats. The vegan trend is geared towards organic food innovations and is likely to continue growing globally in the coming years. According to Allied Market Research, the global vegan industry was valued at US$14.2 billion in 2018 and is expected to reach US$31.4 billion by 2026. With changing diet preferences in Europe, coconut milk produced in low- or lower–middle-income countries could take the lead in capturing the vegan market.

14.4.2 New Products on the Horizon

There are a number of emerging products from coconut entering the global or niche markets. These include coconut aminos (a dark-coloured

Fig. 14.12. Export volume (Mt) of coconut-shell charcoal-based activated carbon from the Philippines and Sri Lanka. From International Coconut Community (ICC) Session Documents, 2022, reproduced by kind permission of the ICC.

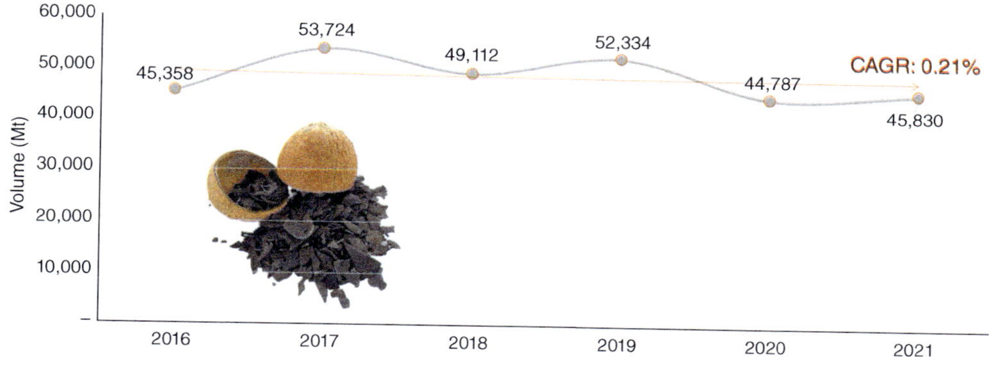

Fig. 14.13. US imports of coconut shell charcoal-based activated carbon (Mt), 2016–2021. From US Census Bureau Source: International Coconut Community (ICC) Session Documents, 2022, reproduced by kind permission of the ICC.

sauce made from the sap of the coconut plant), biofuel, coconut fibreboard, coconut lambanog (fermented toddy), coconut sap syrup, tannins from coconut husks and cocopeat, coconut veneer and waste coconut-leaf-based straw.

Coconut syrup

Coconut syrup is a natural sweetener derived from organically grown palm sap (Andiluhung *et al.*, 2019). The pure, fresh sap is tapped and then evaporated to the right consistency by simmering over a low to moderate heat until it becomes sticky. It has a rich, sweet taste, with a distinct fruity coconut flavour. It contains no additives or artificial flavourings, has a low glycaemic index, and offers a rich source of vitamins and minerals, especially vitamin C and potassium. A coconut palm can produce 50–100 l of sap month^{-1}, depending on the type (Novarianto *et al.*, 2021). Coconut syrup currently has a USA selling price of around US\$1 l^{-1} and is increasingly used as an alternative sweetener to cane sugar owing to its nutritional properties and natural status, along with it being free from artificial preservatives. This trend is also being driven by rising demand for minimally processed food ingredients (Fortune Business Insights, 2022).

Toddy and palm wine

Although not a new product, the refinement of toddy is relatively new. Coconut toddy is produced from the sweetish, milk-white fluid coming from the tapped young coconut inflorescence tissue. The unfermented sweet sap is called neera (Somashekaraiah *et al.*, 2019). For non-alcoholic neera, a little lime juice is added to the collecting pot to prevent fermentation. Toddy is produced in at least 40 coconut-producing countries. It is made by the natural fermentation of the collecting sap, coming from the spadix. Microflora (bacteria and fungi) are responsible for the fermentation and the toddy is fully fermented in 6–8 h. Locally known in Filipino as lambanog, toddy is distilled from coconut sap with an alcohol content of 40–45% by volume. It is a popular drink in rural areas and is produced by local communities during festive seasons. It is indigenous to the Philippines and is an integral part of the Filipino festive heritage. Currently, high-quality lambanog is being produced branded as lakan with the prestigious reputation in the international niche market of a naturally produced palm wine. Distilled by traditional methods, lakan is surprisingly smooth and subtle and is best served neat. This is an example of how branding can add value to coconut products (see www.lakan.com.ph/about, accessed 5 September 2023). Toddy is similarly popular in most coconut-growing countries, with some also receiving similar growing interest in it as a consistent manufactured product.

Coco aminos

Coco aminos is a savoury seasoning sauce made from the fermented sap of coconut palm, containing a range of amino acids. After the coconut sap is extracted, it gets stored and aged. During this time, it ferments due to the natural sugars and wild yeast present. Coconut aminos tastes like a milder version of soy sauce that has less salt. Although coconut aminos is made from coconut sap, it does not taste anything like coconut. Soy sauce has a richer flavour than coconut aminos, but about three times as much sodium. Coconut aminos is also naturally gluten free as it is made from the sap of the coconut palm, while soy sauce is traditionally made from fermented soybeans and wheat. Coconut aminos may be considered a healthy alternative to soy sauce. There is no soya or gluten in coconut aminos, two common allergens, and soy sauce contains anywhere from 280–320 mg of sodium per teaspoon, while coconut aminos has 90 mg of sodium per teaspoon. Amino acids are thought to help cure mental exhaustion. There is increasing demand from the food, pharmaceutical, and nutraceutical industries for this product. According to FMI, 2023, the global amino acid market is expected to grow from US$24.2 billion in 2021 to US$62.1 billion by 2032, although the coco aminos-like proportion of this market is undocumented and needs further assessment.

Coconut veneer

Coconut veneer, used for a range of wood products, can now be produced by rotary peeling of the coconut trunk. The production of veneer increases log volume recovery by 60–65% compared with more traditional wood preparation methods. Senile or diseased palms can be converted into quality veneer products using a new spindle-less lathe technique, at the same time incentivizing coconut plantation regeneration. Representing almost a decade of research, McGavin *et al.* (2019) produced a useful guide to coconut palm rotary veneer processing. However, establishing quality criteria (i.e. colour, visual characteristics and hardness) for architectural veneers from coconut wood still needs to be developed. Work also needs to be undertaken to explore the emerging coconut veneer value chain to develop a viable business model for existing veneer producers and potentially additional small-scale processors.

Coconut fibreboard

Studies indicate that the commercial use of coconut husks as a raw material (coir) for coconut fibreboard could substantially increase coconut profitability and boost bio-economies of coconut-producing countries. Coconut-producing countries importing wood-based building materials would welcome a local alternative (Snijder, *et al.*, 2006). Production of coconut fibreboard could contribute to national low-cost housing programmes, such as in the Philippines. Coconut fibreboard could replace traditional construction materials such as tiles, bricks, plywood, asbestos and cement hollow blocks. It could be used for internal and exterior walls, partitions and ceilings, or for furniture. One ton of coconut fibreboard can be produced from 3000 fruit husks. Although a study on husk-based medium- and high-density fibre confirmed the high potential in terms of environmental performance, coconut fibreboard production still requires more efficient production, processing and transportation to successfully compete with wood-based fibreboards (Freire *et al.*, 2017).

Biofuels

There is interest emerging regarding the use of coconut oil in jet aircraft fuels to reduce carbon dioxide emissions. Such reductions can only be achieved when using jet fuels derived from alternative sources such as blends of coconut oil with kerosene (Eller, *et al.*, 2019). Because coconut oil is produced sustainably, it is likely to be a preferred choice in aviation fuels compared with other less sustainably produced vegetable oils. Its use would help meet carbon-offsetting and reduction-scheme criteria for international aviation-eligible fuels.

Tannins from coconut husks and cocopeat

Coconut husks and cocopeat offer a potential way to supply naturally produced tannins for manufacturing adhesives or natural dyes (Morbeck *et al.*, 2019). Tannins can easily be extracted from green coconut husks and can be part of farmers' solid waste management plan. Another emerging product is waste coconut leaf-based single-use items such as drinking straws and plates. The commercial sector now features the use of coconut leaf-based fibres for the manufacture of these products as an alternative to using plastic, but further market intelligence and development are required.

14.5 Conclusion

This chapter has outlined the main market trends for current coconut products and briefly described some that are emerging. The economic outlook for the coconut industry is extremely positive. Front-runner products include coconut water, coconut sugar, coconut milk, VCO and shell-charcoal-based activated carbon products. Coconut is a key livelihood crop, contributing to the nutrition, health, environment, industrial sectors and national economies in all coconut-growing countries. Coconut can offer an increasingly relevant, climate-resilient component to a sustainable agricultural production system, but there are challenges, the largest arguably being that coconut farmers remain some of the poorest of the poor.

Ensuring an economically viable future for coconut must include the further development of breeding programmes and pathways for the micropropagation of elite, clean planting materials and their use in affordable and resilient replanting programmes, improved production practices and undertaking stricter market compliance. Producers need to be empowered to produce high-performing crops that meet market needs but also provide a viable and sustainable livelihood. This is likely to involve applying innovative technologies, policies, and agronomic and commercial approaches to stabilize product supply chains and maximize coconut's commercial competitiveness. Farmers must be integrated across the coconut value chain. Perhaps most importantly, producers are likely to need to adopt new coconut-product and business models, as recommended by Turner *et al.* (2022), to improve coconut livelihoods and boost exports, tourism and industry investments.

One of the key drivers for industry growth is consumer burgeoning awareness of coconut's health, nutritional and environmental benefits. As a result, there is increasing interest in coconut as a component of many foods, snacks, beverages, and cosmetic and personal-care products. Producers will need to be better supported to work with key coconut-product market players to respond to these growth opportunities. This may include coconut-growing countries repurposing land for specific outputs, such as for coconut water and sap production, or for developing certain kinds of intercropping with high-demand food crops or to undertake other agricultural endeavours to increase land productivity.

The industry will almost certainly gain value from reducing production waste, a trend already being seen in many countries; for example the industry must use coconut water effectively and efficiently for concentrate, considering extraction, consolidation and processing, so that farmers can also benefit from coconut water's added value rather than them just being paid for copra and the water being wasted. This approach would also allow full use of the fruit shell for high-value products (Aguilar *et al.*, 2022). New breeding and planting programmes should aim for varieties with high water and sap production capacity from high-yielding and early-bearing dwarf or hybrid varieties to help reduce waste even further.

With prudent and innovative management of the genetic, agronomic and commercial opportunities, the timing is right for coconut stakeholders to work together to create a sustainable industry, with viable economic initiatives to achieve prosperity for all, particularly smallholder farmers. Without improving their prosperity, the industry is unlikely to thrive.

14.6 References

Adeyemi, W.J., Olayaki, L.A., Abdussalam, T.A., Toriola, A.P., Olowu, A.B. *et al.* (2020) Investigation of the effects of dietary modification in experimental obesity: low dose of virgin coconut oil has a potent therapeutic value. *Biomedicine and Pharmacotherapy* 126: 110110.

Aguilar, E.A., Lozada, E.P. and Aragon, C.T. (2022) *Philippines Coconut Industry Roadmap 2021–2040*. Philippine Coconut Authority, Department of Agriculture, Quezon, Philippines.

Alouw, J.C. and Wulandari, S. (2020) Present status and outlook of coconut development in Indonesia. *IOP Conference Series: Earth and Environmental Science* 418: 012035.

AMRA Farms (2020) *Coconut Farming and Cultivation in India*. AMRA Farms, Palakkad, India. Available at: https://amrafarms.com/coconut-farming-and-cultivation-in-india.html (accessed 4 September 2023).

Andiluhung, A., Mela, E., Widjonarko, G. and Sitoresmi, I. (2019) Products of coconut nira to develop medium and small micro enterprise in Banyumas Regency. *Agrin Journal* 23, 85–102.

Andoh-Mensah, E. and Ofosu-Budu, G.K. (2012) Evaluation of coconut–citrus intercropping systems in the context of lethal yellowing disease of coconut in Ghana. *African Journal of Food, Agriculture, Nutrition and Development* 12, 6945–6962.

Angeles-Agdeppa, I., Nacis, J.S., Capanzana, M.V., Dayrit, F.M. and Tanda, K.V. (2021) Virgin coconut oil is effective in lowering C-reactive protein levels among suspect and probable cases of COVID-19. *Journal of Functional Foods* 83: 104557.

Arshad, S., Rehman, T., Saif, S., Rajoka, M.S.R., Ranjha, M.M.A.N. *et al.* (2022) Replacement of refined sugar by natural sweeteners: focus on potential health benefits. *Heliyon* 8: e10711.

Asghar, M.T., Yusof, Y.A., Mokhtar, M.N., Ya'acob, M.E., Ghazali, H. *et al.* (2020) Coconut (*Cocos nucifera* L.) sap as a potential source of sugar: antioxidant and nutritional properties. *Food Science and Nutrition* 8, 1777–1787.

Astrup, A., Magkos, F., Bier, D.M., Brenna, J.T., de Oliveira Otto, M.C. *et al.* (2020) Saturated fats and health: a reassessment and proposal for food-based recommendations. *Journal of the American College of Cardiology* 76, 844–857.

Beroe (2018) Oleochemicals market supplier. Risk and competitive intelligence. Available at: www.beroeinc.com/category-intelligence/oleochemicals-market/ (accessed 4 September 2023).

Bourdeix, R. and Sourisseau, J.M. (2021) *Coconut Risk Management and Mitigation Manual for the Pacific Region*. Land Resources Division, The Pacific Community, Suva, Fiji.

CBI (Centre for the Promotion of Imports) (2020) *The European Market Potential for Coconut Water*. Centre for the Promotion of Imports, Ministry of Foreign Affairs. Available at: www.cbi.eu/market-information/processed-fruit-vegetables-edible-nuts/coconut-water/market-potential (accessed 4 September 2023).

Costales, C. (2019) *Three NTP Value Chain Analysis and Development of VC Models*. Excoll Travera Enterprise Study.

Dehghan, M., Mente, A., Zhang, X., Swaminathan, S., Li, W. *et al.* (2017) Associations of fats and carbohydrate intake with cardiovascular disease and mortality in 18 countries from five continents (PURE): a prospective cohort study. *Lancet* 390, 2050–2062.

de Vasconcelos, M.H.A., Tavares, R.L., Torres Junior, E.U., Dorand, V.A.M., Batista, K.S. *et al.* (2022) Extra virgin coconut oil (*Cocos nucifera* L.) exerts anti-obesity effect by modulating adiposity and improves hepatic lipid metabolism, leptin and insulin resistance in diet-induced obese rats. *Journal of Functional Foods* 94: 105122.

Eller, Z., Varga, Z. and Hancsók, J. (2019) Renewable jet fuel from kerosene/coconut oil mixtures with catalytic hydrogenation. *Energy and Fuels* 33, 6444–6453.

Espaldon, M.V.O., Lansigan, F.P., Salazar, A.M., Khan, C.L., Faustino-Eslava, D.V. *et al.* (2017). Smarter approaches to reinvigorate agriculture as an industry in the Philippines (project SARAi). Available at: https://sarai.ph (accessed 5 September 2023).

Eyres, L., Eyres, M.F., Chisholm, A. and Brown, R.C. (2016) Coconut oil consumption and cardiovascular risk factors in humans. *Nutrition Reviews* 74, 267–280.

FAO (Food and Agriculture Organization of the United Nations) (2022) *FAOSTAT: Crops and Livestock Products*. FAO, Rome. Available at: www.fao.org/faostat/en/#data/QCL (accessed 4 September 2023).

Fernando, W.M.A.D.B., Martins, I.J., Goozee, K.G., Brennan, C.S., Jayasena, V. *et al.* (2015) The role of dietary coconut for the prevention and treatment of Alzheimer's disease: potential mechanisms of action. *British Journal of Nutrition* 4, 1–14.

Fortune Business Insights (2022) *Virgin Coconut Oil Global Market Report: Market Size Share and COVID-19 Impact Analysis by Type (Organic and Conventional), and by End-User (Food and Beverages; Pharmaceutical, and Cosmetics & Personal Care), and Regional Forecast 2021–2028)*. Fortune Business Insights, Maharashtra, India. Available at: www.fortunebusinessinsights.com/virgin-coconut-oil-market-106554 (accessed 4 September 2023).

Freire, A.L.F., de Araújo Júnior, C.P., de Freitas Rosa, M., de Almeida Neto, J.A. and de Figueirêdo, M.C.B. (2017) Environmental assessment of bioproducts in development stage: the case of fiberboards made from coconut residues. *Journal of Cleaner Production* 153, 230–241.

Gallagher, K. (2021) How environmentally friendly is coconut oil? TreeHugger. Available at: www.treehugger.com/how-environmentally-friendly-is-coconut-oil-5206943 (accessed 4 September 2023).

Gliessman, S. (2011) Transforming food systems to sustainability with agroecology. *Journal of Sustainable Agriculture* 35, 823–825.

Global Oil Market Research Report (2022) *Global Coconut Oil Market by Product Type, Source, Application, Nature, and Region – Forecast to 2027*. Research and Markets, Dublin. Available at: www.researchandmarkets.com/reports/5546307/global-coconut-oil-market-by-product-type (accessed 4 September 2023).

Gratuito, M.K.B., Panyathanmaporn, T., Chumnanklang, R.A., Sirinuntawittaya, N.B. and Dutta, A. (2008) Production of activated carbon from coconut shell: optimization using response surface methodology. *Bioresource Technology* 99, 4887–4895.

Haron, U.A., Abllah, Z. and Nasir, N.A.M.M. (2018) The comparative antimicrobial effect of activated virgin coconut oil (AVCO) and virgin coconut oil (VCO) against dental caries-related pathogens. In: *International Dental Conference of Sumatera Utara 2017 (IDCSU 2017)*. Atlantis Press, Amsterdam, pp. 312–317.

Hawkins, D. (2019) *Worlds Apart: A Coconut Industry Report*. Agronomy Capital Advisors, Uckfield, UK. Available at: www.agronomycapitaladvisors.com/agribusiness-value-chain/coconuts (accessed 4 September 2023).

Hewlings, S. (2020) Coconuts and health: different chain lengths of saturated fats require different consideration. *Journal of Cardiovascular Development and Disease* 7: 59.

ICC (International Coconut Community) (2022) Global Scenario of Coconut. Presented at the *58th ICC Session and Ministerial Meeting*. ICC, Jakarta, India, and Government of Kingdom of Tonga.

JICA (Japan International Cooperation Agency) (2017) *The Urgent Development Study on the Project on Rehabilitation and Recovery from Typhoon Yolanda in the Philippines*. Available at: https://openjicareport.jica.go.jp/pdf/12283438_01.pdf (accessed 4 September 2023).

Joseph, G.H., Kindangen, J.G., Paat, P.C. and Taulaby, D. (2022) Opportunities for the development of the oleochemical industry of coconut products. *E3S Web of Conferences* 361: 02024.

Kawau, D.S., Pakasi, C.B., Sondakh, M.L. and Rengkung, L.R. (2015) Kajian Pendapatan Usahatani Kelapa Dengan Diversifikasi Horizontal Pada Gapoktan Petani Jaya Di Desa Poigar 1 Kecamatan Sinonsayang Kabupaten Minahasa Selatan. *Agri-Sosioekonomi* 11, 41–52.

Khaeruddin, A.A. (2022) IoT based smart irrigation system in management of coconut nurseries. Presented at the *50th International COCOTECH Conference & Exhibition,* 7–11 November 2022, Kuala Lumpur, Malaysia. Department of Agriculture Malaysia, Agriculture Engineering Division.

Khosravi, R., Azizi, A., Ghaedrahmati, R., Gupta, V.K. and Agarwal, S. (2017) Adsorption of gold from cyanide leaching solution onto activated carbon originating from coconut shell – optimization, kinetics and equilibrium studies. *Journal of Industrial and Engineering Chemistry* 54, 464–471.

Kumar, K.N. and Maheswarappa, H.P. (2019) Carbon sequestration potential of coconut based cropping systems under integrated nutrient management practices. *Journal of Plantation Crops* 47, 107–114.

Lasco, R.D., Lales, J.S., Arnuevo, M.T., Guillermo, I.Q., de Jesus, A.C. *et al.* (2000) Carbon dioxide (CO_2) storage and sequestration in the Leyte Geothermal Reservation, Philippines. In: *Proceedings of the World Geothermal Congress,* 28 May–10 June 2000, Kyushu-Tohoku, Japan, pp. 639–644.

Liyanage, M.D.S., Tejwani, K.G. and Nair, P.K.R. (1984) Intercropping under coconuts in Sri Lanka. *Agroforestry Systems* 2, 215–228.

Loshali, M. (2022) *Global Market Study on Coir: Increasing Popularity of Geotextile Products Leading to Demand Surge*. Persistence Market Research, New York. Available at: www.persistencemarketresearch.com/market-research/coir-market.asp (accessed 4 September 2023).

Louarrat, M., Enaime, G., Baçaoui, A., Yaacoubi, A., Blin, J. *et al.* (2019) Optimization of conditions for the preparation of activated carbon from olive stones for application in gold recovery. *Journal of the Southern African Institute of Mining and Metallurgy* 119, 297–306.

Market Research Future (2019) Coconut oil market report insights and industry analysis by type (virgin coconut oil, refined coconut oil, end-user (industry, food) & region competitive market size, share, trends & forecast, 2018–2025. Available at: www.globenewswire.com/en/news-release/2019/05/07/1818394/0/en/Coconut-Oil-Market-Size-to-Reach-USD-8-403-3-Million-by-2025-at-6-37-CAGR-Predicts-Market-Research-Future.html (accessed 5 September 2023).

McGavin, R., Leggate, W., Bailleres, H., Hopewell, G. and Fitzgerald, C. (2019) *A Guide to the Rotary Veneer Processing of Coconut Palms*. Forest Product Innovations, Salisbury Research Facility, Department of Agriculture and Fisheries, Queensland, Australia. Available at: www.aciar.gov.au/publication/books-and-manuals/a-guide-rotary-veneer-processing-coconut-palms (accessed 5 September 2023).

Mendoza, T.C., Teves, M.R.G. and Miciano, M.N.R. (2018) Value chain assessment of coconut-intercropping systems of smallholders in southern Mindanao, Philippines: options for improvement. Available at: www.researchgate.net/publication/326835420_Value_Chain_Assessment_of_Coconut-Intercropping_Systems_of_Smallholders_in_Southern_Mindanao_Philippines_Options_for_Improvement (accessed 5 September 2023).

Morbeck, F.L., Lelis, R C.C., Schueler, M.V.E., Santos, W.A., Sampaio, D.A. *et al.* (2019) Extraction and evaluation of tannin from green coconut mesocarp. *Matéria* 24: 0748.

Murthy, I.K., Gupta, M., Tomar, S., Munsi, M., Tiwari, R. *et al.* (2013) Carbon sequestration potential of agroforestry systems in India. *Journal of Earth Science and Climatic Change* 4, 1–7.

Novarianto, H., Tulalo, M.A. and Mawardi, S. (2021) Coconut sugar production of dwarf coconut varieties. *IOP Conference Series: Earth and Environmental Science* 800: 012024.

OFI (Oil and Fats International) (2021) *Oleochemicals: Supply and Demand Drivers*. OFI, Redhill, UK. Available at: www.ofimagazine.com/content-images/news/Oleochemicals-supply-and-demand-drivers.pdf (accessed 5 September 2023).

Pestaño, L.D.B. and Wifredo, I.J. (2016) Engineering an improved coconut processing system in the Philippines at the farm-level. *Journal of Advanced Agricultural Technologies* 3, 58–62.

Phukon, R.M., Nath, J.C., Sumitha, S., Maheswarappa, H.P. and Elain Apshara, S. (2021) Early evaluation of cocoa (*Theobroma cacao* L.) clones as intercrop in coconut gardens under red river region of Assam state, India. *Journal of Plantation Crops* 49, 77–82.

Praagman, J., Beulens, J.W.J., Alssema, M., Zock, P.L., Wanders, A.J. *et al.* (2016) The association between dietary saturated fatty acids and ischemic heart disease depends on the type and source of fatty acid in the European prospective investigation into cancer and nutrition – Netherlands cohort. *American Journal of Clinical Nutrition* 103, 356–365.

Pruseth, B., Banerjee, S. and Ghosh, A. (2020) Integration of *in silico* and *in vitro* approach to reveal the anticancer efficacy of virgin coconut oil. *CORD* 36, 1–9.

Purbopuspito, J. (2022) Coconut farming practices and their contribution on improving biodiversity, farm productivity, climate change mitigation, and farmer's livelihood. Presented at the *50th ICC COCOTECH,* 7–11 November 2022, Kuala Lumpur, Malaysia.

Report Linker (2020) An AI data platform. Available at: www.reportlinker.com/market-report/search?keywords=coconut (accessed 5 September 2023).

Samriddhi, C. and Roshan, D. (2022) *Coconut Sugar Market by Form (Organic, Conventional), by End User (Commercial, Residential), by Application (Food and Beverage, Cosmetics, Personal Care), by Sales Channel (Hypermarkets, Specialty Stores, E-Commerce, Business to Business): Global Opportunity Analysis and Industry Forecast, 2021–2031*. Allied Market Research, Maharashtra, India. Available at: www.alliedmarketresearch.com/coconut-sugar-market-A16949 (accessed 4 September 2023).

Sangamithra, A., Swamy G.J., Sorna Prema, R., Chandrasekar, V., Sasikala S. *et al.* (2013) Coconut: an extensive review on value added products. *Indian Food Industry Magazine* 32, 29–36.

Septiana, L.A.E.S., Hidayah, A.N., Laurencia, A. and Agustine, E. (2019) Utilization of activated carbon from coconut shell in remediating heavy metal waste in soil around mining area. *IOP Conference Series: Earth and Environmental Science* 311: 12012.

Shankar, B. (2019) *Coconut Products Market by Type (Coconut Water, Coconut Oil, Coconut Milk, Dried Coconut Products, and Others), Application (Food, Beverage, Cosmetics, and Others), and Form (Solid and Liquid): Global Opportunity Analysis and Industry Forecast, 2019–2026*. Allied Market Research, Maharashtra, India. Available at: www.alliedmarketresearch.com/coconut-products-market (accessed 4 September 2023).

Snijder, M.H., Keijsers, E.R., van den Oever, M.J. and van Dam, J.E. (2006) *Coir Based Building and Packaging Materials: Final Report of Project CFC/FIGHF/11 (No. 592)*. Agrotechnology and Food Innovations, Wageningen, Netherlands.

Somashekaraiah, R., Shruthi, B., Deepthi, B.V. and Sreenivasa, M.Y. (2019) Probiotic properties of lactic acid bacteria isolated from neera: a naturally fermenting coconut palm nectar. *Frontiers in Microbiology* 10: 1382.

Sushant, M., Amit, N. and Eswara, P. (2020) *Oleochemicals Market by Type (Fatty Acid, Fatty Alcohol, Glycerol, and Others) and Application (Pharmaceuticals, Personal Care & Cosmetics, Food & Beverages, Soap & Detergents, Polymers, and Others): Global Opportunity Analysis and Industry Forecast, 2021–2030*. Allied Market Research, Maharashtra, India. Available at: www.alliedmarketresearch.com/oleochemicals-market (accessed 4 September 2023).

Takeishi, J., Tatewaki, Y., Nakase, T., Takano, Y., Tomita, N. *et al.* (2021) Alzheimer's disease and type 2 diabetes mellitus: the use of MCT oil and a ketogenic diet. *International Journal of Molecular Sciences* 22: 12310.

Turner, C., Andrews, J., Kumar, S., Iuvale, S.J., Matatumua, T. *et al.* (2022) *Building a Business Case for Investment in a Coconut Industry in the Pacific*. Australian Centre for International Agricultural Research, Canberra.

Vijayakumar, M., Vasudevan, D., Sundaram, K., Krishnan, S., Vaidyanathan, K. *et al.* (2016) A randomized study of coconut oil versus sunflower oil on cardiovascular risk factors in patients with stable coronary heart disease. *Indian Heart Journal* 68, 498–506.

Wezel, A., Casagrande, M., Celette, F., Vian, J.F., Ferrer, A. *et al.* (2014) Agroecological practices for sustainable agriculture. *A review. Agronomy Sustainable Development* 34, 1–20.

Wezel, A., Herren, B.G., Kerr, R.B., Barrios, E., Gonçalves, A. *et al.* (2020) Agroecological principles and elements and their implications for transitioning to sustainable food systems. A review. *Agronomy for Sustainable Development* 40: 40.

WHO (World Health Organisation) (2022) *Diabetes*. WHO, Geneva. Available at: www.who.int/health-topics/diabetes#tab=tab_1 (accessed 5 September 2023).

Index

Note: page numbers in italic denote tables and figures.